FUZZY IF-THEN RULES IN COMPUTATIONAL INTELLIGENCE
Theory and Applications

THE KLUWER INTERNATIONAL SERIES IN ENGINEERING AND COMPUTER SCIENCE

FUZZY IF-THEN RULES IN COMPUTATIONAL INTELLIGENCE
Theory and Applications

Edited by
DA RUAN
Belgian Nuclear Research Centre (SCK•CEN)

ETIENNE E. KERRE
University of Ghent, Belgium

Springer Science+Business Media, LLC

Library of Congress Cataloging-in-Publication

Fuzzy if-then rules in computational intelligence : theory and applications / edited by Da Ruan, Etienne E. Kerre.

p. cm. -- (The Kluwer international series in engineering and computer science ; SECS 553)

Includes bibliographical references and index.

ISBN 978-1-4613-7035-2 ISBN 978-1-4615-4513-2 (eBook)

DOI 10.1007/978-1-4615-4513-2

1. Expert systems (Computer science) 2. Fuzzy systems. 3.Computational intelligence. I. Ruan, Da. II. Kerre, Etienne E. III. Series.

QA76.76.E95 F8825 2000

006.3'3--dc21

00-022047

Originally published by Kluwer Academic Publishers in 2000

Softcover reprint of the hardcover 1st edition 2000

Printed on acid-free paper.

TABLE OF CONTENTS

Preface

Expert systems have been one of the first commercial products resulting from research in artificial intelligence. Expert systems are computer programs that model the knowledge of experts and that are able to solve concrete problems where knowledge of experts is needed. Most of the expert knowledge is pervaded with imprecision (i.e., using linguistic instead of numerical values for the variables involved) and uncertainty (i.e., using specifications such as possible, probable, plausible, more-or-less certain etc.). Therefore techniques that are able to represent and to process incomplete information would certainly improve the knowledge-based expert systems.

During the last 35 years many new theories have been initiated to model imprecise and uncertain information: fuzzy set theory of Zadeh (1965), L-fuzzy set theory of Goguen (1967), flou set theory of Gentilhomme (1968), rough set theory of Pawlak (1982), intuitionistic fuzzy set theory of Gargov and Atanassov (1983), and fuzzy rough sets of Dubois and Prade (1990). Undeniably, fuzzy set theory with more than 27,000 papers with fuzzy in title and/or abstract as mentioned by the INSPEC database in 1999 may be considered as the most successful theory to treat incomplete information.

This edited volume is dedicated to the application of fuzzy set theory or fuzzy logic in a broad sense to one of the key issues in knowledge-based systems namely IF-THEN rules. The main part of an expert system, i.e., the so-called kernel section consists of a knowledge base and an inference engine. The knowledge base consists of a database and a rule base. More particularly the rule base is composed of IF-THEN rules where the antecedent (the IF-part) and the consequent (the THEN-part) mostly contain imprecise predicates that can be modelled using fuzzy set theory. A fuzzy rule of Zadeh (1973) is a conditional statement involving fuzzy predicates. The use of fuzzy rules, expanding from theory to applications in computational intelligence, is abundantly demonstrated in this volume.

IF-THEN statements and questions have given rise to difficulties not only in fuzzy set theory, but also in traditional 2-valued logic. These difficulties disappear in the *chain-set logic* described in Chapter 1 by Hisdal. A single IF-THEN chain-set table can represent a complicated structure of intertwined trees. The chain-set logic uses probabilities for the representation of *information supply*, and possibilities or truth-values for the representation of non-IF-THEN questions. Chapter 1 discusses the stepwise building-up and updating of a knowledge base that has the form of a non-IF-THEN and an IF-THEN chain-set. The chain-set logic is built on top of 2-valued logic. It makes use of both probabilities and possibilities, and it solves the IF-THEN problem of propositional calculus.

Chapter 2 deals with fuzzy predicate calculus and fuzzy rules. Hájek presents the basic many-sorted fuzzy predicate calculus to express and prove logic properties of fuzzy IF-THEN rules.

Starting from an input fuzzy set and an IF-THEN rule, implementations of the Generalized Modus Ponens (GMP) in a fuzzy set theoretical framework allow the derivation of an output fuzzy set. If the GMP is implemented by means of the compositional rule of inference in general this derivation involves a lot of computational efforts, and the shape of the resulting membership function often seems quite arbitrary compared to that of the input membership function. In Chapter 3, Cornelis, De Cock and Kerre present a review of techniques, generating as their output a fuzzy set belonging to a predefined class, that are not afflicted with these disadvantages.

The compositional rule of inference as a generalization of the classical modus ponens in the framework of approximate reasoning is discussed in Chapter 4. Kolesárová and Kerre show the relationship with the special fuzzy quantities and their calculus under various triangular norms.

Chapter 5 approaches the theories of approximate reasoning in the IF-THEN type approximate-reasoning mode based on the lattice-valued proposition logic. The IF-THEN rule is the rule in which both antecedent and consequence are L-type lattice-valued proposition sets. Xu *et al.* propose two approximate reasoning methods that are reasonable due to their strict semantic interpretation and strict syntactical deduction in the lattice-valued propositional logic.

In Chapter 6, Savinov describes the problem of mining set-valued rules in large relational tables containing categorical attributes taking a finite number of values. Such rules allow for an interval of possible values to be selected for each attribute in condition instead of a single value for association rules, while conclusion contains a projection of the data restricted by the condition onto a target attribute. The author further describes an efficient algorithm, which carries out the generalized dual transformation from possibilistic disjunctive normal form representing data into conjunctive normal form representing knowledge and thus generates all the most interesting prime disjunction. Once obtained they can be used for building different forms of rules as well as for other purposes such as prediction, clustering etc.).

Though fuzzy techniques are popular engineering tools, their utilization is being restricted by their exponential complexity property. The objectives of Chapter 7 by Baranyi and Yam are twofold: one is to find a general form for fuzzy system output covering the widest possible areas of applications, and the other is to present a complexity reduction algorithm for the general form.

Generalized fuzzy rules provide the essential means for computational representation of equations and inequality constraints over fuzzy granular structures. In Chapter 8, Kohout and Kim are concerned with the ways the rules can be acquired from the real world data. They also provide a method for elicitation of knowledge of experts and link this to Zadeh's computational theory of perceptions. The algorithms for extracting and consistency checking of the rules explore the Bandler-Kohout (BK)-products of relations. This makes it possible to treat the numerical and linguistic computations in a unified framework with the methods for extrapolation of knowledge and data.

Nowadays, linguistic modeling is considered as one of the most important applications of fuzzy set theory, along with fuzzy control. Linguistic models have the advantage of providing a human-readable description of the system modeled in the form of a set of linguistic rules. In Chapter 9, Cordón *et al.* analyze several approaches to improve the accuracy of linguistic models while maintaining their descriptive power. All these approaches share the common idea of improving the cooperation between the rules in the linguistic model knowledge base.

In Chapter 10, Novák and Kovář describe a large-scale application of fuzzy control in the Czech Republic. They have used the linguistically oriented fuzzy logic controller, for which it is specific to interpret the IF-THEN rules as linguistically characterized logical implications and the inference is logical deduction based on the formal fuzzy logic in broader sense.

Autonomous robots, which achieve tasks without human operators, are required in many fields. They have to be intelligent to determine their own motions in unknown environments based on both sensory data (numerical) and expert knowledge (linguistic), that is, the linguistic-numerical heterogeneous data. Chapter 11 looks at the recent development in the area of data fusion approach in conjunction with the concept of fuzzy IF-THEN rules for intelligent robotic control. Zhou and Ruan also address current problems and future research directions in this area.

Chapter 12 illustrates how fuzzy IF-THEN rules can be used for pattern classification problems. Ishibuchi *et al.* describe a heuristic method for automatically generating fuzzy IF-THEN rules for pattern classification problems from training patterns and propose a fuzzy rule selection method for designing compact fuzzy rule-based systems with high classification ability. They finally present two genetics-based machine learning algorithms for designing fuzzy rule-based systems for high-dimensional pattern classification problems with successful simulation results.

Experiments are demonstrated where a newly developed fuzzy logic system based, indirect adaptive control algorithm is applied to trajectory control of a flexible link manipulator. All the experimental results, obtained in Chapter 13 by Lee and Vukovich, are quite satisfactory, indicating that the fuzzy logic systems based adaptive control algorithm is effective in dealing with complex, nonlinear systems, and can be useful in practical applications.

This volume pays special attention to the fuzzy IF-THEN rules in today's computational intelligence. The 13 chapters contributed by 27 authors and coauthors from 14 countries made this volume possible. Special thanks are due to all these authors and coauthors for their fine contributions to this book, and to Alexander Greene (Publisher, Kluwer Academic Publishers) for his kind advice and help to publish this volume. We hope this volume will benefit many students, researchers, and engineers around the intelligent computational world.

D. RUAN AND E.E. KERRE
Mol and Ghent.

Chapter 1

THE IF-THEN LOGIC OF NATURAL LANGUAGES AND NATURALLY CORRECT INFERENCES

Ellen Hisdal

Institute of Informatics

University of Oslo

Box 1080 Blindern

0316 Oslo

Norway

Abstract *IF-THEN* statements and questions have given rise to difficulties not only in fuzzy set theory, but also in traditional 2-valued logic. These difficulties disappear in the 'chain-set logic' described here. A complicated structure of intertwined trees can be represented by a single *IF-THEN* chain-set table. The chain-set logic uses probabilities P for the representation of *information supply,* and possibilities or truth values for the representation of non-*IF-THEN questions.* The declarative sentence *IF A THEN C* is interpreted in the sense of the information supply '$P(C|A)=1$'. The question *IF A THEN C*? is interpreted either in the sense of 'Is $P(C|A)=1$?', to be answered by 'yes' or 'no'; or in the sense of 'What is the value of $P(C|A)$?' to be answered by 0, or m, or 1, or $0m$, or $m1$, or $0m1$. m is a value in the open real interval $(0,1)$ (excluding the points 0 and 1). It signifies uncertainty and can be replaced by its numerical fractional value if known. The last three 'interval-valued' answers signify partial or complete 'ignorance' or insufficient information in the knowledge base concerning the question being directed at it. Quantification sentences like 'All dogs are animals' or 'Some animals are dogs' are interpreted as the *IF-THEN* sentences '*IF* x is an instance of a dog *THEN* x is an instance of an animal' and '*IF* x is an instance of an animal *THEN* x may be an instance of a dog' respectively. As a result, the necessity for the separate and very-difficult-to-program predicate calculus of 1st order logic disappears.

1. INTRODUCTION

1.1 OVERVIEW

This chapter discusses the stepwise building-up and updating of a knowledge base from sequentially supplied information. The knowledge base has the form of a non-*IF-THEN* and an *IF-THEN* chain-set. A chain-set is a two-dimensional table which has some external similarities to a transposed truth-table of propositional calculus. However, inferences in the chain-set logic are, in contrast to inferences in traditional 1st order logic, always 'naturally correct' (see Sect. 1.2).

The extreme importance of a correct representation of the implication, i.e., of a conditional or *IF-THEN* sentence, is, to a great extent, due to the role of the implication in inferences in all the logical systems mentioned in this chapter. A wrong representation of *IF-THEN* can result in a wrong inference.

Sections 1.3-1.5 show examples of *IF-THEN* and inference difficulties in 1st order logic (propositional calculus+predicate calculus), in fuzzy set theory and in '*is a*' trees of AI. For each naturally strange result, the corresponding result in the chain-set logic is also stated. Sections 2-6 show the procedures of the chain-set logic which give these results. For a fuller and explanatory description of the chain-set procedures, see [17].

A chain-set table for the representation of a declarative (information supply) sentence consists of *chains (columns)* of 1's and 0's (corresponding to *rows* of 'true' and 'false' values in a truth table) to each of which is attached a probability value. *IF-THEN* and non-*IF-THEN* questions can then be directed at the chain-set-structured knowledge base. A non-*IF-THEN* question is also represented by a chain-set. However, to each chain of such a *question* chain-set there is attached a possibility- instead of a probability-value. Possibility values have a probabilistic interpretation which differs from the interpretation of the probability values in the chain-set for the corresponding information supply sentence.

All non-*IF-THEN* information can, in principle, be stored in a single chain-set which is updated for each new information supply item. Similarly, all *IF-THEN* information can be stored in a single *IF-THEN* chain-set which is updated for each new *IF-THEN* information supply. The two chain-set structures must be kept apart because the updating rule for *IF-THEN* information by new *IF-THEN* information is not the same as the updating by non-*IF-THEN* information. The reason for this is that *IF-THEN* information, i.e. a conditional statement, specifies a *conditional* probability. This is in contrast to a non-*IF-THEN* statement, such as an affirmed or negated atomic sentence, or a sentence with an *OR* or *AND* connective, which specifies the *joint* probability of the truth or falsity of the components. However, the rule for the construction of an *IF-THEN* chain-set sees to it that the entries of the probability row of this chain set are also joint probabilities.

An inference is looked upon in the chain-set logic as an answer to a question *qu-kb* directed at the information *info-kb* contained in the *knowledge base.* The answer is the value of $answ{=}P(qu\text{-}kb\,|\,info\text{-}kb)$. For a non-*IF-THEN* question it is obtained as the 'scalar' or 'dot' product of the probability row of the non-*IF-THEN* information chain-set by the possibility row of the question chain-set (see Sect. 3.1). *answ* can have the values 1 (yes), 0 (no) or $m \in (0, 1)$ (a true fraction, signifying uncertainty; m is a mnemonic for 'medium' or 'maybe'). E.g., when the information supply is 'Margy will be at home on Sunday OR Monday', then this sentence tells us that according to the present state of the information supply, there is a probability m, $0 < m < 1$, that she will be at home on Sunday BUT NOT on Monday, and a probability m that she will be at home on Monday BUT NOT on Sunday (see Fig. 1.1 (b4), (b5)).

When the knowledge base contains insufficient information to supply a '0', or 'm', or '1' answer, then we say that we have a case of partial or complete ignorance. The answer can then be 'interval-valued', such as $0m$ (signifying that the event specified in the question is not certain to occur), or $m1$ (signifying that it is not certain that the event specified by the question will not occur), or $0m1$ (signifying complete ignorance in the knowledge base concerning the event specified in the question).

A conditional *question qu=IF A THEN C* does not have its own question chain-set representation. It is answered by 'multiplying' (taking the *AND* composition of) the chain-set stored in the knowledge base *kb,* by the chain-set for A. The question *qu=C* is then directed at this product chain-set (see Sect. 3.2).

1.2 NATURAL LOGIC

In Sections 1.3-1.5 we discuss three logical systems, all of which give rise to difficulties in connection with the representation of-, and procedures for-, *IF-THEN* sentences. Difficulties in connection with *IF-THEN* can again give rise to wrong inferences. For each of the three logical systems we show one or more examples in which the logical procedures of the system fail compared with the logic of all natural languages or 'natural logic'. The corresponding chain-set result (BP-, and the more correct M-logic) are also shown.

Natural logic refers to the logical inferences which every teacher of mathematics or science expects of her or his pupils. The Paris-Rome Example 1..5 of Sect. 1.3 demonstrates a bizarre inference of propositional calculus as compared with that of natural logic. A math teacher would give a 'flunk' grade to a pupil who would infer Eq. (1.3) from Eq. (1.2).

The attitude of the present author towards logic is that of an 'ideal physicist' in the sense of Popper who says *'The very refutation of a theory is always a step forward that takes us nearer to the truth'* ([19], p. VII). She thus looks upon the logic of natural language as a natural phenomenon, and upon mathematical logic as a theory whose procedures are able to simulate correct natural language reasoning. A logical theory which fails in this respect cannot be accepted. This attitude is analogous to the attitude of Chomsky and Pinker concerning innate human grammar structures (see [17], p. 12). The procedures used by humans to generate and to analyze grammatically correct sentences are performed subconsciously in everyday life. It is then the task of science to find out how they work, or to devise equivalent ones.

The confusion of the tower of Babel of the old testament [2] concerns not only phonology (the coding of concepts into sound patterns), but also grammar. Even related languages can have completely different grammatical constructions.[1]

In this respect, the field of logic is superior to the subfields of phonology and grammar of natural language. Logical procedures are by their nature a sort of international and natural logical Esperanto. Probably all languages have formulations which are equivalent to '*IF* Bobby is a dog *THEN* Bobby is an animal'; or, more generally, formulations such as Aristoteles' 'All B is A', 'No C is A', 'Some D is A', 'E may (may not) be A' (see pp. 128, 34 of [1]). Aristotelian syllogisms ([1] p. 4) are procedures for drawing conclusions or inferences from two or more such sentences. The inferences are independent of the particular natural language used.

Our emphasis on the importance of natural logic presents no hindrance to the building up of higher logical systems on top of the logic of natural languages. But we must not forget that natural language is our basic metalanguage. Without the logic of natural language our whole house of cards would collapse due to the lack of a foundation in and on which it can be expressed. We would then be unable to formulate any logical system whatsoever.

1.3 IF-THEN DIFFICULTIES IN FIRST ORDER LOGIC

Practically every introductory textbook on traditional mathematical logic (see, e.g. [18] or [20] or [7]) starts with the so-called propositional calculus. This calculus operates with propositions or sentences (or 'formulas' according to the strange terminology of propositional calculus, see, e.g., [20, p. 52, 54] for more precise definitions) which can be either true (t) or false (f). The next

[1]Consider, e.g., the sentence 'The dog bites the man', versus 'The man bites the dog'. In English, it is the word order which determines who bites whom. In contrast, in the related language German, the sentences 'Der Hund beisst den Mann' and 'Den Mann beisst der Hund' have the same meaning, in spite of the reversed order of Hund and Mann in the two sentences. It is the 'case' of 'Mann' and 'Hund' (indicated by the form of the definite article ('der' versus 'den') which determines who bites whom.

step is to define 'truth tables' for the negation (*NOT* or $\neg$), the conjunction (*AND* or $\wedge$), the inclusive disjunction (*ORA* or $\vee$), the exclusive disjunction (*ORE* or $\vee_{ex}$) and the implication (*IF-THEN* or $\rightarrow$ or $\supset$). These are also called connectives because they connect two sentences such as A and B. E.g., in the composite, conjunctive sentence '*A AND B*', *AND* is a connective which connects A and B.

The (a) tables of Figs. 1.1, 1.2 show the basic truth tables of propositional calculus. Note that the truth tables of the negation and the *AND* and *OR* connectives are in agreement with their meanings in natural language. E.g., the statement 'λ=Margy is at home on Sunday and Monday' implies that she must be at home on both days. This agrees with the truth table (a3) for *AND* which has only one row for which λ=*A AND B* is true, assuming that both A and B are atomic sentences. In contrast, a statement with an *ORA* connective (*ORA*=inclusive *OR* or *AND/OR*), Fig. 1.1 (a4), leaves open three combinations of truth values of A and B for which *A ORA B* is true. We will soon see that there is no such satisfactory agreement between natural language logic and the truth table for the implication.

After the postulation of the truth tables, propositional calculus defines a tautology. This is a composite sentence (a sentence with a connective) whose truth table contains only 't' values in its last column; this being the column whose entries denote the truth of the composite sentence as a whole for the corresponding combination of truth values of the atomic component sentences in the given row of the truth table. For example, the sentence 'A *ORA* NOT A' is a tautology (this follows from the last column of tables (a1), (a2) and rows 2, 3 of (a4)). In contrast, a contradiction is a composite sentence whose truth table contains only 'f' values in its last column.

A composite sentence is said to be a contingency when the last column of its truth table contains one or more 't' values, as well as one or more 'f' values.

The importance of the concepts of tautology, contingency and contradiction is due to their use in connection with inferences.

The procedure of propositional calculus for the purpose of determining whether a sentence C can be inferred (or is a Conclusion) from a sentence (or premiss or Antecedent) A is to set up the truth table for the conditional or *IF-THEN* statement *IF A THEN C*. Each lowest-level atomic component of A and C must have its own column in this truth table. The number of rows in the truth table is then equal to 2^G, the number of different possible assignments of t and f values to the G atomic components. Iff the truth table for *IF A THEN C* is a tautology, then C is said to be a conclusion, i.e., a correct and necessary inference from A. Iff *IF A THEN C* is a contradiction, then C can never be inferred from A. The truth table of the implication or *IF A THEN C* sentence is thus an essential part of the inference procedure of propositional calculus. (Iff=If and only If).

The inferences of propositional calculus and of the chain-set logic are equivalent in many cases. Seven such equivalences, among them the two de Morgan laws, are enumerated in Sect. 6.4.5 of [17].

However, the truth table of propositional calculus for the implication (Fig. 1.2 (a6)) does not always agree with the meaning of *IF-THEN* in natural languages. The cases of disagreement occur for the last two rows of the truth table, i.e., when A is false. The natural language statement *IF A THEN C* does not tell us, as the truth table does, that *IF A THEN C* is true whenever A is false. Actually the *IF-THEN* statement gives us no information whatsoever in this case. As a result, there exist cases of inferences in propositional calculus which disagree completely with correct natural inferences.

Examples 1..1-1..6 below illustrate some such difficulties in propositional calculus. For each example we give also the answer according to the chain-set logic of Sections 2.-6.. Both the more elementary BP logic and the M logic are parts of the chain-set logic. They are discussed in detail in parts I and II of [17] respectively. The M-logic is the more advanced and correct one according to the criterion of agreement with natural logic.

In the examples below, *info* denotes an information supply sentence, and *qu* denotes a question (interrogative sentence) directed at *info*. $answ(qu \mid info) = P(qu \mid info)$ is the probability that *qu* is true, given that *info* is true.

$\neg, \wedge, \vee, \vee_{ex}$ stand for *NOT*, *AND*, *ORA* (inclusive *OR*), and *ORE* (exclusive *OR*), respectively. $A{\rightarrow}C$ stands for *IF A THEN C*.

Example 1..1 $\quad info = B, \quad qu? = \neg B? = NOT\ B?$.

BP and M logic: $\quad answ(qu \mid info) = P(\neg B \mid B) = 0$ *(no)* ,

Propositional calculus: $\quad B{\rightarrow}\neg B$ *is a contingency, not a contradiction* .

For the formal derivation of these inferences of propositional calculus and of the chain-set logic see [17], Fig. 6.6.

Example 1..2 $info = (A{\rightarrow}C), \quad qu? = \neg(A{\rightarrow}C)? = NOT\,(A{\rightarrow}C)?$.

BP and M logic: $answ(qu \mid info) = 0$ *(no)* ,

Prop. calc. : $\quad [(A{\rightarrow}C) \rightarrow \neg(A{\rightarrow}C)]$ *is a contingency, not a contradiction* .

The answer 0 of the chain-set logic is found in [17], chapter 19, solution to exercise 7.2.

Example 1..3 $info = (A{\rightarrow}C), \quad qu? = (A{\rightarrow}\neg C)? = (A{\rightarrow}NOT\ C)?$.

BP and M logic: $answ(qu \mid info) = 0$ *(no)* ,

Prop. calc. : $\quad (A{\rightarrow}C) \rightarrow (A{\rightarrow}\neg C)$ *is a contingency, not a contradiction* .

For the formal derivation of the answer 0 according to the procedures of the chain-set logic see [17], chapter 6, exercise 6.3.

Note how self-evident the answer 0 in the last example is in a probability logic. The given *info* $=(A{\rightarrow}C)$ is interpreted in the chain-set logic as the assertion '$P(1C|1A){=}1$', where $1C$ denotes the outcome corresponding to the affirmation of C, and $0C$ denotes the outcome corresponding to the negation of C. It then follows immediately from the theory of probability that $P(0C|1A){=}0$, using the 2-element probability space $\{1C,\,0C\}$, conditioned on the occurrence of $1A$. Similarly in Example 1..2, the chain-set answer is interpreted as $P[(P(1A|1C) \neq 1)) \mid (P(1A|1C){=}1))]$.[2] This quantity *must* be 0 because $P(1A|1C)$ cannot be both different from 1 and equal to 1.

The following example illustrates a conjunctive sentence which is not contradictory according to both propositional calculus and the BP logic; although it should have been contradictory according to the use of '$\rightarrow$' or '*IF-THEN*' in natural language. It turns out that the sentence *is* contradictory in the more advanced M logic of part II of [17] which uses 'type 1 updating' of *IF-THEN* information by *IF-THEN* information.

Example 1..4 *Consider the label*

$$\lambda = (A{\rightarrow}C)\ AND\ (A{\rightarrow}\neg C)\,. \qquad (1.1)$$

It can be shown that the following results hold in the different logics,

Prop. calculus and BP logic:	*λ is not a contradiction.*
M logic:	*λ has a contradictory chain-set with an empty set of probability rows.*

The proof of the M logic result is shown in Fig. 11.1 of [17]; see also Eq.(14.21) of [17] in this connection.

A hairraising breakdown of the inference procedure of propositional calculus is illustrated by the following 'Paris-Rome problem'.[3]

[2]Frege ([6], p. 11) was probably the first to comment on the ambiguity of natural language sentences concerning affirmation and negation. A sentence A may expresss merely an idea which can be asserted or negated. More often it expresses the assertion of this idea. The chain-set logic resolves this notational ambiguity of natural languages (see [17], Sect. 2.4.2 and Eq. (8.12)). Frege's idea of A is interpreted as the yes-no sequence $< 1A, 0A >$ of A. $1A$ is the outcome corresponding to the affirmation of A, and its assertion is denoted by the specification $P(1A){=}1$, $P(0A){=}0$. $0A$ is the outcome corresponding to the negation of A. and its assertion is denoted by the specification $P(0A){=}1$, $P(1A){=}0$. The idea of *NOT A* is interpreted as the yes-no sequence $< 0A, 1A >$ of *NOT* $A{=}\neg A$. When the context is quite clear, then we may deviate from this unambiguous notation. E.g., $P(qu\,|info)$ should actually have been written as $P(1qu\,|1info)$, and $P(C|A)$ as $P(1C|1A)$.

[3]The Paris-Rome problem is taken from a textbook on logic [5, p. 15] by Fenstad and Normann.

$\lambda_1 = A$ (a1)

A	λ_1
t	t
f	f

$\lambda_2 = \neg A$ (a2)

A	λ_2
t	f
f	t

$\lambda_3 = A \wedge B$ (a3)

A	B	λ_3
t	t	t
t	f	f
f	t	f
f	f	f

$\lambda_4 = A \vee B$ (a4)

A	B	λ_4
t	t	t
t	f	t
f	t	t
f	f	f

$\lambda_5 = A \vee_{ex} B$ (a5)

A	B	λ_5
t	t	f
t	f	t
f	t	t
f	f	f

(b1)

A	1
prob	1
poss	1

(b2)

A	0
prob	1
poss	1

(b3)

A	1
B	1
prob	1
poss	1

(b4)

A	1	1	0
B	1	0	1
prob	m	m	m
poss	1	1	1

(b5)

A	1	0
B	0	1
prob	m	m
poss	1	1

Figure 1.1 Postulated Truth Table of Propositional Calculus for a label λ ((a) frames), and the Chain Set for λ ((b) frames underneath). λ is an atomic or composite sentence. A, B, . . . are atomic sentences (without connectives). The set of elements in the upper left hand margin is called the ground universe GU of the chain-set. For (b3)-(b5) we have GU=$\{A, B\}$. Each element of the ground universe can be affirmed (1 entry in a given chain (column) or negated (0 entry)). $\neg$, $\wedge$, $\vee$, $\vee_{ex}$ stand for NOT, AND, ORA (inclusive OR or OR/AND), and ORE (exclusive OR) respectively.

Truth tables should be read one row at a time, chain-sets should be read one chain (column) at a time.

There exists an external similarity between transposed truth tables and chain sets. Each row *of a truth table with a 't' entry in the last column corresponds to a chain* (column) *of the chain-set. The 0-1 entries of a chain correspond to f-t entries of the truth table, excepting the last column. This column corresponds to the possibility row of the chain-set. Chains with probability and possibility 0 can, by default, be left out in the chain-set representation.*

A chain-set can have a probability row and/or a possibility row. The probability row is used for a declarative sentence λ, representing information supply. It is interpreted as $P(ch|\lambda)$, the probability that the given 0-1 chain (column) of the chain set will occur for the information supply λ. The entries of the probability row must add up to 1. m is a proper fraction.

The possibility row is used for interrogative sentences (questions directed at the knowledge base). It has the same interpretation as the possibilities of fuzzy set theory according to the TEE model, namely $P(yes{-}\lambda \mid ch)$ (see Sect. 2. here and [16], [17, Sect. 4.4.5]. For more details, see [12], [13],) [14], [15]). In the terminology of the theory of probability, the set of 0-1 chains with possibility 1 (considered as a subset of all possible 0-1 chains of length G, G=number of rows in the chain-set above the probability row)) represents the event *'yes-λ'. The fuzzy case of possibility values between 0 and 1 is discussed in [17, 93-98].*

Note that the probability row(s) of a chain-set has (have) no equivalent in a truth table whose last column is used for both statements and questions.
For the implication (IF-THEN) see Fig. 1.2 (a6), (b6).

$\lambda_6 = A \rightarrow C$, truth table

(a6)

A	C	$\lambda_6 = A \rightarrow C$
t	t	t
t	f	f
f	t	t
f	f	t

$\lambda_6 = A \rightarrow C$, information chain set

(b6)

A	1	1	0	0
C	1	0	1	0
prob	m	0	m	m
prob	m	0	0	m

Figure 1.2 (a6): Postulated truth table of the implication IF A THEN C. (b6): Information chain-set of the implication. In contrast to the last column of the truth table *for the implication which is also used for the representation of IF-THEN* questions *(conclusions) in propositional calculus, the IF-THEN* chain-set *has no possibility row. One reason for this is that the procedure for answering the question IF A THEN C? (Sect. 3.2) has no need for such a row.*

However, the IF A THEN C information chain-set (b6) has two probability rows. The second of these corresponds to the case that $A \equiv C$. The conjunction of several IF-THEN statements can have more than two probability rows, see Eq. (1.20).

In contrast to propositional calculus, a universal or existential quantification (classification) structure can always be represented as an IF-THEN chain-set. For a given probability row of the chain-set, each chain with a nonzero probability represents a rootpath in the classification tree. E.g., (b6) could represent 'All dogs are animals', or $A \rightarrow C$=(IF x is an instance of a dog THEN x is an instance of an animal). For the first probability row of (b6), the first chain represents the rootpath [dog, animal], the third chain represents the rootpath [nondog-animal, animal] (e.g. [cat, animal] or [fish, animal]). The last chain represents [nondog, nonanimal], e.g. [tree, plant]. The root of the tree (e.g. 'organism') is not included in the ground universe of an 'is a' chain-set. It should be specified separately for a given classification structure. A more complicated classification chain-set is shown in Eq. (1.20). It summarizes five classification trees, all of which are possible ones according to the (insufficient) information supplied to the knowledge base.

For the chain-set of an existential quantification structure, expressed, e.g., as 'Some animals are dogs', see [17], Fig. 10.1 and chapters 15-17.

Example 1..5 *If one uses the truth table of the implication for IF-THEN, then the premiss*

$$\begin{array}{c} \textit{IF I am in Paris THEN I am in France,} \\ \textit{AND} \\ \textit{IF I am in Rome THEN I am in Italy ,} \end{array} \qquad (1.2)$$

results in the following true conclusion in propositional calculus,

$$\begin{array}{c} \textit{IF I am in Paris THEN I am in Italy} \\ \textit{ORA (AND/OR)} \\ \textit{IF I am in Rome THEN I am in France .} \end{array} \qquad (1.3)$$

The proof of this absurd inference of propositional calculus is given in [17], Fig. 7.4.

In the BP logic the answer to the question of Eq. (1.3), based on the information of Eq. (1.2), is '0' or 'no'. ([17], Figs. 7.8, 7.9). In the more correct M logic the answer is $0m1$ *or the whole real interval [0,1]; in other words 'don't know' (Sect. 14.6.3 of [17]). This is indeed the correct answer as long as the knowledge base has no information as to whether there is an overlap between France and Italy, and where Paris and Rome are located with respect to such a possible overlap. Indeed the information chain-set representation of Eq. (1.2) ([17], Eq.(14.65)) has as many probability rows as the number of different such overlap and location situations. We have here a typical case of partial ignorance because the knowledge base has not been supplied with information as to which of these situations is the correct one in our world.*

The 'cat-mammal' Example 1..6 below shows what could happen in propositional calculus if we would allow the treatment of quantification problems as *IF-THEN* sentences (as we do in the chain-set logic).

Example 1..6 *Let* p *and* q *be the sentences*

$$p = x \textit{ is an instance of a cat} , \qquad q = x \textit{ is an instance of a mammal} , \quad (1.4)$$

and consider the following knowledge base information,

$$\begin{aligned} \textit{info-kb} &= (p \rightarrow q) \\ &= (\textit{IF} \quad x \textit{ is an instance of a cat} \\ &\quad \textit{THEN } x \textit{ is an instance of a mammal}). \end{aligned} \qquad (1.5)$$

Let us now direct, in turn, the following two questions at this knowledge base,[4]

$$qu\text{-}kb_1? = (p \rightarrow \neg q)?$$
$$= (IF \quad x \text{ is an instance of a cat} \qquad (1.6)$$
$$THEN \quad x \text{ is NOT an instance of a mammal})?$$

$$qu\text{-}kb_2? = (q \rightarrow \neg p)?$$
$$= (IF \quad x \text{ is an instance of a mammal} \qquad (1.7)$$
$$THEN \quad x \text{ is an NOT instance of a cat})?$$

According to the logic of natural language, and also according to correct mathematical reasoning, the answer to qu-kb$_1$ *is 'no' and that to qu-kb*$_2$ *is 'sometimes'. However, in traditional propositional calculus the answers to both questions are the same. The implication info-kb* $\rightarrow$*qu-kb*$_1$ *as well as the implication info-kb* $\rightarrow$*qu-kb*$_2$ *are both contingencies (i.e., neither tautologies nor contradictions).*

In contrast, in the M chain-set logic the answer to qu-kb$_1$ *is answ(qu-kb*$_1$*| info-kb)=0, corresponding to 'no'. And the answer to qu-kb*$_2$ *is answ(qu-kb*$_2$*| info)=*$0m$ *corresponding to ' 'no' or 'sometimes' '. The 'no' part of this answer takes care of the case in which 'cat' and 'mammal' should have the same meaning. Such an equality of meaning is possible according to info-kb.*
For the chain-set treatment of the cat-mammal example, see Sect. 9.4 of [17].

After introducing propositional calculus, traditional first order logic introduces the completely new field of predicate calculus. This calculus adds the symbols $\forall$ (for all) and $\exists$ (exists) to the logical theory, together with the use of variables. Variables are not allowed in the sentences (propositions) of propositional calculus, probably because of the wrong inferences which this could lead to in quantification problems, such as the cat-mammal Example 1..6 here.

However, Examples 1.1.5, 1.1.6 of the intelligent dolphins, (pp. 5-7 of [17]) show that predicate calculus is not always completely adequate, besides being complicated and very difficult to program. The chain-set logic treats quantification problems (also called 'is a' problems in AI) simply as *IF-THEN* sentences with variables (see abstract and Eqs. (1.9), (1.20)). The same holds for the structurally equivalent 'is in' and 'is a part of' structures. The Paris-Rome problem of Eqs. (1.2), (1.3) is actually an 'is in' problem.

One of the reasons mentioned in textbooks for the introduction of predicate calculus in addition to propositional calculus is that propositional calculus alone is not accepted as being able to handle sentences with variables, such

[4] *'kb'* stands for 'knowledge base'. For *IF-THEN* questions we must distinguish between *info* and *info-kb*, and between *qu* and *qu-kb*, see Sect. 3.2 here and Sect. 6.3 of [17].

as quantification sentences (see abstract). However, the Paris-Rome Example 1..5, as well as Examples 1..1-1..3, show that *IF-THEN* sentences without variables, which are treated by propositional calculus alone in first order logic, can also give rise to unreasonable inferences. It is therefore quite possible that it is the representation of the *IF-THEN* connective, not a possible, although at present forbidden, use of variables in propositional calculus which gives rise to the strange inferences of first order logic.

1.4 IF-THEN DIFFICULTIES IN FUZZY SET THEORY

Also fuzzy set theory has its share of difficulties in connection with *IF-THEN* sentences, and with inferences from such sentences.

According to Zadeh, [21] Section V, a fuzzy *IF-THEN* statement such as '*IF* x is small *THEN* y is big' is represented by a fuzzy *IF-THEN* relation R. Assuming that x and y are fuzzy subsets of the universes U and V respectively, the *IF-THEN* relation R is a fuzzy subset of the universe $U \times V$. The 2-dimensional relation R is a matrix whose elements are found by a pointwise operation on the grades of membership of A=small and C=big respectively.

Assigning a new fuzzy value to x, such as x=very small, one can then infer a new fuzzy value of y, using Zadeh's compositional rule of inference [21], Eqs. (5.16), (5.17). This rule is a 'max-min matrix multiplication' of x by R.

In later papers, Zadeh leaves open the choice between the above-mentioned formula for R and a possible new one (see [22], Eqs. (4.38), (4.39), or [10], Eqs. (6.2.1), (6.2.2)). The existence of Zadeh's two *IF-THEN* relations indicates that he is not completely satisfied with either of them.

The treatment of the *IF-THEN* statement in fuzzy logic has been discussed by many authors. A list of some of these up to 1991, can be found in papers by Hellendoorn [9], [8]. Hellendoorn discusses the modus ponens problem in detail. Later Dujet and Vincent [4] suggested still another *IF-THEN* relation.

In [10], Sect. 6.2, Example 2 and Fig. 15, Hisdal shows that both of Zadeh's *IF-THEN* relations, combined with his rule of inference, can give rise to dificulties. As an illustration, consider the universes $U = \{u_1, u_2, u_3, u_4\}$, $V = \{v_1, v_2, v_3, v_4\}$, and the *IF-THEN* statement '*IF*x is $.25/u_1$ *THEN* y is $1/v_2$'. Using the the modus ponens specification 'x is $.25/u_1$', it is then shown that both of Zadeh's *IF-THEN* relations result in the inference '$y = .25/v_1 + .25/v_2 + .25/v_3 + .25/v_4$', instead of the originally specified '$y = 1/v_2$'.

In [10] Sect. 6.3 and Fig, 18, Hisdal shows how this problem can be solved in the framework of the max-min fuzzy set theory by starting out with an interval-valued representation of the state of ignorance for an *IF-THEN* relation.

Hisdal's later TEE model interpretation of grades of membership $\mu_\lambda(u)$ as an estimate of the probability of assignment (to an object of exact attribute value u) of a label λ from a label set $\Lambda=\{\lambda_1, \ldots \lambda_L\}$, (see [16], and in more detail [11]-[15]) has, in part, been the inspiration for the probability logic of [17] which is summarized in Sects. (2.)-(7.) here. This logic also makes use of the state of ignorance for *IF-THEN* statements (Sect. 6., and [17] Fig. 12.4(b)), and it distinguishes even more clearly between uncertainty and ignorance.

1.5 IF-THEN DIFFICULTIES IN 'IS A' STRUCTURES OF AI

In artificial intelligence (see, e.g., [3], pp. 26, 342) quantification problems are often called *is a* problems (such as '*IF* Pooh **is a** bear *THEN* Pooh **is an** animal') and represented as tree structures. *is in* and *is a part of* problems are structurally equivalent to *is a* problems.

The Alex '*is a*' man-machine computer dialog system is described shortly in [17], pp. 380-386, including URL's for software. The system works well for pure tree structures, and it can also draw negative inferences such as '*IF* x is a dog, *THEN* x is *NOT* a cat' (see [17], Figs. 18.1, 18.3). Not all *is a* systems can draw negative inferences from *is a* trees.

The Alex system can also handle multiple partitions of a node; e.g., the partitioning of the node 'animal' into 'mammal', 'non-mammal' as well as into 'sea animal', 'land animal'. However, the dialog, which has to be preprogrammed in detail, becomes quite complicated in this case. Furthermore, information supply by the 'man' which is insufficient to pinpoint the exact location of a new node in the tree structure is thrown away; unless the informant is prepared to answer detailed preprogrammed questions asked by the system concerning the precise position of the node. The only way to avoid such throwing away of partial information in a tree structured *is a* system is to build up several trees. The number of trees grows exponentially with additional partial information supply concerning new nodes.

In contrast, in a corresponding chain-set representation of an *is a* system there is no difficulty in representing interwoven trees, such as possible multiple partitions of a node, if one follows the rules for building up an *is a* chain-set in the form of an *IF-THEN structure*. Insufficient information results in more than one probability row of the *is a* information chain-set. The man-machine dialog program can then devise its own questions to decide between the various probability rows. Alternatively the multiple probability rows are left untouched, although new information supply may erase one or more of them. Answers to questions must be computed separately for each remaining probability row. It can then happen that the answer is 'interval-valued', such as $0m$ ('no' or 'maybe'), or $m1$ ('maybe' or 'yes'), or $0m1$ ('don't know'). E.g., when the answer is 0 for one probability row of the *info* chain-set, and m for another, then we say that we have the interval-valued answer $0m$.

2. NON-(IF-THEN) CHAIN SETS

The chain-sets for five different labels λ are shown in Fig. 1.1, (b1)-(b5). Note that in contrast to a truth table, a chain-set should be read one chain (*column*) at a time.

The set of elements in the top left margin of a chain-set (above the probability row) is called the ground universe *GU* of the chain-set. E.g., in (b3)-(b5), we have $GU = \{A, B\}$. The elements of the ground universe *GU* are the non-composite components of the (possibly composite) label λ. They represent complete sentences which may contain variables. Thus we could have, e.g.,

$$A = \text{Margy is at home}, \quad B = \text{John is at home}, \tag{1.8}$$

or

$$A = x \text{ is an instance of a cat}, \quad B = x \text{ is an instance of a dog}. \tag{1.9}$$

In principle there need not be any structural similarity between the sentences A and B. We could have a chain-set whose first ground universe element is A of the first equation, and B of the second.

A '1' entry in a chain denotes the affirmed form of the element of the ground universe in the left margin, a '0' entry denotes its negated form. The elements of a given chain are, per definition, connected by an *AND* connective. E.g., the last chain in (b4) of Fig. 1.1 denotes the outcome *NOT A AND B* or $(0A, 1B)$ for short (see footnote 2).

The probability entry of a given chain *ch* denotes $P(ch|\lambda)$, the probability of occurrence of the chain *ch*, given the assertion of λ. E.g., the last probability entry of (b4) tells us that when λ=*A ORA B* is asserted, then there is a probability m, $0 < m < 1$, that $0A$, the negation of A, and $1B$, the affirmation of B will occur at the same time. In the terminology of the theory of probability, the *joint* probability of the outcome $(0A,\ 1B)$ is equal to m. The entries of the probability row of a chain-set must always add up to 1 because one of the chains *must* occur.

When a reliable informant supplies the information λ_1 = A *ORE* B to a new knowledge base, then the chain-set (b5) must be stored in the knowledge base. 0-1 chains which are left out in a chain-set are, per default, assigned the probability and possibility 0.

A new information supply λ_2 results in a knowledge base chain-set which is the conjunction (*AND* combination) of λ_1 and λ_2.

The universe or 'space' of the probability entries is the set of all possible 1-0 chains of length G, where G is the number of elements in the ground universe. This set has 2^G elements. G=1 for (b1)-(b2) in Fig. 1.1, for (b3)-(b5) we have G=2.

The meaning of an entry of the possibility row is $P(\text{yes-}\lambda|ch)$. E.g., the meaning of the last entry of the possibility row of (b5) is that when the information chain-set of the knowledge base consists of the single chain $\substack{0\\1}$, then the question *qu?*=λ?=*A ORE B?* directed at the knowledge base will be answered by 1 or 'yes'. In the terminology of the theory of probability, the set of chains with possibility entries 1 denotes an event, i.e. a subset of the set of all 0-1 chains. Here we consider only the case in which this subset is nonfuzzy. The case of a fuzzy event, in which the possibility entries may lie between 0 and 1, is discussed in in [17], Sect. 4.1.5.

Note that there exists a significant difference between the way of thinking about the truth table for λ, versus the chain-set table for the information supply λ. An element of the last column of a truth table tells us whether λ is true for a given combination of truth values of the components of λ. In the chain-set way of thinking, we wish to represent the statement λ in the knowledge base, *where λ is assumed to be true.* We are not interested in the cases in which λ is false. We could, of course, complete the chain-set for any label λ by adding all the missing 0-1 chains, and assigning the probability and possibility 0 to each of them. This would seem to make a chain *(column)* of a chain-set for a composite label λ (including its possibility value), completely equivalent to a *row* of the truth table (including the entry of the last column). However, such an equivalence does not hold for the implication, and therefore the inferences in propositional calculus and in the chain-set logic are not always equivalent.

In the following we summarize how the operations of conjunction (*AND*), disjunction (*ORA*) and negation (*NOT*) are performed in the chain-set logic. For details see chapter 3 of [17].

Consider the label λ=λ_1 *AND* λ_2, where at least one of the labels λ_1, λ_2 is a non-*IF-THEN* one. If the ground universes of λ_1 and λ_2 are not identical, then the ground universe and chains of these labels must be 'prolonged' into the ground universe $GU = GU_1 \cup GU_2$. The chain and chain-set of each label is then expanded into the new ground universe *GU*. E.g., when λ_1=A (see the chain-set (b1) of Fig. 1.1), then its expansion into the ground universe $\{A, B\}$ consists of the two chains $\substack{1\\0}$ and $\substack{1\\1}$ with probabilities m (see [17], Sects 3.6 and 12.2).

The chain-set for λ=λ_1 *AND* λ_2 then consists of the intersection of the set of 0-1 chains of λ_1 (with nonzero probability) with the set of 0-1 chains of λ_2. Each chain is then assigned the probability m (or 1 if it is the only chain in the chain-set for λ). When the intersection of chains is empty, then the label λ_1 *AND* λ_2 is contradictory ([17], Sect 4.2). Similarly the chain-set for λ_1 *ORA* λ_2 consists of the union of the sets of chains of λ_1 and λ_2.

As a simple example, we set λ_1=A, λ_2=B. The chain-set for λ_1 is shown in (b1) of Fig. 1.1. Its prolongation and expansion over the ground universe $\{A, B\}$ consists of the two chains $\substack{1\\1}$, and $\substack{1\\0}$. The chain-set of B over the same ground universe consists of the two chains $\substack{1\\1}$ and $\substack{0\\1}$. (b3), (b4) of Fig. 1.1 then show the chain-sets for the conjunction and disjunction of A and B according to the above intersection and union rules.

No prolongation and expansion is needed for finding the chain-set of *NOT* λ. It consists of all chains which are missing (have probability and possibility 0) in the chain-set of λ.

3. INFERENCES OR THE ANSWERING OF QUESTIONS

3.1 NON-(IF-THEN) QUESTIONS

Let *qu-kb* be a question directed at *info-kb*, the information stored in the knowledge base in the form of a chain-set. Both *qu-kb* and *info-kb* must be complete sentences which may, however, contain variables. Sequentially supplied information to the knowledge base can be considered as a single sentence which is the conjunction (*AND* connection) of all the supplied sentences.

The chain-set procedure for answering the question is to find the value of

$$answ = P(qu\text{-}kb\,|info\text{-}kb\,) = \sum_{ch} P(qu\text{-}kb\,|ch) \cdot P(ch|info\text{-}kb\,)\,. \tag{1.10}$$

This is the probability that *qu-kb* is true, given that *info-kb* is true.

Eq. (1.10) tells us that in order to find the answer value we must, for each chain *ch*, multiply the possibility value of that chain in the question chain-set by its probability value in the information chain-set; The answer value is then the sum of these products over all chains. The right hand side of Eq. (1.10) is called a possibility-probability dot product. Its value lies always in the real interval [0,1].

When the ground universes of *qu-kb* and *info-kb* are not identical, then one must shorten or prolong-and-expand *info-kb* before (1.10) is computed (see Sect. 6.2 of [17]).

When *answ*=1, then it can be replaced by 'yes'. When *answ*=0, then it can be replaced by 'no'. And when *answ*=m, then there exists uncertainty as to whether *qu-kb* is true.

As an example of the answer procedure of (1.10), let *info-kb*=*A ORA B*, and *qu-kb* =*A ORE B*. (b4) and (b5) of Fig. 1.1 show the chain-sets of these two. Taking the dot product of Eq. (1.10) we obtain

$$answ = P(qu\text{-}kb\,|info\text{-}kb\,) = 1 \cdot m + 1 \cdot m = m. \tag{1.11}$$

There exists uncertainty whether *A ORE B* occurs when *A ORA B* has occurred.

The two m's of the middle expression of the last equation are the probabilities of the two last chains of (b4). Their sum is smaller than 1 because (b4) contains still another chain with probability m, and the probabilities of all its three chains must add up to 1.

Applying the analogous procedure to the case in which *info-kb* =*A ORE B*, and *qu-kb* =*A ORA B*, we find that

$$answ = P(qu\text{-}kb\,|info\text{-}kb\,) = m \cdot 1 + m \cdot 1 = 1, \tag{1.12}$$

where the two m's are the probabilities of the only two chains of (b5). They must therefore add up to 1. The answer to *qu-kb* ?=*A ORA B*?, directed at *info-kb* =*A ORE B*, is therefore 1 or 'yes'.

3.2 IF-THEN QUESTIONS

Let *qu-kb* ?=*IF A THEN C*? be a conditional question directed at an information chain-set *info-kb*. *info-kb* can be a non-*IF-THEN* or an *IF-THEN* chain-set. According to natural logic, this question inquires about the probability that C is true in all those cases in which A is true.

To answer the question, we must therefore find the chain-set for

$$info = info\text{-}kb\ AND\ A \tag{1.13}$$

and direct the question *qu*=*C* at *info*. This gives the answer value

$$answ = P\,[\,C \mid (info\text{-}kb\ \wedge A)]. \tag{1.14}$$

Note that in contrast to non-*IF-THEN* questions, and also in contrast to the usual procedure of propositional calculus for *IF-THEN* questions, we have no use for an *IF-THEN* question chain-set, i.e. for a possibility row in an *IF-THEN* chain-set. All we need is the information chain-set for the conjunction *info*, Eq. (1.13) (constructed according to the intersection rule for the sets of prolonged and expanded chains, Sect. 2.), and the question chain-set (i.e. the possibility row) of *qu* =C. We then make use of Eq. (1.10), replacing *info-kb* by *info* of Eq. (1.13) and *qu-kb* by *qu*=*C*.

The entries of the following table show the answers to three different *IF-THEN* questions directed at two different information chain-sets according to this rule,

	qu-kb = $A \to C$	$A \to \neg C$	$\neg A \to C$
info-kb =			
$A \to C$	$1 \equiv$ 'yes'	$0 \equiv$ 'no'	$0m \equiv$ 'no' or 'sometimes'
$A \wedge C$	$1 \equiv$ 'yes'	$0 \equiv$ 'no'	contradictory *info* Eq. (1.13)

(1.15)

The information chain-set for *info-kb* =$(A \to C)$ is shown in Fig. 1.2, (b6). It has no possibility row, but two possible probability rows. For both of these, the answer to *qu-kb* $=(A\to\neg C)$ is 0 or 'no'. This is in contrast to propositional calculus in which $[(A\to C)\to(A\to\neg C)]$ is a contingency, not a contradiction. The answer to *qu-kb* $=(\neg A\to C)$ is m according to the first probability row of (b6), and 0 according to the second, resulting in the 'interval-valued' answer $0m$. This answer can be narrowed down by additional information supply as to whether $C\to A$ is valid in addition to $A\to C$. If it is, then the answer is 0.

An interesting case occurs for *info-kb* $=(A\wedge C)$ and *qu-kb* =*IF NOT A THEN C*?$=(\neg A\to C)$?. The *info* chain-set of Eq. (1.13) turns out to be contradictory in this case, it is a chain-set without chains (see [17] sect. 4.2.2 and fig. 4.2). In such a case, an automatic computer system for the storage of supplied information, and for the answering of questions directed at the stored information, must give a warning that the *IF-THEN* question *qu-kb* is nonsensical with respect to *info-kb* $=(A\wedge C)$. *info-kb* tells us that A is certain to occur. It is then nonsensical to ask about the probability of C when A does not occur. No such warning concerning a nonsensical *IF-THEN* question (with respect to *info-kb*) is given in propositional calculus in which $[(A\wedge C)\to(\neg A\to C)]$ is a tautology!

4. THE IF-THEN INFORMATION CHAIN SET

In this section we present and discuss the *IF-THEN* information chain-set. Its updating by new *IF-THEN* information is discussed both here and in Sect. 5..

The interpretation of the information supply

$$\textit{info} = \textit{IF A THEN C} \tag{1.16}$$

is, according to natural logic, equivalent to the statement that $1C$ occurs with certainty whenever $1A$ occurs, or, (see footnote 2 concerning $1A$, $0A$ etc.)

$$P(1C \mid 1A) = 1\,. \tag{1.17}$$

Since C can only occur or not occur,[5] it then follows immediately that,

$$P(0C \mid 1A) = 0\,. \tag{1.18}$$

The information chain-set of λ_6=*IF A THEN C* is shown in (b6), Fig. 1.2. The reason for the choice of this chain-set is given below.

We note that the *IF-THEN* chain-set has no possibility row, it can therefore not be used for answering the question '*IF A THEN C*?'. We have already seen in Sect. 3.2 that a chain-set representation of an *IF-THEN question* is not needed for answering it.

[5]It is frequently claimed that the law of the excluded middle does not hold in fuzzy set theory. See [16], Sect. 5.1, concerning this claim.

An important point in connection with an *IF-THEN* statement versus a non-*IF-THEN* one is the following.

A non-*IF-THEN* statement, such as λ_3, λ_4 or λ_5 of Fig. 1.1, defines, through the probability row of its chain-set, a probability distribution (in terms of 0, m, 1 probability values) over the 4-element chain universe $\{(1A, 1B), (1A, 0B), (0A, 1B), (0A, 0B)\}$. The elements of the probability row are the joint probabilities of A and B. When these are known, then it is possible to compute all the conditional, and all the marginal probabilities in $A \times B$. E.g., for (b4) $P(1B|1A|)$=m because the (b4) chain-set has two chains with with a 1 entry in the A row, and only one of them has a 1 entry in the B row. For (b5) $P(1B|1A)$=0, and for (b3) $P(1B|1A)$=1.

The marginal probability of, e.g., $1A$ is defined as the probability of occurrence of the outcome $1A$, irrespective of the outcome of B. For (b3), (b4), (b5) $P(1A)$ is equal to 1, m, m respectively, and $P(0A)$ is equal to 0, m, m.

An *IF A THEN C* statement specifies merely the value 1 for the single conditional probability $P(1C|1A)$, and consequently the value 0 for $P(0C|1A)$. Both probability rows of (b6), Fig. 1.2, satisfy these values.

Thus, the mere specification $P(1C|1A)$=1 does not allow us to compute all the other conditional probabilities. Neither does it allow us to compute the marginal probabilities $P(1A)$, $P(0A)$, $P(1C)$, $P(0C)$. Consequently it does not allow us to compute unique values for all the *four* entries of the joint probability row of the *IF-THEN* information chain-set.

There exist two determining reasons why all the marginal probabilities of an *IF-THEN* chain-set must be equal to m, not to 0 or 1. These are explained in Sects. 14.2.2, 14.2.3 of [17]. With this additional requirement, it turns out that the requirement $P(1C|1A)$=1 leaves open exactly the two probability rows of (b6). This result is derived in two different ways in Sects. 14.3.2, 14.3.3 of [17]. A single *IF-THEN* statement represents therefore a case of partial ignorance, or insufficient information supply. With the additional information $A \equiv C$ or $C \rightarrow A$, it turns out that only the second probability row applies (see [17], Fig. 9.1). When it is known that A and C have different meanings, then only the first row is left.

5. THE UPDATING OF PROBABILITIES AND OF IF-THEN BY IF-THEN

In chapter 11 of [17] it is demonstrated that there exist two types of updating of probabilities by new information supply. We have already encountered 'type 2 updating of probabilities'. It corresponds to the intersection-of-sets-of-(nonzero-probability)-chains rule of Sect. 2.. This type of updating can change an 'm' value of the probability of a specific chain to 0, and sometimes to 1. It can thus alter a given probability distribution.

Generalized modus ponens updating (see [17], p.290) is an example of updating of type 2. Thus, suppose that the knowledge base has been supplied with the information *IF A THEN C* where 'A=Margy is at home', and 'C=John is at home'. According to (b6), Fig. 1.2, $P(1A)=P(1C)=m$. The additional information supply 'Margy is at home' or '$P(1A)=1$' changes both probability rows of (b6) to the certainty distribution (1 0 0 0).

This alteration of the probability row may be acceptable for the specific example of the *IF A THEN C* statement concerning Margy and John, assuming that it refers to a given point of time only. If it is to be valid for any point of time, then updating of type 2 can result in loss of *IF-THEN* information. This is all the more so when A and C are quantification sentences with variables for which generalized modus ponens type 2 updating of *IF-THEN* information can lead to a loss of analytic (meaning-related) information. (See Sects. 9.2.1 and 15.1 of [17] where it is shown that type 2 updating of '*info* $_1$=All dinosaurs are reptiles' by '*info* $_2$=There exist no dinosaurs' erases the analytical information *info* $_1$.)

To avoid the loss of analytic information we must keep *IF-THEN* information in a special knowledge base chain-set called the *IF-THEN structure* of the knowledge base. This *IF-THEN* structure may be updated by new *IF-THEN* information to a new *IF-THEN* structure. In contrast, the result of updating the *IF-THEN*-structure by non-*IF-THEN* information can be used to update the non-*IF-THEN* chain-set of the knowledge base. But non-*IF-THEN* information is not allowed to update the *IF-THEN* structure itself. [6]

The updating of *IF-THEN* by *IF-THEN* must be of type 1. Type 1 updating consists of the intersection of the sets of probability rows of the two chain-sets. It can erase one or more probability rows, but those which are left, are not altered. In this way *IF-THEN* chain-set structures always retain marginal probabilities m for all elements of the ground universe. When the intersection of the two sets of probability rows is empty, then the labels of the two *IF-THEN* chain-sets are contradictory, and the information supply cannot be accepted.

As an example of updating of type 1, consider the updating of $\lambda_6=(A\rightarrow C)$ by $\lambda_7=(C\rightarrow A)$. In Fig 9.1 of [17] it is shown that the chain-sets for these two contain only one common probability row. This is the last row of (b6), Fig. 1.2. According to this row, the only nonzero probability chains of the chain-set for

[6] In praxis, one will probably use several *IF-THEN* structures. One for classification or *is a* information, one for *is in*, and one for *is a part of*; in addition to the *IF-THEN* structure for a sentence without variables. For each of the first three one can define one or more variables, e.g., x, just as one defines an integer or a real number in many programming languages. A variable is something which may be replaced by an element from some predefined set with the condition that in a specific application we must always replace x (or y or whatever) by the same element. Thereby one avoids strange definitions such as 'By an individual variable we mean the small italic "x"', to be found in many textbooks on logic (e.g., [7], p. 19).

λ=(A→C) *AND* (C→A) are $\substack{1\\1}$ and $\substack{0\\0}$. The chain-set of λ is therefore identical with the chain-set of $A \equiv C$.

6. THE STATES OF IGNORANCE AND THEIR UPDATING

We saw at the end of Sect. 2. that when a chain-set for a label λ_1 is *type 2* updated by the chain-set for a label λ_2 with a different ground universe, (by constructing the chain-set for 'λ_1 *AND* λ_2') then one or both chain-sets must first be prolonged into the union of the two ground universes and expanded. The operation of expansion for *type 1* updating, including the updating of an *IF-THEN* structure, is described on pp. 260, 340 of [17]. Such updating will, in general, lead to a chain-set with more than one probability row. A simple, though perhaps computationally inefficient way to look upon an expansion for type 1 updating is the following.

Suppose that we start out with a knowledge base to which no information has yet been supplied, A seemingly natural way to look upon the initial state of the knowledge base is to say that the storage space for its information is empty.

An alternative way to look at the initial state of the knowledge base is to say that it represents the state of complete ignorance over a given ground universe. Its chain-set consists of all possible probability rows over its universe of 0-1 chains. Every truly new information supply erases one or more of these rows. Seen from this point of view, information supply has the effect of narrowing down the set of possible outcomes, each outcome being a probability *distribution* (in terms of 0, m, 1 probability values). This point of view agrees with that of Shannon's information theory according to which information supply decreases the informational entropy (the uncertainty) of the system.

As an example, consider again an information supply λ_1 over GU_1, followed by an information supply λ_2 over GU_2. The initial state of complete ignorance with respect to these ground universes is then a chain-set over $GU=GU_1 \cup GU_2$ with all possible 0, m, 1 probability rows for its 0-1 chains. There exist 3^I ways to assign one of the three values to each of the $I=2^G$ 0-1 chains (G=number of elements in *GU*). However, the condition that each probability row must sum up to 1 reduces this number. E.g., a probability row cannot contain a 1 entry, and another nonzero entry. Neither can it contain a single m entry, all the other entries being 0. And it cannot contain only 0 entries.

It can then be shown that the state of complete ignorance for a non-*IF-THEN* chain-set contains $2^I - 1$ probability rows. This number is further reduced for an *IF-THEN* chain-set which has the additional requirement that the two marginal probabilities of each element of the ground universe must be m for each probability row.

The non-*IF-THEN* and *IF-THEN* states of complete ignorance for a 2-element ground universe are shown in [17], Fig. 12.4 (a) and (b) respectively. Erasing from (b) all probability rows for which $P(1B|1A) \neq 1$, we get the 2-probability-row *IF-THEN* chain-set of Fig. 1.2.

The following quantification example illustrates another case of partial ignorance. It is treated in more detail in Sect. 14.6.2 of [17]. Let

$$\begin{aligned} \textit{info} &= \text{All } A\text{'s are } C\text{'s } \textit{AND} \text{ all } B\text{'s are } C\text{'s} \\ &= (A \overset{\text{proper}}{\rightarrow} C) \wedge (B \overset{\text{proper}}{\rightarrow} C)\,, \end{aligned} \tag{1.19}$$

The italic A of the last row stands for 'x is an instance of class A', and similarly for B and C. The superscript *'proper'* above the *IF-THEN* arrow in $A{\rightarrow}C$ indicates that $A \equiv C$ is not valid.

The *IF-THEN* structure for *info* turns out to have 5 probability rows,

$$\textit{info} = (A \overset{\text{proper}}{\rightarrow} C) \wedge (B \overset{\text{proper}}{\rightarrow} C)$$

A	1	1	1	1	0	0	0	0
B	1	1	0	0	1	1	0	0
C	1	0	1	0	1	0	1	0
prob row 1)	0	$\boxed{0}$	m	$\boxed{0}$	m	$\boxed{0}$	0	m
prob row 2)	m	$\boxed{0}$	m	$\boxed{0}$	0	$\boxed{0}$	m	m
prob row 3)	m	$\boxed{0}$	0	$\boxed{0}$	m	$\boxed{0}$	m	m
prob row 4)	m	$\boxed{0}$	m	$\boxed{0}$	m	$\boxed{0}$	m	m
prob row 5)	m	$\boxed{0}$	0	$\boxed{0}$	0	$\boxed{0}$	m	m

. (1.20)

Each row satisfies both $(A \overset{\text{proper}}{\rightarrow} C)$, and $(B \overset{\text{proper}}{\rightarrow} C)$. The five different rows correspond to the five different classification trees which *info* of Eq. (1.19) can generate. E.g., row 1) corresponds to the case in which A and B are disjoint, such as for A=dog, B=cat, C=mammal. Row 2) corresponds to the case of $B \subset A \subset C$, such as for A=dog, B=terrier, C=mammal. Row 4) corresponds to a partial overlap between A and B, such as for A=mammal, B=sea animal, C=animal. *info* contains insufficient information to determine which of the five rows (5 classification trees) applies to the information supplied by the informant. However, *info* does supply *some* information which is not lost when the information supply $(A \overset{\text{proper}}{\rightarrow} C)$ is followed by $(B \overset{\text{proper}}{\rightarrow} C)$. In the Alex '*is a*' system the $(B \overset{\text{proper}}{\rightarrow} C)$ information supply is lost unless the informant is ready to answer questions posed by the system (see Sect. 1.5).

7. CONCLUSION

Many-valued logics with intermediate truth values, including fuzzy set theory, usually try to make their formulas agree with those of classical 2-valued logic in the limiting case of truth (or membership) values of 1 and 0.

However, we demonstrated in Sect. 1.3 that classical logic has its flaws in connection with *IF-THEN* sentences, and in certain cases of inferences. If classical logic does not work for special cases of inferences, then its interpolation to the much more complicated case which operates with imprecise values of variables (such as the height value 'tall man') cannot be expected to work in all cases either.

The chain-set logic of this chapter is built on top of 2-valued logic. It makes use of both probabilities and possibilities, and it solves the *IF-THEN* problem of propositional calculus. As a result, its *IF-THEN* chain-set can also be applied to quantification problems, no 'predicate calculus' is needed. Furthermore the chain-set logic distinguishes between uncertainty (symbolized by a probability value m) on the one hand, versus ignorance or insufficient information supply on the other. Ignorance is symbolized by a chain-set with more than one probability row, and by the possibility of interval-valued answers, to questions; such as $m1$, in the case when two different probability rows of the information chain-set give the answers m and 1 respectively. An example is the case '*info* =All dogs are animals' (without specifying whether the words'dog' and 'animal' have the same meaning), and the question 'Is an animal a dog?'. This case differs from an uncertainty answer m, such as in the case of the information 'Some animals are dogs' (see Fig 10.1 of [17]) and the question 'Is an animal a dog?'.

References

[1] Aristotle. *Prior and Posterior Analytics.* Everyman's Library, Dutton:New York, 1964; original ≈344 B.C. Edited and translated by John Warrington.

[2] Bible. *The Holy Bible.* Collins' Clear Type Press. King James Translation, Genesis, chapter 11.

[3] Charniak, Eugene and McDermott, Drew. *Introduction to Artificial Intelligence.* Addison-Wesley, 1985.

[4] Dujet, Ch. and Vincent, N. Force implication: A new approach to human reasoning. *Fuzzy Sets and System*, 69:53–63, 1995.

[5] Fenstad, Jens Erik and Normann, Dag. *Innfring i Matematisk Logikk.* Institute of Mathematics, University of Oslo, 1988. In Norwegian.

[6] Frege, Gottlob. *Translations from the Philosophical Writings of Gottlob Frege.* Basil Blackwell, Oxford, third edition, 1980. Edited by Peter Geach and E.D. Black.

[7] Hatcher, William S. *The Logical Foundations of Mathematics.* PWS-Kent Publishing Company, Boston, 1989.

[8] Hellendoorn, H. Fuzzy logic and the generalized modus ponens. *Int. J. of Approximate Reasoning*, 1990 or 1991.

[9] Hellendoorn, H. The generalized modus ponens considered as a fuzzy relation. *Fuzzy Sets and Systems*, 46:29–48, 1992.

[10] Hisdal, E. The IF THEN ELSE statement and interval-valued fuzzy sets of higher type. *Int. J. Man-Machine Studies*, 15:385–455, 1981.

[11] Hisdal, E. A theory of logic based on probability. Research Report 64, Institute of Informatics, University of Oslo, Box 1080 Blindern, 0316 Oslo 3, Norway, 1984. ISBN 82-90230-60-5.

[12] Hisdal, E. Infinite-valued logic based on two-valued logic and probability, part 1.1. Difficulties with present-day fuzzy set theory and their resolution in the TEE model. *Int. J. Man-Machine Studies*, 25:89–111, 1986.

[13] Hisdal, E. Infinite-valued logic based on two-valued logic and probability, part 1.2. Different sources of fuzziness and uncertainty. *Int. J. Man-Machine Studies*, 25:113–138, 1986.

[14] Hisdal, E. Infinite-valued logic based on two-valued logic and probability, part 1.3. Reference experiments and label sets. Research Report 147, Institute of Informatics, University of Oslo, Box 1080 Blindern, 0316 Oslo 3, Norway, 1988,1990. ISBN 82-7368-053-3. Can also be found on http://www.ifi.uio.no/~ftp/publications/research-reports/Hisdal-3.ps.

[15] Hisdal, E. Infinite-valued logic based on two-valued logic and probability, part 1.4. The TEE model. Research Report 148, Institute of Informatics, University of Oslo, Box 1080 Blindern, 0316 Oslo 3, Norway, 1988,1990. ISBN 82-7368-054-1. Can also be found on http://www.ifi.uio.no/~ftp/publications/research-reports/Hisdal-4.ps.

[16] Hisdal, E. Open-mindedness and probabilities versus possibilities. In Ruan, Da, editor, *Fuzzy Logic Foundations and Industrial Applications*, pages 27–55. Kluwer Academic Publishers, Boston, 1996.

[17] Hisdal, Ellen. *Logical Structures for Representation of Knowledge and Uncertainty*. Physica Verlag, A Springer-Verlag Company, 1998. ISBN 3-7908-1056-8 Physica Verlag Heidelberg New York.

[18] Kleene, S.C. *Mathematical Logic*. John Wiley, New York, London, 1968.

[19] Popper, K.R. *Conjectures and Refutations*. Routledge and Kegan Paul, London, 3rd edition, 1969.

[20] Suppes, Patrick. *Introduction to Logic*. D. Van Nostrand Company, 1957.

[21] Zadeh, L.A. Outline of a new approach to the analysis of complex systems and decision processes. *IEEE Transactions on Systems, Man and Cybernetics*, SMC-3:28–44, 1973.

[22] Zadeh, L.A. Pruf–a meaning representation language for natural languages. *Int. J. Man-Machine Studies*, 10:395–460, 1978.

Chapter 2

FUZZY PREDICATE CALCULUS AND FUZZY RULES

Petr Hájek
Institute of Computer Science
Academy of Sciences
182 07 Prague, Czech Republic
hajek@uivt.cas.cz

Abstract The basic many-sorted fuzzy predicate calculus $BL\forall$ is presented and used to express and prove logical properties of "fuzzy IF–THEN rules".

1. FUZZY PREDICATE CALCULI

[1] The notion of a *t-norm* (triangular norm) as a possible semantics of fuzzy conjunction is well known in the fuzzy community: it is a binary operation $*$ on the real unit interval $[0,1]$ which is commutative, associative, non-decreasing in both arguments and has 1 as its unit element. If it is a continuous function we speak of a continuous t-norm. The following three examples are particularly important:

$$x * y = \max(0, x + y - 1) \quad \text{(Łukasiewicz } t\text{-norm)},$$
$$x * y = \min(x, y) \quad \text{(Gödel } t\text{-norm)},$$
$$x * y = x \cdot y \quad \text{(product } t\text{-norm)}.$$

Each continuous t-norm defines its *residuum* $\Rightarrow$ which is a unique operation defined as follows:

$$x \Rightarrow y = \max\{z | x * z \leq y\}.$$

---

[1]This chapter is a revised version of the paper Hájek, 1998a.Thanks are due to Professor S. Gottwald for his permission to publish the material here.

Note that $x \Rightarrow y = 1$ iff $x \leq y$; for $x > y$ the residua of the above t-norms are

$$\begin{array}{ll} x \Rightarrow y = 1 - x + y & \text{(Łukasiewicz)}, \\ x \Rightarrow y = y & \text{(Gödel)}, \\ x \Rightarrow y = y/x & \text{(product)}. \end{array}$$

If you choose a continuous t-norm for the semantics of your conjunction, its residuum is (should be) your preferred choice of semantics of your implication. More than that: you can define the semantics $(-)$ of your negation by $(-)x = x \Rightarrow 0$ (x implies falsity); this gives $(-)x = 1 - x$ for Łukasiewicz, but for both Gödel and product we get $(-)0 = 1, (-)x = 0$ for x positive.

In my book Hájek, 1998b I systematically develop propositional and predicate calculi based on continuous t-norms and apply them to an analysis of fuzzy approximate reasoning, fuzzy quantifiers etc. Here I sketch directly the many-sorted fuzzy predicate calculus and its use for understanding "fuzzy rules". (Some knowledge of classical predicate calculus will be helpful.)

The language. We want our language to be able to express properties of objects of various sorts (pressure, temperature,..); let us have finitely many *sorts* $s, \ldots, t$. A *type* is a finite sequence of sorts. We have *predicates* $P, Q, \ldots$, each having a type. (A predicate of type (s) will describe a property of objects of type s; a predicate of the type (s, t) will describe a relation among objects of type s and objects of type t etc.) We shall have *variables* and *constants*; each variable and constant has a sort. *Atomic formulas* have the form $P(x_1, \ldots, x_n)$ where P is a predicate of type $(s_1, \ldots s_n)$ and each x_i is a variable or a constant of the sort s_i. Also $\bar{0}$ is considered to be an atomic formula (FALSE). We have one truth constant $\bar{0}$ (for falsity), two binary connectives $\&, \rightarrow$ (conjunction and implication) and two quantifiers: $\forall$ (for all), $\exists$ (there exists). Formulas are built from atomic formulas using connectives and quantifiers, i.e., if φ, ψ are formulas and x is a variable (of any sort) then $\varphi \rightarrow \psi, \varphi \& \psi, (\forall x)\varphi, (\exists x)\varphi$ are formulas. We write

$$\neg\varphi \text{ for } \varphi \rightarrow \bar{0},$$

$$\varphi \wedge \psi \text{ for } \varphi \& (\varphi \rightarrow \psi)$$

$$\varphi \vee \psi \text{ for } ((\varphi \rightarrow \psi) \rightarrow \psi) \wedge ((\psi \rightarrow \varphi) \rightarrow \varphi)$$

$$\varphi \equiv \psi \text{ for } (\varphi \rightarrow \psi) \& (\psi \rightarrow \varphi).$$

An *interpretation* of a language is given by the following things:

- for each sort s, a non-empty set M_s of objects of sort s,
- for each predicate P of type $(s_1, \ldots, s_n)$ a fuzzy relation r_P on $M_1 \times \ldots \times M_n$, i.e., a mapping assigning the each n-tuple $(a_1, \ldots, a_n)$ of objects of respective sorts the degree $r_P(a_1, \ldots, a_n)$ of membership to the relation r_P.
- for each constant c of sort s, an element $m_c \in M_s$. We write

$$\mathbf{M} = ((M_s)_{s \text{ sort}}, (r_P)_{P \text{ predicate}}, (M_c)_{c \text{ constant}}).$$

Let us pause for a trivial example: s is the sort of concert goers and t the sort of music composers. $M_s = \{u, v, w\}$, $M_t = \{b, d, h\}$. We have a predicate *likes* of the type (s, t); the following is the relation r_{likes}

	b	d	h
u	1.0	0.7	0.9
v	0.3	0.9	0.1
w	0.4	0.5	0.8

Finally, we shall have a constant *Bach* for b. The *truth value* of a formula φ depends in general on three things: the interpretation $\mathbf{M}$, the t-norm $*$ giving meaning to connectives, and an evaluation e of variables and constants by objects of the corresponding sort. If x_i is a constant c then e must satisfy $e(x_i) = m_c$ (the meaning is fixed by $\mathbf{M}$). The truth value of φ determined by given $*, \mathbf{M}, e$ is denoted by $\|\varphi\|^*_{\mathbf{M},e}$.

The following defines $\|\varphi\|^*_{\mathbf{M},e}$:

$$\|P(x_1, \ldots, x_n)\|^*_{\mathbf{M},e} = r_P(e(x), \ldots e(x_n));$$

$$\|\varphi \& \psi\|^*_{\mathbf{M},e} = \|\varphi\|^*_{\mathbf{M},e} * \|\varphi\|^*_{\mathbf{M},e}$$

$$\|\varphi \to \psi\|^*_{\mathbf{M},e} = \|\varphi\|^*_{\mathbf{M},e} \Rightarrow \|\psi\|^*_{\mathbf{M},e}$$

$$\|(\forall x)\varphi\|^*_{\mathbf{M},e} = \inf_{e_x} \|\varphi\|^*_{\mathbf{M},e_x}$$

$$\|(\exists x)\varphi\|^*_{\mathbf{M},e} = \sup_{e_x} \|\varphi\|^*_{\mathbf{M},e_x}$$

where e_x runs over all evaluations differing from e at most in the value for the argument x. The example helps: let x be of the sort s, and y, z of the sort t, each $e(x) = u, e(y) = h, e(z) = d$. Then

$$\|likes(x, Bach)\|^*_{\mathbf{M},e} = 1,\ \|likes(x, y)\|^*_{M,e} = 0.9,\ \|likes(x, z)\|^*_{M,e} = 0.7$$

independently of $*$; if $*$ is the Łukasiewicz t-norm then

$$\|likes(x,y) \to likes(x,z)\|^*_{M,e} = 1 - 0.9 + 0.7 = 0.8;$$
$$\|(\exists y)likes(x,y)\|^*_{M,e} = \sup(1, 0.7, 0.9) = 1;$$
$$\|(\forall x)(\exists y)likes(x,y)\|^*_{M,e} = \inf(1, 0.9, 0.8) = 0.8;$$
$$\|(\forall x)likes(x, Bach)\|^*_{\mathbf{M},e} = \inf(1, 0.3, 0.4) = 0.3$$

The truth value of the last two formulas is independent of e (all variables are bound by quantifiers) and of $*$ (no connectives occur).

Note also that for each continuous t-norm $*$,

$$\|\varphi \wedge \psi\|^* = \min(\|\varphi\|^*, \|\psi\|^*),$$
$$\|\varphi \vee \psi\|^* = \max(\|\varphi\|^*, \|\psi\|^*),$$

(indices $\mathbf{M}, e$ deleted); furthermore, $\|\varphi \equiv \psi\|^* = 1$ iff $\|\varphi\|^* = \|\psi\|^*$.

A formula φ is a $*$-*tautology* if $\|\varphi\|^*_{\mathbf{M},e} = 1$ for each $\mathbf{M}, e$; φ is a *tautology* (or: a t-tautology, t-norm-tautology) if it is a $*$-tautology for each continuous t-norm $*$. The following formulas are tautologies; we take them for the axioms of the basic (many-sorted) fuzzy predicate logic $BL\forall$:

Axioms for connectives:

(A1) $(\varphi \to \psi) \to ((\psi \to \chi) \to (\varphi \to \chi))$
(A2) $(\varphi \& \psi) \to \varphi$
(A3) $(\varphi \& \psi) \to (\psi \& \varphi)$
(A4) $(\varphi \& (\varphi \to \psi)) \to (\psi \& (\psi \to \varphi))$
(A5a) $(\varphi \to (\psi \to \chi)) \to ((\varphi \& \psi) \to \chi)$
(A5b) $((\varphi \& \psi) \to \chi) \to (\varphi \to (\psi \to \chi))$
(A6) $((\varphi \to \psi) \to \chi) \to (((\psi \to \varphi) \to \chi) \to \chi)$
(A7) $\bar{0} \to \varphi$

Axioms for quantifiers:

($\forall$1) $(\forall x)\varphi(x) \to \varphi(y)$
($\exists$1) $\varphi(y) \to (\forall x)\varphi(x)$
($\forall$2) $(\forall x)(\chi \to \psi) \to (\chi \to (\forall x)\psi)$
($\exists$2) $(\forall x)(\varphi \to \chi) \to ((\exists x)\varphi \to \chi)$
($\forall$3) $(\forall x)(\varphi \vee \chi) \to ((\forall x)\varphi \vee \chi$

where y is a constant or a variable substitutable for x in φ and the formula χ does not contain free occurrences of x. (If the reader does not understand these conditions he/she just accepts that they just eliminate some "pathological" cases.)

Example of $(\forall 2)$:

$$(\forall x)((\forall y)P(y) \to Q(x)) \to ((\forall y)P(y) \to (\forall x)Q(x)).$$

There are two *deduction rules:* modus ponens – from φ and $(\varphi \to \psi)$ derive ψ; and generalization – from φ derive $(\forall x)\varphi$. A *proof* in $BL\forall$ is a sequence $\varphi_1, \ldots, \varphi_n$ of formulas such that each φ_i is an axiom of $BL\forall$ or is derived from some previous members of the sequence. A formula φ is *provable* in $BL\forall$ if it is the last member of a proof.

If these definitions are entirely new for the reader, he/she should just observe that we take some tautologies for axioms and define some deduction rules; it can be shown that they are *truth-preserving,* i.e., – produce tautologies from tautologies. It follows immediately that each formula φ provable in $BL\forall$ is a tautology, i.e., the truth value of φ is always 1 (for each $*, \mathbf{M}, e$). To be able to construct interesting proofs in $BL\forall$ one needs some training; let us only state that very many interesting tautologies are provable in $BL\forall$.[2]

2. THEORIES AND MODELS; SIMILARITIES AND FUZZY FUNCTIONS

A *theory* is given by its set of axioms – here are some particular formulas. A *proof* in a theory T (over the logic $BL\forall$) is a sequence $\varphi_1, \ldots, \varphi_n$ of formulas such that each φ_i is a logical axiom *or an axiom of* T or it is derived from some previous members of the sequence by a deduction rule. An interpretation $\mathbf{M}$ is a $*$-*model* of a theory T if each axiom α of T is $*$-*true* in $\mathbf{M}$, i.e., $\|\alpha\|^*_{\mathbf{M},e} = 1$ for each e. (We write $\|\alpha\|^*_{\mathbf{M}} = 1$.) If $\mathbf{M}$ is a $*$-*model* of T and φ is provable in T (notion: $T \vdash \varphi$) then $\|\varphi\|^*_{\mathbf{M}} = 1$: each formula provable in T is $*$-true in each $*$-model of T.

We shall study some important theories. First, let $\approx$ be a binary predicate of type (s, s), let x, y, z be variables of the sort s. The following are *axioms of similarity* for $\approx$ and the sort s:

$$x \approx x \text{ (reflexivity)}$$

$$x \approx y \to y \approx x \text{ (symmetry)}$$

$$(x \approx y \& y \approx z) \to x \approx z \text{ (transitivity)}.$$

[2] It can be shown that $BL\forall$ does not prove all tautologies (in the present meaning of the word "tautology"); moreover, tautologies are not recursively axiomatizable. (Not yet published.) But one can prove completeness of $BL\forall$ will respect to a modified notion of a tautology, admitting besides truth functions given by continuous t-norms more general truth functions (not necessary on $[0, 1]$), given by so-called BL-algebras. (See Hájek, 1998b for details.)

A fuzzy binary relation r on M (i.e., $r : M^2 \rightarrow [0,1]$) is a $*$-similarity on M if (M, r) is a $*$-model of the above axioms. For example let $*$ be Łukasiewicz t-norm, let $M = [0,1]$ and $r_{\approx}(u,v) = \max(0, 1 - k|u - v|)$ (for some positive k); then $r_{\approx}$ is an Ł-similarity on $[0,1]$.

Note that if we have various sorts then similarity predicates for different sorts should be, pedantically, denoted by different symbols; but for simplicity we use just one $\approx$ and the sort is apparent from the variables used.

Let P be a predicate of type $(s, \ldots, t)$ and let T contain axioms of similarity for $\approx$. The *congruence axiom* for $P, \approx$ is

$$(x \approx x' \& \ldots \& y \approx y') \rightarrow (P(x, \ldots, y) \equiv P(x', \ldots, y')).$$

(Thus P *respects* the similarity.)

Now we come to our main notion of a *fuzzy function.* Let F be a predicate of type (s, t), let $\approx$ be a similarity predicate in T. The *axioms of a fuzzy function* for F (with respect to $\approx$) are the congruence axiom for $F, \approx$ (F respects the similarity) and

$$(F(x,y) \& F(x,y')) \rightarrow y \approx y'$$

(any two images of x are similar).

Figure 1 shows an Ł-fuzzy function – a fuzzy relation $r_F : [0,1] \times [0,1] \rightarrow [0,1]$ such that $(([0,1],[0,1]), r_F, r_{\approx})$ is an Ł-model of similarity and of a fuzzy function, $r_{\approx}$ being *as above* (with $k = 4$ on the first copy of [0,1] and $k = 2$ on the second). Note that if $G(x,y)$ means $y = x^2$ (crisp relation!) then our F satisfies $F(x,y) \equiv (\exists x', y')(G(x',y') \& x \approx x' \& y \approx y')$.

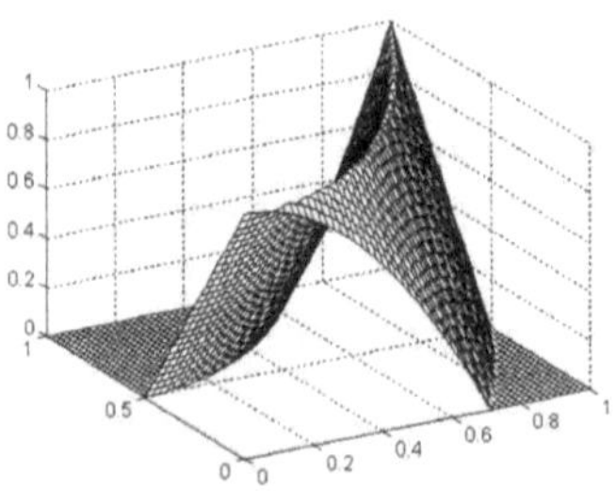

Figure 1.1 Fuzzy function.

3. ON FUZZY RULES

Imagine now we consider a fuzzy structure **M**, interpretation of our predicate language with a similarity named $\approx$ and a fuzzy function named F (the type of F being (s,t), where s,t are two sorts). Thus we know that the above axioms are true in **M** (more precisely, $*$-true for a chosen t-norm $*$). We do not know the meaning of F precisely; but we know some *examples*: let $c_1,\dots,c_n$ name some elements of the sort s, let $d_1,\dots,d_n$ name some elements of the sort t and assume $F(c_1,d_1),\dots,F(c_n,d_n)$ be true in **M**. (Thus: it is true in degree 1 that F maps c_i to d_i.) Add these formulas

$$F(c_1,d_1),\dots,F(c_n,d_n)$$

to our axioms (for $\approx$ and F) and define

$$A_i(x) \equiv x \approx c_i,\ B_i(x) \equiv y \approx d_i$$

(thus $A_1(x)$ says x is similar to the first example of an argument etc.). Now we may prove (in $BL\forall$ from the axiom accepted) that

$$(A_i(x)\&B_i(y)) \to F(x,y)$$

and

$$F(x,y) \to (A_i(x) \to B_i(y))$$

(in words: if x is similar to c_i and y is similar to d_i then $F(x,y)$; if $F(x,y)$ then if x is similar to c_i then y is similar to d_i). Using provable properties of implication we can prove the following in our theory:

$$\bigvee_{i=1}^{n} (A_i(x)\&B_i(y)) \to F(x,y),$$

$$F(x,y) \to \bigwedge_{i=1}^{n} (A_i(x) \to B_i(y)).$$

The formula $\bigvee_{i=1}^{n}(A_i(x)\&B_i(y))$ will be called MAMD(x,y) (Mamdani's formula) and the formula $\bigwedge_{i=1}^{n}(A_i(x) \to B_i(y))$ will be called RULES(x,y). Observe that MAMD is a disjunction of &-conjunctions (and recall that disjunction is evaluated by maximum) and RULES is a min-conjunction of implications.

In figure 2 find the picture[3] of MAMD(x,y) and RULES(x,y) for 1, 3, 5 and 9 examples.

[3]Thanks are due to David Coufal for the preparation of the figures in this chapter.

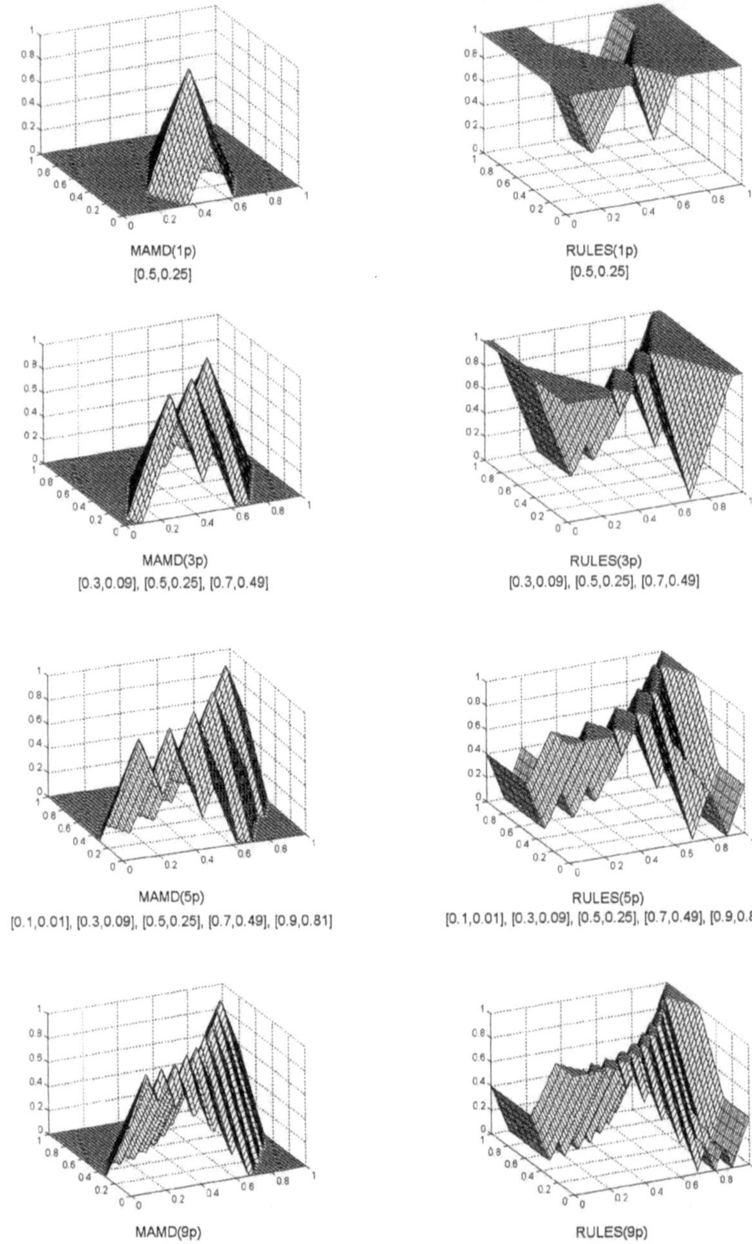

Figure 1.2 MAMD, RULES.

Thus the formula MAMD(x, y) defines a lower approximation of the fuzzy function F and RULES(x, y) defines an upper approximation. How tight are they?

Our axioms allow to prove (over $BL\forall$) the following:

$$\bigvee_i A_i^2(x) \rightarrow [\text{MAMD}(x, y) \equiv \text{RULES}(x, y)].$$

Here $A_i^2(x)$ stands for $A_i(x) \& A_i(x)$ and may be read "x is *very* A_i". Thus the above formula is $*$-true in each $*$-model of our axioms, which means that for each meaning of x, the truth degree of $\bigvee_i A_i^2(x)$ is the lower bound of the truth degree of the equivalence MAMD$(x, y) \equiv$ RULES(x, y). Coming back to our definition of $A_i(x)$, we may read the formula as follows: the more it is true that x is very similar to one of our examples, still more true is (for each y) that MAMD$(x, y) \equiv$ RULES(x, y).

Proofs of all the above formulas (from the chosen axioms, over $BL\forall$) are fully presented in Hájek, 1998b, where various further logical properties of "fuzzy rules" are proved. We hope to have contributed to a better understanding of logical properties of "approximate fuzzy inference". Note that our approach is largely inspired by Kruse et al., 1993 (even if the notion of a fuzzy function seems to occur only in Hájek, 1998b). For references to other analyses of fuzzy rules see Hájek, 1998b p. 177.

References

Hájek, P. (1998a). Fuzzy predicate calculus and fuzzy rules. In *Fuzzy-Neuro Systems 99 (Ed.: Brewka G. et. all)*, pages 1–8. Leipziger Universitätsverlag, Leipzig.

Hájek, P. (1998b). *Metamathematics of fuzzy logic.* Kluwer Academic Publishers, Dordrecht.

Kruse, R., Gebhart, J., and Klawonn, F. (1993). *Fuzzy-Systeme.* B. E. Teubner, Stuttgart.

Chapter 3

THE GENERALIZED MODUS PONENS IN A FUZZY SET THEORETICAL FRAMEWORK

Chris Cornelis, Martine De Cock* and Etienne E. Kerre
Department of Applied Mathematics and Computer Science
Ghent University
Krijgslaan 281 (S9), B-9000 Ghent, Belgium
{Chris.Cornelis, Martine.DeCock, Etienne.Kerre}@rug.ac.be

Abstract Starting from an input fuzzy set and an IF–THEN rule, implementations of the Generalized Modus Ponens (GMP) in a fuzzy set theoretical framework allow the derivation of an output fuzzy set. If the GMP is implemented by means of the Compositional Rule of Inference (CRI) in general this derivation involves a lot of computational efforts, and the shape of the resulting membership function often seems quite arbitrary compared to that of the input membership function. In this chapter we present a review of techniques, generating as their output a fuzzy set belonging to a predefined class, that are not afflicted with these disadvantages.

Keywords: generalized modus ponens, compositional rule of inference, closed system, linguistic hedge

1. INTRODUCTION

The modus ponens (MP)

$$\begin{array}{l} \text{x is } A \\ \text{IF x is } A \text{ THEN y is } B \\ \hline \text{y is } B \end{array} \qquad \text{(SCHEME–1)}$$

*Supported by the Fund for Scientific Research Flanders (FWO) is a well–known deduction rule in (boolean) logic. From the fact "x is A" and the IF–THEN rule "IF x is A THEN y is B," we can derive a new fact, namely "y is B." However, if we do not exactly know that "x is A," we cannot make any deduction concerning y, even if we would have tons of other information on x. This implies that if we want to develop a useful derivation system (e.g., for a computer–controlled car) based on this MP, we have to provide an IF–THEN rule for each possible A. Needless to say this would be highly inefficient, if not impossible.

A better solution would be to use an alternative derivation rule, called the generalized modus ponens (GMP):

$$\frac{\begin{array}{l}\text{x is } A' \\ \text{IF x is } A \text{ THEN y is } B\end{array}}{\text{y is } B'} \qquad \text{(SCHEME–2)}$$

In this scheme it is possible to make a deduction concerning y even if the fact we know about x (namely "x is A'") does not exactly match the premise of the rule (namely, "x is A"). B' will of course depend on A', A and B. We postpone the discussion of how to derive B' to the following sections.

If x is a variable taking its values in a universe X, and y a variable taking its values in a universe Y, then in a fuzzy set theoretical framework, A' and A are usually represented by fuzzy sets on X, while B and B' are modelled by fuzzy sets on Y.

Definition 1 (fuzzy set) *A fuzzy set A on a universe X is a $X - [0,1]$ mapping. This mapping is often referred to as the membership function of A. For all x in X, $A(x)$ is called the degree of membership of x in A. The class of all fuzzy sets on X is denoted $\mathcal{F}(X)$. A fuzzy relation from a universe X to a universe Y is a fuzzy set on $X \times Y$, i.e., an element of $\mathcal{F}(X \times Y)$. Furthermore for A and B in $\mathcal{F}(X)$:*

$$A \subseteq B \Leftrightarrow (\forall x \in X)(A(x) \leq B(x))$$

$$support(A) = \{x | x \in X \text{ and } 0 < A(x)\}$$

$$shell(A) = \{x | x \in X \text{ and } 0 < A(x) < 1\}$$

$$(\forall x \in X)((A \cup B)(x) = max(A(x), B(x)))$$

$$(\forall x \in X)((A \cap B)(x) = min(A(x), B(x)))$$

In applications A' corresponds either to a crisp object a of the universe (e.g.,a temperature, a speed,...) or to a linguistic term <term> (e.g., *"warm," "slow,"*...). In the first case $A' = \{a\}$, while in the second case for each x in X, $A'(x)$ is the degree to which x satisfies <term>. Likewise A and B represent objects or terms (the latter being much more frequent). However, while A', A and B are *constructed* to model objects or terms, B' is a fuzzy set *derived* during the inference process (in a way we still have to explain). B' is the representation of "some" object or linguistic term, but there is often no obvious way in determining which one.

If the output of the inference process is meant to control a machine, it is preferable that B' is associated to a crisp object of the universe (most often a crisp number). This association process is called defuzzification (see e.g., [10]). However, if the output is meant to guide a person, B' would better be interpreted linguistically. To do this, there is often a need for some process of linguistic approximation (see e.g., [12]). In this chapter we will only focus on implementations of the GMP generating an output that can be interpreted as a linguistic term without further need for linguistic approximation.

2. THE COMPOSITIONAL RULE OF INFERENCE

The most popular way of deriving B' from A', A and B is no doubt the compositional rule of inference (CRI) introduced by Zadeh [12]. First we recall the notions of triangular norm and implicator.

Definition 2 (triangular norm) *A triangular norm $\mathcal{T}$ (or shortly t-norm) is an associative and commutative increasing*[1] $[0,1]^2 - [0,1]$ *mapping that satisfies the boundary conditions* $\mathcal{T}(1,1) = 1$, $\mathcal{T}(1,0) = \mathcal{T}(0,1) = \mathcal{T}(0,0) = 0$ *and* $(\forall x \in [0,1])$ $(\mathcal{T}(1,x) = \mathcal{T}(x,1) = x)$.

Definition 3 (implicator) *An implicator $\mathcal{I}$ is a hybrid monotonic* $[0,1]^2 - [0,1]$ *mapping (i.e.,* $(\forall x \in [0,1])$ $(\mathcal{I}(.,x)$ *is decreasing and* $\mathcal{I}(x,.)$ *is increasing)) that satisfies the boundary conditions* $\mathcal{I}(1,0) = 0$, $\mathcal{I}(1,1) = \mathcal{I}(0,1) = \mathcal{I}(0,0) = 1$.

For a particular t–norm (e.g., Table 1.1) , we can associate with each fact "x and y have relation R" an $\mathcal{F}(X) - \mathcal{F}(Y)$ mapping.

---

[1] We recall that for a poset $(X, \leq)$ and f a $X - [0,1]$ mapping :

$$f \text{ is increasing} \Leftrightarrow (\forall (x,y) \in X^2)(x \leq y \Rightarrow f(x) \leq f(y))$$

$$f \text{ is decreasing} \Leftrightarrow (\forall (x,y) \in X^2)(x \leq y \Rightarrow f(x) \geq f(y))$$

The minimum operator $\mathcal{T}_M$	$\mathcal{T}_M(x,y) = min(x,y)$
The algebraic product $\mathcal{T}_P$	$\mathcal{T}_P(x,y) = x.y$
The Łukasiewicz t-norm $\mathcal{T}_W$	$\mathcal{T}_W(x,y) = max(0, x+y-1)$

Table 1.1 Examples of popular t-norms ($(x,y) \in [0,1]^2$).

The Kleene–Dienes implicator $\mathcal{I}_{KD}$	$\mathcal{I}_{KD}(x,y) = max(1-x, y)$
The Lukasiewicz implicator $\mathcal{I}_W$	$\mathcal{I}_W(x,y) = min(1, 1-x+y)$
The Reichenbach implicator $\mathcal{I}_R$	$\mathcal{I}_R(x,y) = 1-x+xy$
The Standard Strict implicator $\mathcal{I}_s$	$\mathcal{I}_s(x,y) = \begin{cases} 1 & \text{if } x \leq y \\ 0 & elsewhere \end{cases}$
The Standard Star implicator $\mathcal{I}_g$	$\mathcal{I}_g(x,y) = \begin{cases} 1 & \text{if } x \leq y \\ y & elsewhere \end{cases}$
The Gaines implicator $\mathcal{I}_\Delta$	$\mathcal{I}_\Delta(x,y) = \begin{cases} 1 & \text{if } x \leq y \\ \frac{y}{x} & elsewhere \end{cases}$

Table 1.2 Examples of popular implicators ($(x,y) \in [0,1]^2$).

Definition 4 (compositional rule of inference) *If $\mathcal{T}$ is a t–norm, $R \in \mathcal{F}(X \times Y)$, then the $\mathcal{F}(X) - \mathcal{F}(Y)$ mapping $cri_R^{\mathcal{T}}$ corresponding to the fact*

$$\mathsf{x} \textit{ and } \mathsf{y} \textit{ have relation } R$$

is defined by for $A' \in \mathcal{F}(X)$, $y \in Y$:

$$cri_R^{\mathcal{T}}(A')(y) = \sup_{x \in X} \mathcal{T}(A'(x), R(x, y))$$

The application of the operator $cri_R^{\mathcal{T}}$ to a fuzzy set A', generating a new fuzzy set, is called the compositional rule of inference.

The rule "IF x is A THEN y is B" states a relation between x and y. This relation can be expressed, by means of an implicator $\mathcal{I}$ (e.g., Table 1.2), by $A \Rightarrow_{\mathcal{I}} B$ which is defined as (for $x \in X$, $y \in Y$):

$$(A \Rightarrow_{\mathcal{I}} B)(x, y) = \mathcal{I}(A(x), B(y))$$

The B' in (SCHEME–2) can now be computed using the *cri*–operator defined above, namely $B' = cri^{\mathcal{T}}_{A \Rightarrow_{\mathcal{I}} B}(A')$.

3. THE IMPORTANCE OF BEING CLOSED

Much effort has already been put in investigating various techniques of inference based on the compositional rule of inference [4], and also on similarity relations [11], on fuzzy truth values [1], on interval-based implications [9], ... resulting in an extensive and detailed literature. In practice, however, most of these often sophisticated methods are being sparsely used because of the high computational demands they pose. The CRI requires that, for each evaluation of the output fuzzy set, the supremum over a possibly large set of values is computed. When the input is crisp (e.g., a crisp temperature) calculating the supremum can be avoided. However, when the input is fuzzy (e.g., a linguistic term), there is no straightforward way to eliminate the calculation of the supremum. As a consequence, most working systems involve simple computations transforming crisp inputs into crisp outputs, a task for which efficient algorithms have been devised. Intelligent systems interacting with humans, on the other hand, should be expected to cope with imprecise information conveyed by a human operator under the form of linguistic statements; to be meaningful, the answer produced by the system should preferably be stated in the user's own natural language (i.e., be of a linguistic nature).

The class of fuzzy sets commonly encountered in computer applications is often restricted to a subclass of fuzzy sets that can all be modelled by the same general shape membership function characterized by a small number of parameters. In that case all membership functions can be kept in memory by storing the parameters instead of all the membership values. If operations are performed on these fuzzy sets, it is desirable that the resulting fuzzy sets can also be characterized by this general shape function. This brings us to the concept of a closed system.

Definition 5 (closed system) *For $U \subseteq \mathcal{F}(X)$, $V \subseteq \mathcal{F}(Y)$ and m a $\mathcal{F}(X) - \mathcal{F}(Y)$ mapping (i.e., an operator transforming a fuzzy set on X into a fuzzy set on Y), (U, V, m) is called a closed system iff $(\forall A \in U)(m(A) \in V)$.*

Applied to inferencing, U and V constitute classes of fuzzy sets characterized by a (possibly but not necessarily different) general shape function, while m corresponds to an implementation of the IF–THEN rule (or more general: of the fact "x and y have relation R"). We begin by remarking that the implementation of the GMP by means of the CRI as described above gives rise to a closed system if all the fuzzy sets on X and on Y are taken into account.

Theorem 1 *For every t–norm $\mathcal{T}$, every implicator $\mathcal{I}$, for all $A \in \mathcal{F}(X)$, $B \in \mathcal{F}(Y)$:*

$$(\mathcal{F}(X), \mathcal{F}(Y), cri^{\mathcal{T}}_{A \Rightarrow_{\mathcal{I}} B}) \text{ is a closed system.}$$

In this chapter, however, we will discuss three closed systems (U, V, m) in which U and V are relatively small subsets of $\mathcal{F}(X)$ and $\mathcal{F}(Y)$ respectively. The third technique is a simplification of the CRI, while the first two involve other schemes to implement the GMP.

Closed systems often expose a number of interesting characteristics:

1. The membership functions involved can be efficiently stored in memory by means of a small number of parameters.

2. The membership function resulting from the inference process can be immediately linguistically interpreted.

3. The computational overhead for the CRI mentioned above can be avoided because the inference is essentially restricted to simple parameter manipulation.

The last two arguments make these techniques extremely suitable for inference applications with linguistic inputs and outputs.

4. HELLENDOORN APPROACH

4.1 GENERAL IDEA

In [6] Hellendoorn presents a technique for implementing the GMP, that can be applied when A, B, A' and B' in (SCHEME–2) are fuzzy sets with increasing membership functions. It is assumed that they all have the same shape. Furthermore the membership function for B' depends on the center and the length of its shell. These can be calculated on the basis of the shells of A, B and A'. In this inference process the CRI is not used.

Before we go into the details, we discuss an older proposal Hellendoorn advanced towards a system closed for inference. As opposed to the newer one, this older approach is based on the CRI. We will show that it is not suitable to model inference using IF–THEN rules.

4.2 CLASS OF MEMBERSHIP FUNCTIONS

The universe will be an interval of $\mathbb{R}$, i.e., $[a, b]$ for some $a \in \mathbb{R}, b \in \mathbb{R}, a \leq b$. For such an interval, we define a particular set of quadruples: $\mathbb{P}_{[a,b]} = \{(\alpha, \beta, \gamma, \delta) | (\alpha, \beta, \gamma, \delta) \in [a, b]^4$ and $\alpha \leq \beta \leq \gamma \leq \delta$ and $(\alpha < \beta$ or $\alpha = \beta = a)$ and $(\gamma < \delta$ or $\gamma = \delta = b)\}$ Each quadruple consists of 4 parameters characterizing a membership function.

Definition 6 (ϕ–generator, ϕ–function, Γ–function) *Let ϕ_1 and ϕ_2 be continuous $[0, 1] - [0, 1]$ mappings, ϕ_1 increasing and ϕ_2 decreasing, with boundary conditions $\phi_1(0) = 0$, $\phi_1(1) = 1$, $\phi_2(0) = 1$ and $\phi_2(1) = 0$. Let $[a, b]$ be an interval of $\mathbb{R}$. For $(\alpha, \beta, \gamma, \delta)$ in $\mathbb{P}_{[a,b]}$, the $[a, b] - [0, 1]$ mapping $\phi_{\phi_1,\phi_2}(.; \alpha, \beta, \gamma, \delta)$ is called a ϕ–function and is defined as (for x in $[a, b]$):*

$$\phi_{\phi_1,\phi_2}(x; \alpha, \beta, \gamma, \delta) = \begin{cases} 0, & \textit{for } x \in [a, \alpha] \quad \textit{if } \alpha < \beta \\ \phi_1(\frac{x-\alpha}{\beta-\alpha}), & \textit{for } x \in [\alpha, \beta] \quad \textit{if } \alpha < \beta \\ 1, & \textit{for } x \in [\beta, \gamma] \\ \phi_2(\frac{x-\gamma}{\delta-\gamma}), & \textit{for } x \in [\gamma, \delta] \quad \textit{if } \gamma < \delta \\ 0, & \textit{for } x \in [\delta, b] \quad \textit{if } \gamma < \delta \end{cases}$$

ϕ_{ϕ_1,ϕ_2} is called a ϕ-generator. $\phi_{\phi_1,\phi_2}(.; \alpha, \beta, b, b)$ is an increasing function. It is called a Γ-function and denoted $\Gamma_{\phi_1}(.; \alpha, \beta)$.

4.3 A CRI BASED CLOSED SYSTEM

For a particular class of fuzzy relations R, the application of $cri_R^{\mathcal{T}_M}$ to a ϕ–function results in another ϕ–function.

Lemma 1 *[5] Let $(\alpha_1, \beta_1, \gamma_1, \delta_1) \in \mathbb{P}_{[a_1,b_1]}$, $(\alpha_2, \beta_2, \gamma_2, \delta_2) \in \mathbb{P}_{[a_2,b_2]}$ and D the $\mathbb{R}^2 - \mathbb{R}$ mapping defined by $(\forall(x, y) \in \mathbb{R}^2)(D(x, y) = y - x)$.*

If $R = \phi_{\phi_1,\phi_2}(.; \alpha_2, \beta_2, \gamma_2, \delta_2) \circ D$
$A' = \phi_{\phi_1,\phi_2}(.; \alpha_1, \beta_1, \gamma_1, \delta_1)$
$B' = cri_R^{\mathcal{T}_M}(A')$

Then $B' = \phi_{\phi_1,\phi_2}(.;\alpha_1+\alpha_2,\beta_1+\beta_2,\gamma_1+\gamma_2,\delta_1+\delta_2)$

For the proof we refer to [5]. We remark that R is a fuzzy relation on $[a_1,b_1]\times[a_3,b_3]$, with $[a_1,b_1]$ being the universe of A' and $[a_3,b_3]$ such that $(\forall(x,y)\in\mathbb{R}^2)\ ((x,y)\in[a_1,b_1]\times[a_3,b_3]\Leftrightarrow y-x\in[a_2,b_2])$, i.e., $a_3=a_2+b_1$ and $b_3=a_1+b_2$. Hence while A' is a fuzzy set on $[a_1,b_1]$, B' will be a fuzzy set on $[a_3,b_3]$.

The following theorem is straightforward from this lemma:

Theorem 2 *Let* $(\alpha_2,\beta_2,\gamma_2,\delta_2)\in\mathbb{P}_{[a_2,b_2]}$,

$$R=\phi_{\phi_1,\phi_2}(.;\alpha_2,\beta_2,\gamma_2,\delta_2)\circ D$$

$$U=\{\phi_{\phi_1,\phi_2}(.;\alpha_1,\beta_1,\gamma_1,\delta_1)|(\alpha_1,\beta_1,\gamma_1,\delta_1)\in\mathbb{P}_{[a_1,b_1]}\}$$

$$V=\{\phi_{\phi_1,\phi_2}(.;\alpha_1+\alpha_2,\beta_1+\beta_2,\gamma_1+\gamma_2,\delta_1+\delta_2)|(\alpha_1,\beta_1,\gamma_1,\delta_1)\in\mathbb{P}_{[a_1,b_1]}\}$$

Then $(U,V,cri_R^{T_M})$ *is a closed system.*

In other words applying the CRI, we get the following inference scheme:

$$\begin{array}{l}\mathsf{x}\text{ is }\phi_{\phi_1,\phi_2}(.;\alpha_1,\beta_1,\gamma_1,\delta_1)\\ \mathsf{x}\text{ and }\mathsf{y}\text{ have relation }\phi_{\phi_1,\phi_2}(.;\alpha_2,\beta_2,\gamma_2,\delta_2)\circ D\\ \hline \mathsf{y}\text{ is }\phi_{\phi_1,\phi_2}(.;\alpha_1+\alpha_2,\beta_1+\beta_2,\gamma_1+\gamma_2,\delta_1+\delta_2)\end{array}\qquad\text{(SCHEME-3)}$$

This works well for relations that are fuzzy comparators like *"is taller than," "is approximately equal to,"*... The following theorem, however, indicates that there is no straightforward correspondence between the usual representation of an IF–THEN rule by means of an implicator $\mathcal{I}$, and the representation of a fuzzy relation by means of a ϕ–function as done above.

Theorem 3 *Let* $a\in]0,+\infty[$. *There is no quadruple* (e,f,g,h) *in* $\mathbb{R}^4$ *such that for all* x *and* y *in* $[a,4a]$ *:*

$$\mathcal{I}(\Gamma_{\phi_1}(x;2a,3a),\Gamma_{\phi_1}(y;2a,3a))=\phi_{\phi_1,\phi_2}(y-x;e,f,g,h)$$

Proof Suppose that we could find such an e,f,g and h, then from $\phi_{\phi_1,\phi_2}(2a-3a;e,f,g,h)=\phi_{\phi_1,\phi_2}((2a+a)-(3a+a);e,f,g,h)$ we would derive $\mathcal{I}(\Gamma_{\phi_1}(3a;2a,3a),\ \Gamma_{\phi_1}(2a;2a,3a))=\mathcal{I}(\Gamma_{\phi_1}(3a+a;2a,3a),\ \Gamma_{\phi_1}(2a+a;2a,3a))$. Hence $\mathcal{I}(1,0)=\mathcal{I}(1,1)$, which conflicts with the definition of implicator. □

4.4 A SHELL–BASED APPROACH

In [6] Hellendoorn presents another implementation of the GMP which can be used if all the membership functions involved in (SCHEME–2) belong to a particular class of increasing functions. In this section we will only consider Γ–functions defined by means of one and the same $[0,1] - [0,1]$ mapping ϕ_1 (defined as above in Definition 6). We will therefore drop ϕ_1 in the notation and denote them as $\Gamma(.;\alpha,\beta)$ for $\alpha \subset X$, $\beta \in X$, $\alpha < \beta$, X a closed interval of the reals.

Let X and Y be closed intervals of the reals. For a, b, e and f in X, $a < b$, $e < f$ and c, d in Y, $c < d$: $\Gamma(.;a,b)$ and $\Gamma(.;e,f)$ are fuzzy sets on X, while $\Gamma(.;c,d)$ is a fuzzy set on Y. We want to make the inference:

$$\begin{array}{l} \mathsf{x} \text{ is } \Gamma(.;e,f) \\ \text{IF } \mathsf{x} \text{ is } \Gamma(.;a,b) \text{ THEN } \mathsf{y} \text{ is } \Gamma(.;c,d) \\ \hline \mathsf{y} \text{ is } B' \end{array} \qquad \text{(SCHEME-4a)}$$

Hellendoorn states that B' is a Γ–function characterized by two parameters g and h ($g < h$), i.e., $B' = \Gamma(.;g,h)$. According to him, g and h can be computed from $a, b, c, d, e,$ and f in the following manner:

1. Compute the length l of the shell of B' by means of the lengths of the shells of the other three membership functions. It is assumed that the length of $shell(\Gamma(.;g,h))$ should be proportional to the length of $shell(\Gamma(.;c,d))$ as the length of $shell(\Gamma(.;e,f))$ is to the length of $shell(\Gamma(.;a,b))$. Therefore

$$l = h - g = \frac{(d-c)(f-e)}{b-a}$$

2. Compute the center z of the shell of B' by means of the measure of difference w between $\Gamma(.;a,b)$ and $\Gamma(.;e,f)$, i.e.,:

$$w = \frac{(a+b)-(e+f)}{(b-a)+(f-e)}$$

   Hellendoorn proposes to calculate the center z as follows:

$$z = \frac{1}{2}((min(Y) - \frac{l}{2} - c)w^2 + (min(Y) - \frac{l}{2} - d)w + c + d)$$

   Remark that $w = 0$ iff the centers of the shells of $\Gamma(.;a,b)$ and $\Gamma(.;e,f)$ coincide. In this case $z = \frac{c+d}{2}$ which means that the centers of the shells of $\Gamma(.;c,d)$ and B' will coincide as well. If $w = 1$ then $a = f$, hence $\Gamma(.;a,b) \subseteq \Gamma(.;e,f)$. In other words

$\Gamma(.;e,f)$ is entirely weaker than $\Gamma(.;a,b)$, which means there is not enough information to make a meaningful derivation. In this case $z = min(Y) - \frac{l}{2}$, hence $(\forall y \in Y)(B'(y) = 1)$ which is usually interpreted as $B' = unknown$.

According to this definition of z, z does not necessarily belong to Y. If we want to develop a closed system, we can fix this by chosing $g = max(min(Y), z - \frac{l}{2})$ and $h = min(max(Y), z + \frac{l}{2})$.

Theorem 4 *Let X and Y be closed intervals of $\mathbb{R}$ and*

$$U = \{\Gamma(.;p,q) | \Gamma(.;p,q) \text{ is a } \Gamma\text{-function on } X\}$$

$$V = \{\Gamma(.;r,s) | \Gamma(.;r,s) \text{ is a } \Gamma\text{-function on } Y\}$$

Let a and b in X, $a < b$ and c and d in Y, $c < d$. Let the $U - V$ mapping $m_{(\Gamma(.;a,b),\Gamma(.;c,d))}$ be defined by (for e and f in X, $e < f$):

$$m_{(\Gamma(.;a,b),\Gamma(.;c,d))}(\Gamma(.;e,f)) = \Gamma(.;g,h)$$

with
$$\begin{aligned} g &= max(min(Y), z - \tfrac{l}{2}) \\ h &= min(max(Y), z + \tfrac{l}{2}) \\ z &= \tfrac{1}{2}((min(Y) - \tfrac{l}{2} - c)w^2 + (min(Y) - \tfrac{l}{2} - d)w + c + d) \\ w &= ((a+b) - (e+f))/((b-a) + (f-e)) \\ l &= ((d-c)(f-e))/(b-a) \end{aligned}$$

Then $(U, V, m_{(\Gamma(.;a,b),\Gamma(.;c,d))})$ is a closed system.

Substituting the proper Γ–functions in (SCHEME–2) we can make the following derivation:

$$\frac{\begin{array}{l} \mathsf{x} \text{ is } \Gamma(.;e,f) \\ \text{IF } \mathsf{x} \text{ is } \Gamma(.;a,b) \text{ THEN } \mathsf{y} \text{ is } \Gamma(.;c,d) \end{array}}{\mathsf{y} \text{ is } \Gamma(.;max(min(Y), z - \frac{l}{2}), min(max(Y), z + \frac{l}{2}))} \qquad \text{(SCHEME-4b)}$$

Example 1 *The $[0,1] - [0,1]$–mapping ϵ is defined by $\epsilon(x) = x$, for all x in $[0,1]$. Let $\phi_1 = \epsilon$. Consider the rule*

"IF speed is high THEN braking distance will be long" *(RULE–1)*

where "speed" and "braking distance" are variables that take values in the respective universes $X = [0,150]$ (expressed in km/h) and $Y = [0,300]$ (expressed in meters). The linguistic concept "high speed" *can be modelled by the fuzzy set $A = \Gamma(.;30,100)$ on X, while* "long braking distance" *can be represented by the fuzzy set $B = \Gamma(.;100,200)$ on Y. If speed is* "more or less high," *which can be modelled by $A' = \Gamma(.;10,90)$, then using (RULE–1) and (SCHEME–4b) we derive that braking distance is $B' = \Gamma(.;64, \frac{352}{7} + 128)$. Figure 1.1 depicts the membership functions for A, A', B and B'.*

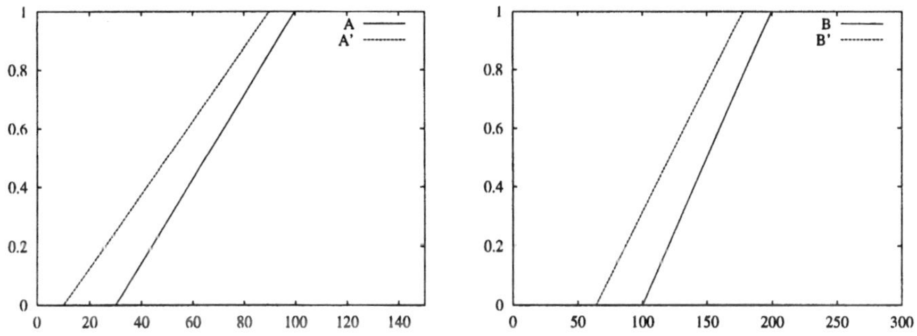

Figure 1.1 Hellendoorn approach a) Fuzzy sets on the universe of speeds b) Fuzzy sets on the universe of braking distances

5. NAFARIEH–KELLER APPROACH

5.1 GENERAL IDEA

The method of inference Nafarieh and Keller [7, 8] have proposed is specifically tailored to reasoning processes where linguistic hedges (i.e., adverbs such as *very, rather, quite, ...*) are modelled by real powers of a given base fuzzy set; we call them **powering hedges**. The general idea is to restate the GMP as (n represents a positive real number)

$$\frac{\begin{array}{l} \mathsf{x} \text{ is } A^n \\ \text{IF } \mathsf{x} \text{ is } A \text{ THEN } \mathsf{y} \text{ is } B \end{array}}{\mathsf{y} \text{ is } B^n} \qquad \text{(SCHEME–5)}$$

This corresponds to a hedge–preserving semantics of a fuzzy rule: when x is $m(A)$, then y should be $m(B)$ (where m represents a hedge). One should however be very cautious when adopting such an interpretation of a rule, as its use is not always justified by the context. Consider, for instance, the rule: "IF it is warm THEN it is pleasant to go out for a walk." Modelling this rule using the above strategy seems unrealistic, for it would lead us to believe that the most pleasant walks can be taken on the surface of the sun.

In [2] it is argued that a requirement like hedge-preservation underlies that the rule expresses more than a simple logical implication, for in classical logic it holds that a proposition inferred by modus ponens can never be more specific than what the consequent of the rule says.

The case where the input fuzzy set A' in (SCHEME–2), is not expressed as a real power of A, may be adequately handled by approximating A' by A^n for a certain n in order that the foregoing scheme can be applied to yield B^n. The approximation step is based on some rather ingenious mathematical machinery which we will discuss further on. To put it in a nutshell, derivations are performed on the basis of a measurement of the compatibility or likeness of the input fuzzy set and the antecedent of an IF–THEN rule, and once this degree of likeness is determined, it gives, in a way to be described, rise to a positive real power which will then act as a modifier on the consequent of the IF–THEN rule, thus yielding a fuzzy set which is a power of the original one in the consequent of the rule.

Nafarieh and Keller have applied this mechanism to control an automatic target recognizer. Their results are described in [8].

5.2 CLASS OF MEMBERSHIP FUNCTIONS

In order to simplify our discussion somewhat, we shall only consider intervals of $\mathbb{R}$ as candidate universes, i.e. $X = [u_1, u_2]$, $u_1 \in \mathbb{R}$, $u_2 \in \mathbb{R}$, $u_1 < u_2$. Within such a universe we will consider the class of convex fuzzy sets with a continuous membership function and an infinite support. We will denote this class of fuzzy sets on $[u_1, u_2]$ for which the Nafarieh–Keller approach is suitable, as $\mathcal{NK}([u_1, u_2])$. The convexity requirement seems quite natural for fuzzy sets representing linguistic concepts as *"old," "very hot," "about five and a half."* Demanding our fuzzy sets to have an infinite support actually rules out crisp numbers, which may seem a bit harsh at first, yet, as was stated above, simpler and more efficient techniques have been conceived in those circumstances where a crisp input is available.

5.3 MODE OF OPERATION

First we introduce a measure of compatibility between two fuzzy sets A and B belonging to the class $\mathcal{NK}([u_1, u_2])$.

Definition 7 (compatibility index) *The compatibility index* comp *is defined as the* $\mathcal{NK}([u_1, u_2]) - [0, 1]$ *mapping defined by (for* A *and* B *in* $\mathcal{NK}([u_1, u_2])$*):*

$$\mathsf{comp}(A, B) = \frac{\int_{u_1}^{u_2} (A \cap B)(x)dx}{\int_{u_1}^{u_2} (A \cup B)(x)dx}$$

Proposition 1 *For* A, A_1 *and* B *in* $\mathcal{NK}([u_1, u_2])$*:*

1. $A \cap B = \emptyset \Rightarrow \mathsf{comp}(A, B) = 0$
2. $\mathsf{comp}(A, A) = 1$ *(reflexivity)*
3. $\mathsf{comp}(A, B) = \mathsf{comp}(B, A)$ *(symmetry)*

4. $A \subseteq A_1 \subseteq B \Rightarrow \mathsf{comp}(A, B) \leq \mathsf{comp}(A_1, B)$
5. $B \subseteq A_1 \subseteq A \Rightarrow \mathsf{comp}(A, B) \leq \mathsf{comp}(A_1, B)$

Proof.
The first property holds because $\emptyset$ is defined as the constant $[u_1, u_2] - [0, 1]$ mapping such that $\emptyset(x) = 0$, for all x in $[u_1, u_2]$ and, hence, the numcrator in definition (7) becomes 0. Properties 2 and 3 are quite obvious, so only the last two statements remain to be proved.

To show that property 4 holds, take A, A_1 and B in $\mathcal{NK}([u_1, u_2])$ such that $A \subseteq A_1 \subseteq B$. This means that $(\forall x \in [u_1, u_2])(A(u) \leq A_1(u) \leq B(u))$, and hence we have, due to the monotonicity of the *min* operator, for all $u \in [u_1, u_2]$ that $min(A(u), B(u)) \leq min(A_1(u), B(u))$. In terms of functions, this is equivalent to saying that $min \circ (A_1, B)$ dominates $min \circ (A, B)$ on $[u_1, u_2]$ and hence due to the monotonicity of integration we obtain $\int_{u_1}^{u_2} (A \cap B)(x)dx \leq \int_{u_1}^{u_2} (A_1 \cap B)(x)dx$. It is also easy to see that $\forall u \in [u_1, u_2], max(A(u), B(u)) = max(A_1(u), B(u)) = B(u)$, so finally we may conclude that $\mathsf{comp}(A, B) \leq \mathsf{comp}(A_1, B)$. Property 5 may be proved analogously. □

Definition 8 (reference fuzzy set) *Given a fuzzy set* $A \in \mathcal{NK}([u_1, u_2])$, *we introduce a reference fuzzy set*[2] *on the unit interval,* ref_A. *To define* ref_A *we first construct the linear mapping h:*

$$\begin{array}{rrcl} h: & [u_1, u_2] & \rightarrow & [0, 1] \\ & x & \mapsto & \frac{x - u_1}{u_2 - u_1}, \quad \forall x \in [u_1, u_2] \end{array}$$

ref_A *is then defined as:*

$$\begin{array}{rrcl} \mathsf{ref}_A: & [0, 1] & \rightarrow & [0, 1] \\ & x & \mapsto & A(h^{-1}(x)), \quad \forall x \in [0, 1] \end{array}$$

Theorem 5 *Let the universe* $[u_1, u_2]$ *be an interval of* $\mathbb{R}$, $u_1 < u_2$, $A \in \mathcal{NK}([u_1, u_2])$ *and* ref_A *its reference set. Then we have the following equality for any* $n \in [0, +\infty[$:

$$\frac{\int_0^1 \mathsf{ref}_A^n(x)dx}{\int_0^1 \mathsf{ref}_A(x)dx} = \frac{\int_{u_1}^{u_2} A^n(x)dx}{\int_{u_1}^{u_2} A(x)dx} \qquad (1.1)$$

[2]Nafarieh and Keller call it the fuzzy truth value "true."

Proof.

The equality in (1.1) can be shown by applying a simple substitution to the integrands in the numerator and the denominator of the left hand side. Let $u = h^{-1}(x) = (u_2 - u_1)x + u_1$, so that $du = (u_2 - u_1)dx$. We easily get

$$\frac{\int_0^1 (A \circ h^{-1})^n(x)dx}{\int_0^1 (A \circ h^{-1})(x)dx} = \frac{\int_{u_1}^{u_2} \frac{A^n(u)}{u_2-u_1} du}{\int_{u_1}^{u_2} \frac{A(u)}{u_2-u_1} du}, \tag{1.2}$$

which is equivalent to the right hand side of (1.1), since $u_1 \neq u_2$. □

Due to this result, we are now in a position to establish a remarkable link between the previously defined compatibility index comp and the reference set ref_A. Indeed, since

$$\frac{\int_{u_1}^{u_2} A^n(x)dx}{\int_{u_1}^{u_2} A(x)dx} = \begin{cases} \frac{\int_{u_1}^{u_2}(A\cap A^n)(x)dx}{\int_{u_1}^{u_2}(A\cup A^n)(x)dx} = \mathsf{comp}(A, A^n) & \text{for } n \geq 1 \\ \frac{\int_{u_1}^{u_2}(A\cup A^n)(x)dx}{\int_{u_1}^{u_2}(A\cap A^n)(x)dx} = \frac{1}{\mathsf{comp}(A,A^n)} & \text{for } n < 1 \end{cases} \tag{1.3}$$

we have

$$\int_0^1 \mathsf{ref}_A^n(x)dx = \begin{cases} \left(\int_0^1 \mathsf{ref}_A(x)dx\right) \mathsf{comp}(A, A^n) & \text{for } n \geq 1 \\ \frac{\int_0^1 \mathsf{ref}_A(x)dx}{\mathsf{comp}(A,A^n)} & \text{for } n < 1 \end{cases} \tag{1.4}$$

In case the input A' in (SCHEME–2) is not expressed as a power of A, then (1.4) can still be used to approximate the degree that A' matches A. For instance, when $A' \subset A$, we could try to find a real value $n > 1$ such that $\mathsf{comp}(A', A)$ is approximately equal to $\mathsf{comp}(A^n, A)$. We do this by changing $\mathsf{comp}(A, A^n)$ in (1.4) by $\mathsf{comp}(A, A')$ and solving the resulting equation in the unknown n. A similar argument can be made when $A \subset A'$, to the effect that an $n < 1$ should be searched such that $\mathsf{comp}(A', A)$ is approximately equal to $\mathsf{comp}(A^n, A)$. In the remaining case (i.e., A' is incomparable to A w.r.t. $\subseteq$), A' can be considered less specific than A for the inference result, so we should still look for an $n < 1$ (we want the result of the inference step to be weaker than the consequent of the rule). From (1.4) one can easily deduce that the minimum value of $\mathsf{comp}(A, A^n)$ when $n < 1$ is arrived at when $n = 0$, namely

$$\mathsf{comp}^* = \mathsf{comp}(A, A^0) = \frac{\int_0^1 \mathsf{ref}_A(x)dx}{\int_0^1 \mathsf{ref}_A^0(x)dx} = \int_0^1 \mathsf{ref}_A(x)dx \tag{1.5}$$

When $\mathsf{comp}(A', A) < \mathsf{comp}^*$ (for example when A' and A are disjoint), there will be no solution to the equation (1.4). Nafarieh and Keller suggest to set $n = 0$ in that case. The derived fuzzy set B^n will correspond to the universe. This can be taken as a represention of total ignorance (B^n = *unknown*), and is in accordance with our intuition, because when the input and antecedent differ too much it should not be possible to infer anything with the GMP.

Theorem 6 *Let $[u_1, u_2]$ and $[v_1, v_2]$ be intervals of $\mathbb{R}$, $u_1 < u_2$, $v_1 < v_2$. Let $A \in \mathcal{NK}([u_1, u_2])$ and $B \in \mathcal{NK}([v_1, v_2])$. Let the $\mathcal{NK}([u_1, u_2]) - \mathcal{NK}([v_1, v_2])$ mapping $m_{(A,B)}$ be defined by (for $A' \in \mathcal{NK}([u_1, u_2])$):*

$$m_{(A,B)}(A') = B^n$$

with n defined as follows :

Let (a) : $\int_0^1 \mathsf{ref}_A^n(x)dx = \mathsf{comp}(A, A') \int_0^1 \mathsf{ref}_A(x)dx$

Let (b) : $\mathsf{comp}(A, A') \int_0^1 \mathsf{ref}_A^n(x)dx = \int_0^1 \mathsf{ref}_A(x)dx$

if $A' \subseteq A$ then
if (a) has a solution n_0 and $n_0 \geq 1$ then $n := n_0$
else $n := 0$
else if (b) has a solution n_0 and $n_0 < 1$ then $n := n_0$
else $n := 0$

Then $(\mathcal{NK}([u_1, u_2]), \{B^n | B \in \mathcal{NK}([v_1, v_2]), n \in [0, +\infty[\}, m_{(A,B)})$ is a closed system.

Example 2 *Let us consider (RULE–1) and fuzzy sets A and B from example 1 again. In this section we will also use the $[0, 1]-[0, 1]$ mapping θ defined by $\theta(x) = 1 - x$, for all x in $[0, 1]$. When we receive an input fuzzy set A', we have to distinguish between two cases. If the input is a positive real power n of the antecedent, then there is no need to perform any calculations at all and we immediately acquire the result, B^n. This* *is especially so when the input exactly matches the antecedent, so at that stage we retrieve the (highly desirable!) coincidence with the modus ponens (SCHEME–1).*

Let us concentrate now on what happens when A' is not expressed as a power of A. Our first step will then be to construct the reference set of A. It can be verified that $\mathsf{ref}_{\mathsf{A}} = \Gamma_\epsilon(.;\frac{1}{5},\frac{2}{3})$. *Next, we have to calculate the integral*

$$\int_0^1 \mathsf{ref}_A^n(x)dx, n \in [0,+\infty[$$

It is worth noticing that these calculations are independent of A' and thus have to be performed just once. We find that

$$\int_0^1 \mathsf{ref}_A^n(x)dx = \int_{\frac{1}{5}}^{\frac{2}{3}} \left(\frac{15x-3}{7}\right)^n dx + \int_{\frac{2}{3}}^1 dx = \frac{7}{15(n+1)} + \frac{1}{3} \qquad (1.6)$$

We now consider some specific inputs A' for the linguistic variable "speed." It's interesting to investigate what happens in case **shifting hedges**, *that modify linguistic terms by applying a shift rather than an involution to the associated membership function, are used. For instance, consider $A' = \Gamma_\epsilon(.;40,110)$. The first step in the inference process is to assess the measure of compatibility* $\mathsf{comp}(A,A')$ *between the antecedent and the input fuzzy set. Using Definition (7), we achieve*

$$\begin{aligned} \mathsf{comp}(A,A') &= \frac{\int_0^{150} min(A(x),A'(x))dx}{\int_0^{150} max(A(x),A'(x))dx} \\ &= \frac{\int_{40}^{110} \frac{x-40}{70}dx + \int_{110}^{150} dx}{\int_{30}^{100} \frac{x-30}{70}dx + \int_{100}^{150} dx} = \frac{15}{17} \end{aligned}$$

Since $A' \subset A$, we have to look for an $n \geq 1$ that satisfies equation (a) in theorem 6. From $\frac{7}{15(n+1)} + \frac{1}{3} = \frac{15}{17}\frac{17}{30}$ we find $n = \frac{9}{5}$. The output fuzzy set $B' = B^n$ is then $(\Gamma_\epsilon(.;100,200))^{\frac{9}{5}}$. The fuzzy sets A, A', B and B' are depicted in Figure 1.2.

We should remark that our result does not fully correspond to the hedge–preserving semantics we have outlined in the previous section, since this would mean that the fuzzy set B should be shifted in exactly the same way (generally undesirable when antecedent and consequent are defined on different universes). On the other hand, the input is somewhat stronger than the antecedent, and likewise the output is somewhat

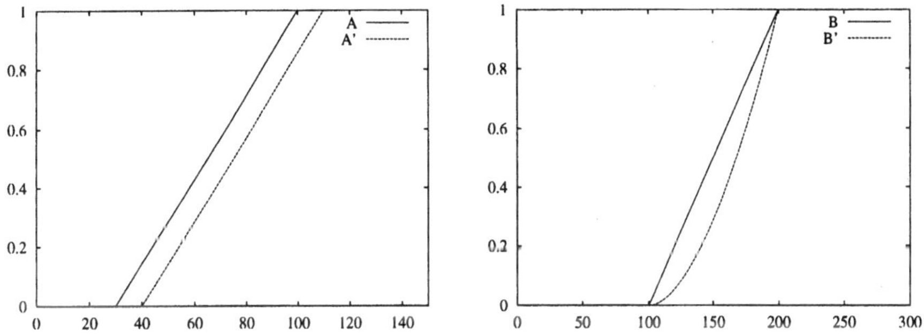

Figure 1.2 Nafarieh–Keller approach a) Fuzzy sets on the universe of speeds b) Fuzzy sets on the universe of braking distances

stronger than the consequent of the rule, which sounds intuitively correct. The hard part in using shifted hedges will be to devise suitable linguistic expressions for characterizing the output.

Finally let us consider the concept "around one hundred" *represented by* $A' = \phi_{\epsilon,\theta}(.;90,100,100,110)$. *Remark that* $A' \subset A$. *We first compute* $\mathsf{comp}(A, A')$:

$$\mathsf{comp}(A, A') = \frac{\int_0^{150} min(A(x), A'(x))dx}{\int_0^{150} max(A(x), A'(x))dx} = \frac{\int_{90}^{100} \frac{x-90}{10}dx + \int_{100}^{110} \frac{110-x}{10}dx}{\int_{30}^{100} \frac{x-30}{70}dx + \int_{100}^{150} dx} = \frac{2}{17}$$

From $\frac{7}{15(n+1)} + \frac{1}{3} = \frac{2}{17}\frac{17}{30}$ *we find* $n = \frac{-11}{4}$. *Hence there is no* $n \geq 1$ *which satisfies equation (a) in theorem 6 and we set* $n = 0$, *which means that the inferred result is* unknown. *The reason for this is that* A *and* A' *are not sufficiently similar.*

6. BOUCHON APPROACH

6.1 GENERAL IDEA

In [3] Bouchon presents a closed system in which the inference is based on the CRI. The key to her approach is the use of operators representing linguistic hedges (approximately, rather, about, almost). Unlike those of other authors, such an adverb representing operator does not act on a fuzzy set on $\mathbb{R}$ (i.e., a $\mathbb{R}-[0,1]$ mapping), but on a $\mathbb{R}-]-\infty,1]$ mapping, which serves as a help mapping. The fuzzy sets representing a <term>, and <rather term>, <approximately term>,... are all constructed from this help mapping following the same process: the application of a suitable operation to the help function.

A disadvantage of this approach is the lack of an operator suitable to represent an adverb like very. To overcome this problem, we will use $\mathbb{R} - [-2, 2]$ mappings as help functions.

6.2 MODE OF OPERATION

Definition 9 *For α and β in $\mathbb{R}$ and f a $\mathbb{R} - \mathbb{R}$ mapping, $f_{(\alpha,\beta)}$ is the fuzzy set on $\mathbb{R}$ defined by (for $x \in \mathbb{R}$):*

$$f_{(\alpha,\beta)}(x) = min(1, max(0, \alpha f(x) + \beta))$$

Proposition 2 *For $\beta \in [0, 1]$, f a $\mathbb{R} - \mathbb{R}$ mapping:*

(1) $f_{(1,-\beta)} \subseteq f_{(1,0)}$
(2) $f_{(1,0)} \subseteq f_{(1-\beta,\beta)}$
(3) $f_{(1,0)} \subseteq f_{(1,\beta)}$

Proof Let $x \in \mathbb{R}$. The inequalities $f(x) - \beta \leq f(x)$ and $f(x) \leq f(x) + \beta$ are quite trivial. Adding the fact that min and max are increasing proves the first and the last inclusion. To prove the middle one, we remark that for $f(x) \leq 1$, $f(x) \leq (1 - \beta)f(x) + \beta$. If $f(x) > 1$ then both $f_{(1,0)}(x)$ and $f_{(1-\beta,\beta)}(x)$ are 1. □

If the fuzzy set $f_{(1,0)}$ represents a <term>, then for a particular choice of $\beta \in [0, 1]$, $f_{(1,-\beta)}$ can represent <very term>, while $f_{(1-\beta,\beta)}$ and $f_{(1,\beta)}$ can model <rather term> and <more or less term> respectively.

We first formulate a lemma that will help us to state under which conditions this generalized Bouchon approach for modelling hedges, gives rise to a closed system for inference.

Lemma 2 *For f a $\mathbb{R}-\mathbb{R}$ mapping, $wd(f) = [-2, 2]$, $b \in [0, 1]$, $\beta \in [0, 1]$:*

(i) $\sup_{x \in \mathbb{R}} \mathcal{T}_W(f_{(1,\beta)}(x), \mathcal{I}_R(f_{(1,0)}(x), b)) = (1 - \beta)b + \beta$
(ii) $\sup_{x \in \mathbb{R}} \mathcal{T}_W(f_{(1,-\beta)}(x), \mathcal{I}_R(f_{(1,0)}(x), b)) = b$
(iii) $\sup_{x \in \mathbb{R}} \mathcal{T}_W(f_{(1-\beta,\beta)}(x), \mathcal{I}_R(f_{(1,0)}(x), b)) = max(b, \beta)$

Proof As an example we prove *(i)*.

$$
\begin{aligned}
&\sup_{x\in\mathbb{R}} \mathcal{T}_W(f_{(1,\beta)}(x), \mathcal{I}_R(f_{(1,0)}(x), b))\\
&= \sup_{x\in\mathbb{R}} \mathcal{T}_W(min(1, max(0, f(x)+\beta)), \mathcal{I}_R(min(1, max(0, f(x))), b))\\
&= \sup_{x\in\mathbb{R}} max(0, min(1, max(0, f(x)+\beta)) + (b-1)min(1, max(0, f(x))))\\
&= \sup_{z\in wd(f)} max(0, min(1, max(0, z+\beta)) + (b-1)min(1, max(0, z)))
\end{aligned}
$$

We divide $wd(f)$ into 5 intervals and calculate the supremum in each of them :

1. $X_1 = [-2, -\beta[$
$$
\begin{aligned}
&\sup_{z\in X_1} max(0, min(1, max(0, z+\beta)) + (b-1)min(1, max(0, z)))\\
&= \sup_{z\in X_1} max(0, min(1, 0) + (b-1)min(1, 0))) = 0
\end{aligned}
$$

2. $X_2 = [-\beta, 0]$
$$
\begin{aligned}
&\sup_{z\in X_2} max(0, min(1, max(0, z+\beta)) + (b-1)min(1, max(0, z)))\\
&= \sup_{z\in X_2} max(0, min(1, z+\beta) + (b-1)min(1, 0)))\\
&= \sup_{z\in X_2} max(0, min(1, z+\beta))\\
&= \sup_{z\in X_2} min(1, z+\beta) = min(1, \beta) = \beta
\end{aligned}
$$

3. $X_3 =]0, 1-\beta]$
$$
\begin{aligned}
&\sup_{z\in X_3} max(0, min(1, max(0, z+\beta)) + (b-1)min(1, max(0, z)))\\
&= \sup_{z\in X_3} max(0, z+\beta+(b-1)z)\\
&= max(0, (1-\beta)b+\beta) = (1-\beta)b+\beta
\end{aligned}
$$

4. $X_4 =]1-\beta, 1]$
$$
\begin{aligned}
&\sup_{z\in X_4} max(0, min(1, max(0, z+\beta)) + (b-1)min(1, max(0, z)))\\
&= \sup_{z\in X_4} max(0, 1+(b-1)z)\\
&= max(0, 1+(b-1)(1-\beta)\\
&= max(0, (1-\beta)b+\beta) = (1-\beta)b+\beta
\end{aligned}
$$

5. $X_5 =]1, 2]$
$$
\begin{aligned}
&\sup_{z\in X_5} max(0, min(1, max(0, z+\beta)) + (b-1)min(1, max(0, z)))\\
&= max(0, 1+(b-1)) = b
\end{aligned}
$$

Since $b \le (1-\beta)b+\beta$ and $\beta \le (1-\beta)b+\beta$:

$$max(0, \beta, (1-\beta)b+\beta, b) = (1-\beta)b+\beta$$

□

Theorem 7 *Let $\tilde{A}$ and $\tilde{B}$ be $\mathbb{R}$–$\mathbb{R}$ mappings, $wd(\tilde{A}) = wd(\tilde{B}) = [-2, 2]$, $\beta \in [0,1]$, k_β the constant $\mathbb{R} - \{k\}$–mapping,*

$$U = \{\tilde{A}_{(1,0)}, \tilde{A}_{(1,\beta)}, \tilde{A}_{(1,-\beta)}, \tilde{A}_{(1-\beta,\beta)}\}$$

$$V = \{\tilde{B}_{(1,0)}, \tilde{B}_{(1-\beta,\beta)} \cup k_\beta, \tilde{B}_{(1,0)} \cup k_\beta\}$$

$$m = cri^{\mathcal{T}_W}_{\tilde{A}_{(1,0)} \Rightarrow_{\mathcal{I}_R} \tilde{B}_{(1,0)}}$$

Then (U, V, m) is a closed system.

Proof Using (i) from Lemma 2, for all $y \in \mathbb{R}$:

$$\begin{aligned} m(\tilde{A}_{(1,\beta)})(y) &= cri^{\mathcal{T}_W}_{\tilde{A}_{(1,0)} \Rightarrow_{\mathcal{I}_R} \tilde{B}_{(1,0)}}(\tilde{A}_{(1,\beta)})(y) \\ &= \sup_{x \in \mathbb{R}} \mathcal{T}_W(\tilde{A}_{(1,\beta)}(x), \mathcal{I}_R(\tilde{A}_{(1,0)}(x), \tilde{B}_{(1,0)}(y))) \\ &= (1-\beta)\tilde{B}_{(1,0)}(y) + \beta \\ &= (1-\beta)min(1, max(0, B(y))) + \beta \\ &= min(1, max(\beta, (1-\beta)B(y) + \beta)) \\ &= max(min(1,\beta), min(1, (1-\beta)B(y) + \beta)) \\ &= max(\beta, min(1, max(0, (1-\beta)B(y) + \beta))) \\ &= max(\beta, \tilde{B}_{(1-\beta,\beta)}(y)) \end{aligned}$$

Likewise we can deduct:

using (i): $m(\tilde{A}_{(1,\beta)}) = \tilde{B}_{(1,0)}$
using (ii): $m(\tilde{A}_{(1,-\beta)}) = \tilde{B}_{(1,0)}$
using (iii): $m(\tilde{A}_{(1-\beta,\beta)}) = \tilde{B}_{(1,0)} \cup k_\beta$

□

If in (SCHEME–2) $A = \tilde{A}_{(1,0)}$ and $B = \tilde{B}_{(1,0)}$, and the GMP is implemented by means of the CRI (using $\mathcal{T}_W, \mathcal{I}_R$) the following derivations can be made:

x is A IF x is A THEN y is B	x is more or less A IF x is A THEN y is B
y is B	y is rather B with uncertainty β
x is very A IF x is A THEN y is B	x is rather A IF x is A THEN y is B
y is B	y is B with uncertainty β

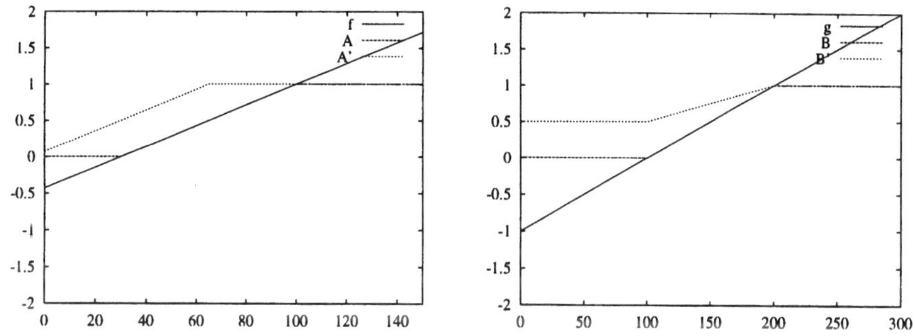

Figure 1.3 Generalized Bouchon approach a) Fuzzy sets on the universe of speeds b) Fuzzy sets on the universe of braking distances

Example 3 *We return to our car control example. Let f and g be defined by:*

$$f: \begin{array}{rcl} [0,150] & \to & [0,1] \\ x & \mapsto & \frac{x-30}{70}, \quad \forall x \in [0,150] \end{array}$$

$$g: \begin{array}{rcl} [0,300] & \to & [0,1] \\ x & \mapsto & \frac{x-100}{100}, \quad \forall x \in [0,300] \end{array}$$

then $A = f_{(1,0)}$ *represents* "high speed" *while* $B = g_{(1,0)}$ *models* "long braking distance." *For* $\beta = \frac{1}{2}$, $A' = f_{(1,\frac{1}{2})}$ *represents* "more or less high speed." *We can make the derivation:*

speed is more or less high (A')
IF speed is high (A) THEN braking distance will be long (B)

---

braking distance is rather long with uncertainty $\frac{1}{2}$ $(B' = g_{(\frac{1}{2},\frac{1}{2})} \cup k_{\frac{1}{2}})$

The mappings f and g and the fuzzy sets A, A', B and B' are depicted in Figure 1.3.

7. CONCLUSION

The generalized modus ponens (GMP) is a useful deduction scheme in case the input "x is A'" does not exactly match the premise of the IF–THEN rule "IF x is A THEN y is B". In a fuzzy set theoretical framework, the GMP is usually implemented by means of the compositional rule of inference (CRI). However, if the input set A' is a fuzzy set (instead of a crisp singleton), the derivation of an output set B' involves large computational efforts. Furthermore the obtained membership function B' often still has to be interpreted by some process of linguistic approximation.

In this chapter we have discussed three methods in which the derived membership function B' belongs to a predefined class of membership functions. The class of possible input fuzzy sets, the class of possible output fuzzy sets and the implementation of the GMP form together a so-called *closed system*. Main advantages of these systems are that the generated fuzzy set can easily be linguistically interpreted and that the computational overhead caused by the CRI, can be avoided.

In the Hellendoorn approach an output Γ–function B' is derived by simple manipulation of the parameters characterizing the input Γ – function A' and the Γ–functions A and B involved in the IF–THEN rule. In the Nafarieh–Keller approach the output fuzzy set B' is a real power B^n of the fuzzy set B in the consequent of the IF–THEN rule. The power n is determined by the degree of compatibility between the input fuzzy set A' and the fuzzy set A in the premise of the rule. Finally the generalized Bouchon approach we have discussed, presents a shortcut to obtain the same results the CRI would give, by means of manipulation of parameters that correspond to linguistic hedges.

References

[1] J. F. Baldwin, "A new approach to approximate reasoning using a fuzzy logic", Fuzzy sets and systems, 2(1979), 309–325

[2] B. Bouchon-Meunier, D. Dubois, L. Godo, H. Prade, "Fuzzy sets and possibility theory in approximate and plausible reasoning", Fuzzy sets in approximate reasoning and information systems, Kluwer Academic Publishers, (1999), 15–190

[3] B. Bouchon–Meunier, "Stability of Linguistic Modifiers Compatible with a Fuzzy Logic", Lecture Notes in Computer Science 313, Proceedings IPMU '88, Springer–Verlag, (1988), 63–70

[4] M. R. Emami, I. B. Türksen, A. A. Goldenberg, "A unified parameterized formulation of reasoning in fuzzy modeling and control", Fuzzy sets and systems, 108(1999), 59–81

[5] H. Hellendoorn, "Closure Properties of the Compositional Rule of Inference", Fuzzy Sets and Systems, 35(1990), 163–183

[6] H. Hellendoorn, "The generalized modus ponens considered as a fuzzy relation", Fuzzy sets and systems, 46(1992), 29–48

[7] A. Nafarieh, J. M. Keller, "A New Approach to Inference in Approximate Reasoning", Fuzzy Sets and Systems, 41(1991), 17–37

[8] A. Nafarieh, J. M. Keller, "A Fuzzy Logic Rule–Based Automatic Target Recognizer", International Journal of Intelligent Systems, 6(1991), 295–312

[9] I. B. Türksen, "Interval valued fuzzy sets based on normal forms", Fuzzy sets and systems, 20(1986), 191–210

[10] W. Van Leeckwijck, E. E. Kerre, "Defuzzification: criteria and classification", Fuzzy Sets and Systems, 108(1999), 159–178

[11] R. R. Yager, "An approach to inference in approximate reasoning", International Journal of Man-Machine Studies, 13(1980), 323–338

[12] L. A. Zadeh, "Calculus of Fuzzy Restrictions", in : "Fuzzy Sets and Their Applications to Cognitive and Decision Processes", L. A. Zadeh, K.-S. Fu, K. Tanaka, M. Shimura (Ed.), Academic Press Inc. New York, (1975), 1–40

Chapter 4

COMPOSITIONAL RULE OF INFERENCE BASED ON TRIANGULAR NORMS

Anna Kolesárová
Department of Mathematics, Fac. of Chemical Technology,
Slovak University of Technology,
Bratislava, Radlinského 9, 812 37 Bratislava, Slovakia
kolesaro@cvt.stuba.sk

Etienne E. Kerre
Etienne.Kerre@rug.ac.be
Department of Applied Mathematics and Computer Science
Ghent University
Krijgslaan 281, 9000 Ghent, Belgium
Etienne.Kerre@rug.ac.be

Abstract The compositional rule of inference as a generalization of the classical modus ponenes in the framework of approximate reasoning is discussed. Rules, observations, as well as conclusions are modeled by fuzzy sets, and the conjunction by triangular norms. The relationship with the special fuzzy quantities and their calculus is shown. In particular, the compositional rule of inference based on the limit triangular norms T_M and T_D, as well as on continuous Archimedean triangular norms is investigated, including the case of linear inputs and outputs. Several examples are given.

Keywords: Compositional rule of inference; Modus ponens; Triangular norm; Fuzzy quantity

1. COMPOSITIONAL RULE OF INFERENCE

In fuzzy logic and approximate reasoning there are several kinds of inference rules. The most important ones are the generalized modus ponens and the compositional rule of inference.

The generalized modus ponens (GMP) is a natural extension of the classical modus ponens, i.e., the inference rule of the form $(a \wedge (a \Rightarrow b)) \Rightarrow b$.

Consider two variables X, Y, whose values are in the universes $\mathbb{X}$ and $\mathbb{Y}$, respectively. Denote by $\mathcal{F}(\mathbb{X})$ and $\mathcal{F}(\mathbb{Y})$ the systems of all fuzzy subsets of $\mathbb{X}$ and $\mathbb{Y}$, respectively.

If A, A' and B, B' are fuzzy sets modeling the values of variables X and Y, i.e., $A, A' \in \mathcal{F}(\mathbb{X})$, $B, B' \in \mathcal{F}(\mathbb{Y})$, then the generalized modus ponens has the following form:

$$\begin{array}{lcccl} \text{Rule} & \text{If } X & \text{is} & A & \text{then } Y \text{ is } B \\ \text{Observation} & X & \text{is} & A' & \\ \text{Conclusion} & Y & \text{is} & B'. & \end{array} \tag{1.1}$$

According to this inference rule, given a fuzzy rule of the form (1) and the fact (observation) that "X is A'," we conclude that "Y is B'."

If there is a fuzzy relation $R \in \mathcal{F}(\mathbb{X} \times \mathbb{Y})$ expressing the relationship between X and Y, then the conclusion can be derived by the compositional rule of inference (CRI) which can be considered to be a special case of the generalized modus ponens. In that case, when it is explicitly given that X is in a fuzzy relation R to Y, instead of the previous scheme we have the following one:

$$\begin{array}{ccc} (X,Y) & \text{is} & R \\ X & \text{is} & A' \\ Y & \text{is} & B', \end{array}$$

and the membership function of the conclusion B' is given by:

$$B'(y) = \sup_{x \in \mathbb{X}} T\left(A'(x), R(x,y)\right), \quad y \in \mathbb{Y}, \tag{1.2}$$

where T is a triangular norm (t–norm for short).

For the definition and more properties of t–norms we refer the reader, e.g., to [17, 20, 33].

Note that we do not make any distinction between a fuzzy set A and its membership function, mostly denoted as μ_A.

This way of inference of the conclusion, when B' is determined by Eq. (2), is precisely called the sup–T compositional rule of inference. We will write

$$B' = A' \circ_T R. \tag{1.3}$$

The fuzzy subset B' representing the value of the variable Y is the image of the fuzzy subset A' in the given fuzzy relation R, and is obtained as the projection of the T-intersection of the cylindrical extension $ce(A')$ of A', $ce(A')(x,y) = A'(x)$, and the fuzzy relation R on the space $\mathbb{Y}$, i.e.,

$$B' = \text{proj}(ce(A') \cap_T R)_{|\mathbb{Y}}.$$

The philosophic background of CRI and GMP is discussed in more detail, e.g., in [6, 10, 12, 13, 31, 35].

If the fuzzy relation R is not explicitly given, it can be determined from the fuzzy statement (1) by means of an appropriate fuzzy implication. Then we write $R = A \to B$, and for each $x \in \mathbb{X}$ and each $y \in \mathbb{Y}$, the membership degree $R(x,y)$ is given by

$$R(x,y) = (A \to_I B)(x,y) = I\left(A(x), B(y)\right),$$

where I is the considered implication operator.
In this case, Eq.(2) has the form

$$B'(y) = \sup_{x \in \mathbb{X}} T\left(A'(x), I\left(A(x), B(y)\right)\right), \quad y \in \mathbb{Y}.$$

Note that the function $I : [0,1]^2 \to [0,1]$ should be an extension of the classical implication from the $\{0,1\}$-valued logic to the domain $[0,1]$ of truth values in fuzzy logic.

In general, to select an appropriate fuzzy implication for calculating the fuzzy relation R, is a difficult problem. Since the generalized modus ponens is an extension of the classical modus ponens, it is reasonable to ask GMP to coincide with the classical one in the case when $A' = A$, i.e., for the observation "X is A," the obtained conclusion should be "Y is B."
Expressed by means of the compositional rule of inference we will ask that

$$B = A \circ_T R, \tag{1.4}$$

or

$$B(y) = \sup_{x \in \mathbb{X}} T\left(A(x), I\left(A(x), B(y)\right)\right), \quad y \in \mathbb{Y}. \tag{1.5}$$

This requirement leads to the requirement of normal inputs. Therefore in the sequel all considered inputs will be normal, i.e., for each considered $A \in \mathcal{F}(\mathbb{X})$ there exists $x_0 \in \mathbb{X}$ such that $A(x_0) = 1$.

Fuzzy implications in the context of approximate reasoning are discussed in more detail, e.g., in [21], compare also [5, 16].

In fuzzy control there are often successfully used fuzzy relations R derived from the input-output information with the membership function defined for each $x \in \mathbb{X}$ and each $y \in \mathbb{Y}$ by

$$R(x, y) = T^* \left(A(x), B(y)\right),$$

where T^* is a t–norm.

The relation R can be understood as the Cartesian product based on the t–norm T^*, $R = A \times_{T^*} B$. Then Eq. (2) has the form

$$B'(y) = \sup_{x \in \mathbb{X}} T\left(A'(x), T^*\left(A(x), B(y)\right)\right), \quad y \in \mathbb{Y}. \tag{1.6}$$

In general, T^* can be different from T. When $T = T^*$ is a left continuous t–norm, due to its associativity we obtain

$$B'(y) = T\left(\sup_{x \in \mathbb{X}} T\left(A'(x), A(x)\right), B(y)\right), \quad y \in \mathbb{Y}. \tag{1.7}$$

If we denote in Eq.(7) the height of the T-intersection of the antecedent A and the observation A' by h, i.e.,

$$h = hgt\left(A \cap_T A'\right) = \sup_{x \in \mathbb{X}} T\left(A'(x), A(x)\right),$$

we obtain

$$B'(y) = T\left(h, B(y)\right), \quad y \in \mathbb{Y},$$

which is a useful formula for determining of the conclusion B'.

Various other cases of CRI are described also in the papers devoted to the analogical reasoning, see [1, 2, 3], compare also [32].

2. T-BASED CRI AND T-SUMS OF FUZZY QUANTITIES

In applications the values of input and output variables are usually modeled by means of appropriate fuzzy subsets of the real line $\mathbb{R}$, i.e., fuzzy quantities. Therefore in the sequel let $\mathbb{X} = \mathbb{Y} = \mathbb{R}$. Following [11, 13], we will use for the representation of linguistic notions the Φ–functions. Note that we will deal with more general types of Φ–functions than they were proposed in [13].

Definition 1 *Any convex, normal unimodal fuzzy quantity with a continuous membership function* $\Phi\colon \mathbb{R} \to [0,1]$ *with the peak in* 0 *(i.e.,* $\Phi(x) = 1$ *if and only if* $x = 0$*), will be called a shape.*

Since we do not differentiate between notation of fuzzy subsets and their membership functions, we will use the symbol Φ also for a shape. Immediately from Definition 1 we obtain the following claim.

Lemma 1 Φ *is a shape if and only if the functions* $\varphi_1, \varphi_2\colon [0,\infty[\to [0,1]$ *defined by*

$$\varphi_1(x) = \Phi(-x), \quad \varphi_2(x) = \Phi(x),$$

are non-increasing, continuous and $\varphi_1^{-1}(\{1\}) = \varphi_2^{-1}(\{1\}) = \{0\}$.

Definition 2 *A mapping* $\varphi : [0,\infty[\to [0,1]$ *which is continuous, non-increasing and* $\varphi(x) = 1$ *if and only if* $x = 0$*, will be called a basic shape. The set of all basic shapes will be denoted by* $\mathcal{S}$.

It is straightforward that for any $\varphi \in \mathcal{S}$ and $\alpha \in]0,\infty[$, the function $\varphi^{(\alpha)} : [0,\infty[\to [0,1]$, defined by

$$\varphi^{(\alpha)}(x) = \varphi\left(\frac{x}{\alpha}\right), \tag{1.8}$$

is also a basic shape.

If Φ is a shape then for its membership function, which will be called a Φ–function, we can write

$$\Phi(x) = \begin{cases} 1 & \text{if } x = 0 \\ \varphi_1(x) & \text{if } x < 0 \\ \varphi_2(x) & \text{if } x > 0, \end{cases}$$

where φ_1, φ_2 are basic shapes.

For the fact that a shape Φ is determined by means of basic shapes φ_1, φ_2 we will use the notation $\Phi = (\varphi_1, \varphi_2)$.

Definition 3 *Let* $a, b \in \mathbb{R}$, $a \le b$, $\alpha, \beta \in]0,\infty[$. *Let* Φ *be a shape. Then the symbol* $\Phi(a,b,\alpha,\beta)$ *will denote a fuzzy quantity with membership function*

$$\Phi(a,b,\alpha,\beta)(x) = \begin{cases} 1 & \textit{if } x \in [a,b] \\ \varphi_1\left(\frac{a-x}{\alpha}\right) & \textit{if } x < a \\ \varphi_2\left(\frac{x-b}{\beta}\right) & \textit{if } x > b. \end{cases} \tag{1.9}$$

Remark 1.
(i) The expression (9) can be also written in the form:

$$\Phi(a,b,\alpha,\beta)(x) = \begin{cases} 1 & \text{if } x \in]a,b[\\ \varphi_1\left(\frac{a-x}{\alpha}\right) & \text{if } x \leq a \\ \varphi_2\left(\frac{x-b}{\beta}\right) & \text{if } x \geq b, \end{cases}$$

which allows us to extend our considerations to fuzzy quantities with unbounded kernels. Namely, we can suppose that $a, b \in [-\infty, \infty]$ and $\alpha, \beta \in]0, \infty[$. Then obviously, if $b = +\infty$, the normality of the discussed quantities forces $a < \infty$, and

$$\Phi(a,+\infty,\alpha,\beta) = \begin{cases} 1 & \text{if } x > a \\ \varphi_1\left(\frac{a-x}{\alpha}\right) & \text{if } x \leq a. \end{cases}$$

Evidently, the kernel $Ker\,\Phi(a,+\infty,\alpha,\beta) = [a,\infty[$, whenever $a > -\infty$ and the basic shape φ_2 as well as the spread β have no influence on the fuzzy quantity $\Phi(a,+\infty,\alpha,\beta)$.

Note that fuzzy quantities of this type are so called LT-fuzzy numbers. The case for $a = -\infty$ is similar to the previous one.

(ii) It is evident that $\Phi(a,b,\alpha,\beta)$ is a normal, convex fuzzy quantity with continuous membership function and the kernel $Ker\,\Phi(a,b,\alpha,\beta) = [a,b] \cap \mathbb{R}$. With this notation we have $\Phi = \Phi(0,0,1,1)$.
Vice versa, any normal, convex fuzzy quantity A with continuous membership function can be represented in the form $\Phi(a,b,\alpha,\beta)$, where $Ker\,A = [a,b] \cap \mathbb{R}$ and if $a, b \in \mathbb{R}$, then

$$\varphi_1(x) = A(a - \alpha x), \quad \varphi_2(x) = A(b + \beta x), \quad x \in [0, \infty[.$$

Note that the choice of α and β in this representation is free, in general.

Example 1.

The fuzzy quantity A with

$$A(x) = \begin{cases} \exp(x-3) & \text{if } x \leq 3 \\ \exp(-\frac{x-3}{2}) & \text{if } x > 3, \end{cases}$$

can be represented by $A = \Phi(3,3,1,2)$, where $\Phi(x) = \exp(-|x|)$, i.e., $\Phi = (\varphi_1, \varphi_2)$, $\varphi_1(x) = \varphi_2(x) = \exp(-x)$, $x \in [0, \infty[$.

Example 2.

The fuzzy quantity (trapezoidal fuzzy interval) B,

$$B(x) = \begin{cases} 1 & \text{if } x \in [3,4] \\ \frac{x-1}{2} & \text{if } x \in [1,3[\\ \frac{7-x}{3} & \text{if } x \in]4,7] \\ 0 & \text{otherwise,} \end{cases}$$

can be represented in the following way: $B = \Phi(3,4,2,3)$, $\Phi(x) = \max(0, 1 - |x|)$, i.e., $\Phi = (\varphi_1, \varphi_2)$, $\varphi_1(x) = \varphi_2(x) = \max(0, 1-x)$, $x \in [0, \infty[$.

Example 3.

Let $C(x) = \frac{1+\exp(-|x|)}{2}$, $x \in \mathbb{R}$. Then

$$C(x) = \Phi(0,0,1,1), \ \Phi = (\varphi_1, \varphi_2),$$

where

$$\varphi_1(x) = \varphi_2(x) = \frac{1+\exp(-x)}{2}, \ x \in [0, \infty[.$$

Note that $\lim_{x \to \pm\infty} \Phi(x) \neq 0$.

In general, LR-fuzzy intervals $U = (a, b, \alpha, \beta)_{\mathrm{LR}}$, see, e.g., [7, 8] can be expressed by Φ–functions in the form $U = \Phi(a, b, \alpha, \beta)$, $\Phi = (\varphi_1, \varphi_2)$, with

$$\varphi_1(x) = \begin{cases} \mathrm{L}(x) & \text{if } x \in [0,1] \\ 0 & \text{if } x > 1, \end{cases}$$

and analogously φ_2 can be expressed by R.

The antecedents and consequences in fuzzy rules and the observations are usually modeled by normal convex fuzzy quantities with continuous membership functions which means that they can be described as $\Phi(a, b, \alpha, \beta)$ with appropriate parameters (see Definition 3 and Remark 1). Moreover, in the next part we will consider that relations R modeling the relationships between variables X and Y are functions of the difference $y - x$ and can be expressed by means of Φ–functions.

Example 4.

If the relationship between two variables X and Y: "X is approximately equal to Y," is modeled by the relation R given by

$$R(x,y) = \begin{cases} 1 - |y-x| & \text{if. } |y-x| \leq 1 \\ 0 & \text{otherwise,} \end{cases}$$

then R can be expressed by the Φ–function $R(x,y) = \Phi(0,0,1,1)(y-x)$, where $\Phi = (\varphi, \varphi)$ and $\varphi(u) = \max(1-u, 0)$ for $u \in [0, \infty[$.

Example 5.

Let the relationship between X and Y: "X is much less than Y," be modeled by the relation R given by

$$R(x,y) = \begin{cases} 0 & \text{if } \; y-x \le 10 \\ \frac{y-x}{10} - 1 & \text{if } \; 10 < y-x \le 20 \\ 1 & \text{if } \; y-x > 20. \end{cases}$$

Then

$$R(x,y) = \Phi(20,+\infty,10,\beta)(y-x) = \begin{cases} 1 & \text{if } y-x > 20 \\ \varphi_1\left(\frac{20-(y-x)}{10}\right) & \text{if } y-x \le 20, \end{cases}$$

where $\varphi_1(u) = \max(1-u,0)$, $u \in [0,\infty[$. As it was explained above, the parameter β and basic shape φ_2 are irrelevant.

In the next part let the observation A' be modeled by a shape $\Phi_1 = (\varphi_{11},\varphi_{21})$, and the relation R by a shape $\Phi_2 = (\varphi_{12},\varphi_{22})$, both with bounded kernels. Namely,

$$A'(x) = \Phi_1(a_1,b_1,\alpha_1,\beta_1)(x) = \begin{cases} 1 & \text{if } x \in [a_1,b_1] \\ \varphi_{11}\left(\frac{a_1-x}{\alpha_1}\right) & \text{if } x < a_1 \\ \varphi_{21}\left(\frac{x-b_1}{\beta_1}\right) & \text{if } x > b_1, \end{cases} \tag{1.10}$$

and

$$R(x,y) = \Phi_2(a_2,b_2,\alpha_2,\beta_2)(y-x) = \begin{cases} 1 & \text{if } y-x \in [a_2,b_2] \\ \varphi_{12}\left(\frac{a_2-(y-x)}{\alpha_2}\right) & \text{if } y-x < a_2 \\ \varphi_{22}\left(\frac{(y-x)-b_2}{\beta_2}\right) & \text{if } y-x > b_2. \end{cases} \tag{1.11}$$

Then the conclusion $B' = A' \circ_T R$ determined by the sup $-T$ CRI is according to (2) given by

$$B'(y) = \sup_{x\in\mathbb{R}} T\left(\Phi_1(a_1,b_1,\alpha_1,\beta_1)(x), \Phi_2(a_2,b_2,\alpha_2,\beta_2)(y-x)\right), \quad y \in \mathbb{R}. \tag{1.12}$$

It is well known that the sum of fuzzy quantities U, V based on a t–norm T is a fuzzy quantity $U \oplus_T V$ with membership function given by

$$U \oplus_T V(y) = \sup_{x\in\mathbb{R}} T\left(U(x), V(y-x)\right), \quad y \in \mathbb{R}. \tag{1.13}$$

Comparing (12) and (13) we obtain

$$B' = A' \circ_T R = \Phi_1(a_1, b_1, \alpha_1, \beta_1) \oplus_T \Phi_2(a_2, b_2, \alpha_2, \beta_2). \qquad (1.14)$$

Therefore for determination of the conclusion B' by means of sup $-T$ *CRI* under assumptions (10) and (11), the results known for T-sums of fuzzy quantities can be applied.
It is easy to see that $Ker\, B'$ of B' determined by means of (14) is always the interval $[a_1 + a_2, b_1 + b_2]$. Moreover, the following theorem shows that it is possible to calculate the outputs B' in Eq. (14) by means of T-sums of special fuzzy quantities on $[0, \infty[$ - basic shapes.

Theorem 1 *The conclusion B' in Eq. (14) can be expressed in the form*

$$B' = \Phi_3(a, b, 1, 1), \qquad (1.15)$$

where

$$a = a_1 + a_2, \quad b = b_1 + b_2, \quad \Phi_3 = (\varphi_{13}, \varphi_{23}),$$

and

$$\varphi_{13} = \varphi_{11}^{(\alpha_1)} \oplus_T \varphi_{12}^{(\alpha_2)}, \qquad (1.16)$$

$$\varphi_{23} = \varphi_{21}^{(\beta_1)} \oplus_T \varphi_{22}^{(\beta_2)}. \qquad (1.17)$$

The proof of this assertion follows from the proof of the independence of the left and right sides of fuzzy intervals (not necessarily LR-fuzzy intervals) which is proved, e.g., in [30] and the definition of a function $\varphi^{(\alpha)}$.

Note that due to Remark 1 (i), Eqs. (16) and (17) can be used also in the case of unbounded kernels of A or R (with possible superfluous calculation of output basic shapes φ_{13} or φ_{23}).

The following lemma will be useful for a more practical formulation and easier applicability of the results.

Lemma 2 *Let $a, b \in \mathbb{R}$. Then*
$\Phi(a, b, \alpha, \beta) = \Phi^*(a, b, 1, 1)$ *if and only if* $\Phi = (\varphi_1, \varphi_2)$ *and* $\Phi^* = (\varphi_1^{(\alpha)}, \varphi_2^{(\beta)})$.

In what follows we focus our attention on the overview of results which can be derived for the sup $-T$ *CRI* from known results for T-sums of fuzzy quantities which can be found in [7, 8, 22-28].

3. *CRI* BASED ON *T*–NORMS T_M AND T_D

Simple results for the sup–T CRI can be obtained when $T = T_D$ or $T = T_M$ and both the observation and relation parts are represented by Φ–functions $\Phi_1 = \Phi_2$.
Note that t–norms T_M (minimum) and T_D (drastic product) are for $(x, y) \in [0,1]^2$ defined by

$$T_M(x,y) = \min(x,y)\ , \qquad T_D(x,y) = \begin{cases} \min(x,y) & \text{if } \max(x,y) = 1 \\ 0 & \text{otherwise.} \end{cases}$$

In that case immediately from results proved, e.g., in [7, 8, 22, 25, 30] it follows that the conclusion B' can be also represented by the same Φ–function.

Proposition 1 *Let Φ be a shape and let membership functions of an observation A' and a relation R are expressed in the form*

$$A'(x) = \Phi(a_1, b_1, \alpha_1, \beta_1)(x) \qquad (1.18)$$

$$R(x,y) = \Phi(a_2, b_2, \alpha_2, \beta_2)(y - x). \qquad (1.19)$$

Then the conclusion $B' = A' \circ_T R$ derived by means of the sup–T CRI is for each $y \in \mathbb{R}$ given by

(i) $B'(y) = \Phi(a_1 + a_2, b_1 + b_2, \alpha_1 + \alpha_2, \beta_1 + \beta_2)(y)$, *if* $T = T_M$,

(ii) $B'(y) = \Phi\left(a_1 + a_2, b_1 + b_2, \max(\alpha_1, \alpha_2), \max(\beta_1, \beta_2)\right)(y)$, *if* $T = T_D$.

Theorem 1 and results from fuzzy arithmetics [25, 30] ensure that B' in (i) of Proposition 1 is described by $\Phi^*(a, b, 1, 1)$ with $a = a_1 + a_2$, $b = b_1 + b_2$ and $\Phi^* = (\varphi_1^{(\alpha_1+\alpha_2)}, \varphi_2^{(\beta_1+\beta_2)})$, whereby φ_1, φ_2 are components of Φ. Using Lemma 2 we obtain B' in the form stated in (i). The result (ii) can be explained in a similar way.

The result corresponding to the claim (i) in Proposition 1 was proved in [13] in the case of a Φ–function describing LR–fuzzy intervals (which have only bounded supports). Note that Φ–functions in [13] are more specific than those considered here.

In the next part we extend the results for CRI based on t–norms T_M and T_D in situations when the observation and relation parts are represented by Φ–functions Φ_1, Φ_2 which are different in general.

To this end we need the notion of the pseudo–inverse of a non–increasing function.

Definition 4 *[19] Let $g : [a,b] \to [c,d]$, where $a,b,c,d \in [-\infty, \infty]$, be a non–increasing function. The function $g^{(-1)} : [c,d] \to [a,b]$ defined by*

$$g^{(-1)}(t) = \sup\{x \in [a,b];\ g(x) > t\}$$

is called the pseudo–inverse of the function g.

In our case, a basic shape φ is a non–increasing function whose domain is the interval $[0,\infty[$ and range $[0,1]$. Each basic shape φ can be uniquely extended into the extended shape $\tilde{\varphi} : [0,\infty] \to [0,1]$ by

$$\tilde{\varphi}(x) = \begin{cases} \varphi(x) & \text{for } x < \infty \\ \lim\limits_{x\to\infty} \varphi(x) & \text{otherwise.} \end{cases}$$

Because of the continuity of the shapes there is a one-to-one correspondence between shapes and extended shapes, and hence we will not differentiate between them.

By Theorem 1, for determining the conclusion B' in Eq. (14) it is important to determine the sum of corresponding basic shapes. The following results for calculation with basic shapes are proved in [30].

Theorem 2 *Let φ, η be basic shapes. Then*

(i) $\varphi \oplus_{T_M} \eta = \left(\varphi^{(-1)} + \eta^{(-1)}\right)^{(-1)}$,

(ii) $\varphi \oplus_{T_D} \eta = \max(\varphi, \eta)$.

The next example illustrates the procedure of calculating the sum in the case described in (i) of this theorem.

Example 6.

Let for $x \in [0,\infty[$

$$\varphi(x) = 2^{-x} \quad \text{and} \quad \eta(x) = \max(0, 2 - 2^x).$$

Then the pseudo–inverses of φ and η are for each $t \in [0,1]$ defined by

$$\varphi^{(-1)}(t) = -\log_2 t \quad \text{and} \quad \eta^{(-1)}(t) = \log_2(2-t).$$

Therefore

$$\left(\varphi^{(-1)} + \eta^{(-1)}\right)^{(-1)}(x) = \frac{2}{1+2^x}.$$

Remark 2.

Proposition 1 describes special cases which are consequences of the previous theorem for $\varphi = \eta$ and Theorem 1.

(i) It can be easily shown that

$$\left(\varphi^{(\alpha)}\right)^{(-1)} = \alpha\varphi^{(-1)}.$$

Using this property and the assertion (i) of Theorem 2 for $\eta = \varphi$ we obtain

$$\begin{aligned} \varphi^{(\alpha)} \oplus_{T_M} \varphi^{(\beta)} &= \left(\left(\varphi^{(\alpha)}\right)^{(-1)} + \left(\varphi^{(\beta)}\right)^{(-1)}\right)^{(-1)} \\ &= \left(\alpha\varphi^{(-1)} + \beta\varphi^{(-1)}\right)^{(-1)} \\ &= \left((\alpha+\beta)\,\varphi^{(-1)}\right)^{(-1)} = \varphi^{(\alpha+\beta)}. \end{aligned}$$

This result and Theorem 1 ensure already the validity of (i) in Proposition 1.

(ii) Directly from the definition of a function $\varphi^{(\alpha)}$ it can be derived that

$$\max\left(\varphi^{(\alpha)}, \varphi^{(\beta)}\right) = \varphi^{(\max(\alpha,\beta))}.$$

Thus, due to Theorem 2,

$$\varphi^{(\alpha)} \oplus_{T_D} \varphi^{(\beta)} = \varphi^{(\max(\alpha,\beta))},$$

which together with Theorem 1 give the claim (ii) of Proposition 1.

4. *CRI* BASED ON CONTINUOUS ARCHIMEDEAN T–NORMS

Following [13], Fullér and Zimmermann derived an exact formula for calculation of B' in the case that Φ–functions representing A' and R have the same components which are twice differentiable and concave and T is a continuous Archimedean t–norm with strictly convex twice differentiable additive generator. The membership function of B' can be determined by this formula only under the assumption that the left and right spreads of A' do not differ from the left and right spreads of R, i.e., A' and R are described by functions $\Phi(a_i, b_i, \alpha, \beta)$, $i = 1, 2$.
Note that conditions concerning the differentiability were removed in [14]. Further generalizations of the Fullér-Zimmermann result can be found, e.g., in [4, 26, 27, 28, 30].

By (14), the conclusion B' determined by the sup $-T$ CRI under assumptions (10) and (11) is given by $B' = \Phi_1 \oplus_T \Phi_2$. Usually, the determination of the membership function of the sum is a very difficult and complicated task. We show several situations when it is possible to determine the sum, and consequently the conclusion B', in a quite simple way. In general, we will consider shapes Φ_1, Φ_2, which need not be necessarily the same. In this section we will ask T to be a continuous Archimedean t–norm [17].

Denote by f an additive generator of the considered continuous Archimedean t–norm T, i.e., a continuous, strictly decreasing function $f : [0,1] \to [0,\infty]$ with $f(0) = 1$, such that

$$T(x,y) = f^{-1}\left(\min\left(f(0), f(x) + f(y)\right)\right), \quad (x,y) \in [0,1]^2.$$

As it has been already shown, for determining the conclusion B' in (14) it is important to determine the sum of corresponding basic shapes. Therefore we will formulate the next results for basic shapes and then show some applications.

Theorem 3 *[30] Let T be a continuous Archimedean t–norm with an additive generator f and let φ, η be basic shapes. Let $f \circ \varphi$ and $f \circ \eta$ be concave functions. Then*

$$\varphi \oplus_T \eta = \max(\varphi, \eta) = \varphi \oplus_{T_D} \eta.$$

Due to last results proved in [15] the concavity of the composed functions $f \circ \varphi$ and $f \circ \eta$ can be relaxed to the subadditivity.

The next rather general theorem was also shown in [30].

Theorem 4 *Let T be a continuous Archimedean t–norm with an additive generator f. Let $K : [0,\infty[\to [0,\infty[$ be a strictly increasing convex function with $K(0) = 0$. Then for each $\alpha \in]0,\infty[$, the function $\varphi_{(\alpha)} : [0,\infty[\to [0,1]$ defined by*

$$\varphi_{(\alpha)}(x) = f^{-1}\left(\min\left(f(0), \alpha K\left(\frac{x}{\alpha}\right)\right)\right),$$

is a basic shape.
Moreover, for any $\alpha, \beta \in]0,\infty[$ we have

$$\varphi_{(\alpha)} \oplus_T \varphi_{(\beta)} = \varphi_{(\alpha+\beta)}.$$

Example 7.

(i) Consider the Φ–function $\Phi = (\varphi, \varphi)$ with the basic shape $\varphi(x) = \max(0, 1 - \sqrt{x})$, $x \in [0, \infty[$ and the Lukasiewicz t–norm T_L given by $T_L(x, y) = \max(0, x + y - 1)$, $x, y \in [0, 1]$, with the additive generator $f(x) = 1 - x$.

Further, let us consider that the observation A' and the relation R are represented by (18) and (19), i.e.,

$$\begin{aligned} A'(x) &= \Phi(a_1, b_1, \alpha_1, \beta_1)(x) \\ R(x, y) &= \Phi(a_2, b_2, \alpha_2, \beta_2)(y - x). \end{aligned}$$

Since $f \circ \varphi(x) = \min(1, \sqrt{x})$ is a concave function, according to Theorem 3, we obtain $\varphi \oplus_{T_L} \varphi = \varphi \oplus_{T_D} \varphi$ and then $B' = A' \circ_{T_L} R = A' \circ_{T_D} R$. Therefore the conclusion B' can be further determined using Proposition 1(ii).

(ii) Let $\Phi = (\varphi, \varphi)$, where $\varphi(x) = \exp(-x^2)$. Consider the product t–norm T_P, given by $T_P(x, y) = xy$, $x, y \in [0, 1]$, with additive generator $f(x) = -\log x$.

Clearly, $f \circ \varphi^{(\alpha)}(x) = -\log \exp\left(-\frac{x^2}{\alpha^2}\right) = \frac{x^2}{\alpha^2}$.

Put $K(x) = x^2$. The function K satisfies all conditions given in Theorem 4. Then

$$\varphi_{(\alpha^2)}(x) = f^{-1}\left(\alpha^2 \left(\frac{x}{\alpha^2}\right)^2\right) = \exp\left(-\frac{x^2}{\alpha^2}\right) = \varphi^{(\alpha)}.$$

Therefore

$$\varphi^{(\alpha)} \oplus_{T_P} \varphi^{(\beta)} = \varphi_{(\alpha^2)} \oplus_{T_P} \varphi_{(\beta^2)} = \varphi_{(\alpha^2+\beta^2)} = \varphi^{(\sqrt{\alpha^2+\beta^2})}.$$

Now, if we consider that the observation A' and the relation R are represented by (18) and (19) as above, then the conclusion $B' = A' \circ_T R$ is given by

$$B'(y) = \Phi\left(a_1 + a_2, b_1 + b_2, \sqrt{\alpha_1^2 + \alpha_2^2}, \sqrt{\beta_1^2 + \beta_2^2}\right)(y), \quad y \in \mathbb{R}.$$

Recall that fuzzy quantities $\Phi(\mu, \mu, \sigma^2, \sigma^2) = G(\mu, \sigma^2)$ are called Gaussian fuzzy numbers. They are given by the formula

$$G(\mu, \sigma^2)(x) = \exp\left(-\frac{(x - \mu)^2}{\sigma^2}\right),$$

and

$$G(\mu_1, \sigma_1^2) \oplus_{T_P} G(\mu_2, \sigma_2^2) = G(\mu_1 + \mu_2, \sigma_1^2 + \sigma_2^2).$$

5. *CRI* IN THE CASE OF LINEAR INPUTS AND OUTPUTS

Very often the values of linguistic variables are modeled by means of linear fuzzy intervals. By a linear fuzzy interval we call a fuzzy set $A\colon \mathbb{R} \to [0,1]$ with the membership function given by

$$A(x) = \begin{cases} 1 & \text{if } x \in [a,b] \\ 1 - \frac{a-x}{\alpha} & \text{if } a-\alpha \le x < a \\ 1 - \frac{x-b}{\beta} & \text{if } b < x \le b+\beta \\ 0 & \text{otherwise,} \end{cases}$$

where $a, b \in \mathbb{R}$, $a \le b$ and $\alpha, \beta > 0$.

Recall that for $a = b$ the linear fuzzy intervals are called triangular fuzzy intervals and for $a < b$ trapezoidal fuzzy intervals.

In the next part, let both the observation and relation parts be represented by the Φ–function with linear components, namely

$$A'(x) = \Phi(a_1, b_1, \alpha_1, \beta_1)(x), \tag{1.20}$$
$$R(x,y) = \Phi(a_2, b_2, \alpha_2, \beta_2)(y-x), \tag{1.21}$$

and

$$\Phi = (\varphi_1, \varphi_2), \quad \varphi_1(x) = \varphi_2(x) = \max(0, 1-x), \quad x \in [0, \infty[. \tag{1.22}$$

It is easy to see that A', R and Φ defined by Eqs. (20), (21) and (22) respectively, are linear fuzzy intervals in the previous sense.

In what follows, using the results for T–sums of linear fuzzy intervals [25, 26] we can express the conditions under which also the conclusion B' derived by means of the sup–T *CRI* will be represented by the same Φ–function, i.e., $B' = \Phi(a, b, \alpha, \beta)$, $\Phi = (\varphi_1, \varphi_2)$, where $\varphi_1(x) = \varphi_2(x) = \max(0, 1-x)$, $x \in [0, \infty[$. Then for determining B' it is enough to find only parameters a, b, α, β.

The first results for this case when $T = T_D$ or $T = T_M$ were already given in Proposition 1 which was formulated for general functions φ_1, φ_2.

Let T_L denote the Lukasiewicz t–norm. Then the following assertion is true [26].

Proposition 2 *Let an observation A' and a relation R be given by Eqs. (20), (21) and (22). Let T be a t–norm such that $T \le T_L$. Then the conclusion B' derived by means of the sup–T CRI is of the form*

$$B'(y) = \Phi(a, b, \alpha, \beta)(y), \quad y \in \mathbb{R}, \tag{1.23}$$

where

$$a = a_1 + a_2, \quad b = b_1 + b_2, \quad \alpha = \max(\alpha_1, \alpha_2), \quad \beta = \max(\beta_1, \beta_2).$$

This means that the conclusion B' is a linear fuzzy interval and that in this case $A' \circ_T R = A' \circ_{T_D} R$ for any considered t–norm $T \leq T_L$.

Corollary 1 *Let an observation A' and a relation R be given by Eqs. (20), (21) and (22). Let T be a continuous Archimedean t–norm with a concave additive generator f. Then the membership function of the conclusion B' derived under the sup–T composition is also given by Eq. (23).*

This assertion follows immediately from Proposition 2 and the fact that a continuous Archimedean t–norm T with the concave additive generator is weaker than the Lukasiewicz t–norm, i.e., $T \leq T_L$ [18], see also [26].

Now, let us consider the system $\{T_s^Y\}_{s \in]0,\infty[}$ of Yager's t–norms which are defined by

$$T_s^Y(x, y) = \max\left(0, 1 - ((1-x)^s + (1-y)^s)^{1/s}\right), \quad (x, y) \in [0, 1]^2$$

and whose normed additive generators are given by $f_s(x) = (1-x)^s$, $x \in [0, 1]$.

It is easily seen that for $s \in]0, 1]$ the generators f_s are concave functions and therefore Corollary 1 for determining B' under the sup-T_s^Y composition can be applied. Moreover, for Yager's t–norm the following proposition is true.

Proposition 3 *Let an observation A' and a relation R be given by (20), (21) and (22). Let $T_s^Y, s > 1$ be the Yager t–norm. Then the conclusion B' derived by means of CRI based on the t–norm T_s^Y is of the form*

$$B'(y) = \Phi(a, b, \alpha, \beta)(y), \quad y \in \mathbb{R}, \tag{1.24}$$

where

$$a = a_1 + a_2, \quad b = b_1 + b_2, \quad \alpha = (\alpha_1^r + \alpha_2^r)^{1/r}, \quad \beta = (\beta_1^r + \beta_2^r)^{1/r}, \tag{1.25}$$

and

$$r = \frac{s}{s-1}. \tag{1.26}$$

Remark 3.

Note that due to the Mesiar transformation principle [29], the results given in Propositions 2, 3 and Corollary 1 for linear shapes can be generalized for arbitrary Φ–functions with components $\varphi_1 = \varphi_2 = \varphi$, which are strictly monotone on their supports and $\lim_{x\to\infty} \varphi(x) = 0$.

Let T be a continuous Archimedean t–norm with additive generator f and $f^{(-1)} = \varphi$. Denote by T^s a t–norm generated by a power f^s, $s \in]0, \infty[$. Let $\Phi = (\varphi, \varphi)$. Then by the generalization of the previous results we obtain

(i) By CRI based on any t–norm $T^* \leq T$ we obtain the same conclusions as by CRI based on the drastic product T_D (compare with Proposition 2).

(ii) The previous assertion (i) also holds for CRI based on any t–norm T^s with $s \leq 1$.

(iii) The conclusion B' derived by CRI based on any t–norm T^s with $s > 1$ is of the form (24) and the parameters α and β are also given by (25) and (26) (compare with Proposition 3).

Let us illustrate the described generalization by the following example.

Example 8.

(i) Consider $\varphi(x) = \exp(-x)$. Then $\varphi^{(-1)}(x) = -\log x$ is just an additive generator of the product t–norm T_P given by $T_P(x, y) = xy$. Therefore CRI based on any t–norm $T^* \leq T_P$ gives the same conclusions as CRI based on T_D, see Remark 3(i), and so, for determination of B' we can use Proposition 1(ii).

(ii) Triangular norms which are generated by s-powers of the generator of the t–norm T_P, i.e., by functions $(-\log x)^s$, $s \in]0, \infty[$, belong to the Aczél–Alsina family of t–norms $\{T_s^{AA}\}_{s\in]0,\infty[}$, see [20].

Then, according to Remark 3(ii), the conclusions B' derived by CRI based on T_s^{AA} for $s \leq 1$ are the same as the conclusions derived by CRI based on T_D, see Proposition 1(ii).

Moreover, according to Remark 3(iii), the conclusion derived by CRI based on T_s^{AA} for $s > 1$ is determined by Eqs. (24)–(26).

Note that it is possible to generalize the results also for different components φ_1, φ_2.

Acknowledgments

This work was supported by the grants VEGA 2/6087/99 and 1/7076/20.

References

[1] B. Bouchon-Meunier, J. Delechamp, C. Marsala and M. Rifqi, "Several forms of fuzzy analogical reasoning," *FUZZ-IEEE'97*, Barcelona, 1997, 45–50.

[2] B. Bouchon-Meunier and L. Valverde, "Analogical reasoning and fuzzy resemblance," *Intelligent Systems* (eds. B. Bouchon-Meunier, L. Valverde, R. Yager), Elsevier Science Pub., 1993, 247-255.

[3] B. Bouchon-Meunier, J. Delechamp, C. Marsala, R. Mesiar and M. Rifqi, "Fuzzy deductive reasoning and analogical scheme," *Proceedings EUSFLAT'99*, Palma de Mallorca, 1999, 139-142.

[4] B. De Baets and A. Marková–Stupňanová, "Analytical expressions for addition of fuzzy intervals," *Fuzzy Sets and Systems* **91** (1997) 203–213.

[5] B. De Baets and E.E. Kerre, "The generalized modus ponens and triangular fuzzy data model," *Fuzzy Sets and Systems* **59** (1993) 305–317.

[6] D. Driankov, H. Hellendoorn and M. Reinfrank, *An Introduction to Fuzzy Control*, Second Edition, Springer-Verlag, Berlin, 1996.

[7] D. Dubois and H. Prade, "Additions of interactive fuzzy numbers," *IEEE Trans. Automat. Control* **26** (1981) 926–936.

[8] D. Dubois and H. Prade, "Fuzzy numbers: An overview," *Analysis of Fuzzy Information* (ed. J.C. Bezdek), CRC Press, Boca Raton, 1988, 3–39.

[9] D. Dubois, E.E. Kerre, R. Mesiar and H. Prade, "Fuzzy Interval Analysis," *Fundamentals of Fuzzy Sets*, The Kluwer handbook series on fuzzy sets (D. Dubois, H. Prade, eds.), Kluwer, Dordrecht, 1999, 483-582.

[10] R. Fullér, *Fuzzy Reasoning and Fuzzy Optimization*, Turku Centre for Computer Science, TUSC General Publication No 9, 1998.

[11] R. Fullér and H.-J. Zimmermann, "On computation of the compositional rule of inference under triangular norms," *Fuzzy Sets and Systems* **51** (1992) 267–275.

[12] S. Gottwald, *Fuzzy Sets and Fuzzy Logic. Foundations of Applications -from a Mathematical Point of View*, Vieweg, Braunschweig/Wiesbaden, 1993.

[13] H. Hellendoorn, "Closure properties of the compositional rule of inference," *Fuzzy Sets and Systems* **35** (1990) 29–48 .

[14] D.H. Hong and S.Y Hwang, "On the compositional rule of inference under triangular norms," *Fuzzy Sets and Systems* **66** (1994) 25–38.

[15] D.H. Hong, "Some results on the additions of fuzzy intervals," *Fuzzy Sets and Systems*, accepted for publication.

[16] E.E. Kerre and D. Ruan, "The power of fuzzy relations for representation of imprecise knowledge," *World Automation Congress'98 Proceedings*, Anchorage, Alaska, USA, ISSCI-054.1-054.6 (CD-rom, ISBN: 1-889335-07-X).

[17] E.P. Klement and R. Mesiar, "Triangular norms," *Tatra Mountains Math. Publ.* **13** (1997) 169–194.

[18] E.P. Klement, R. Mesiar and E.Pap, "A characterization of the ordering of continuous t–norms," *Fuzzy Sets and Systems* **86** (1997) 189–195.

[19] E.P. Klement, R. Mesiar and E. Pap, "Quasi- and pseudo–inverses of monotone functions, and construction of t–norms," *Fuzzy Sets and Systems* **104** (1999) 3–13.

[20] E.P. Klement, R. Mesiar and E. Pap, *Triangular Norms*, Kluwer, 2000 (to appear).

[21] G.J. Klir and B. Yuan, *Fuzzy Sets and Fuzzy Logic: Theory and Applications*, Prentice Hall, 1995.

[22] M. Mareš, *Computation over fuzzy quantities*, CRC Press, Boca Raton, 1994.

[23] A. Marková, "Additions of $L-R$ fuzzy numbers," *Busefal* **63** (1995) 25–29.

[24] A. Marková, "T-sum of $L - R$ fuzzy numbers," *Fuzzy Sets and Systems* **85** (1997) 379–384.

[25] A. Kolesárová, "Triangular norm-based addition of linear fuzzy numbers," *Tatra Mountains Math. Publ.* **6** (1995) 75–81.

[26] A. Kolesárová, "Triangular norm-based addition preserving linearity of T-sums of linear fuzzy intervals," *Mathware and Soft Computing* **1** (1998) 91–98.

[27] A. Kolesárová, "Similarity preserving t–norm-based additions of fuzzy numbers," *Fuzzy Sets and Systems* **91** (1997) 215–229.

[28] R. Mesiar, "A note to the T-sum of $L - R$ fuzzy numbers," *Fuzzy Sets and Systems* **79** (1996) 259–261.

[29] R. Mesiar, "Shape preserving additions of fuzzy intervals," *Fuzzy Sets and Systems* **86** (1997) 73–78.

[30] R. Mesiar, "Triangular norm-based addition of fuzzy intervals," *Fuzzy Sets and Systems* **91** (1997) 231–237.

[31] W. Pedrycz, *Fuzzy Control and Fuzzy Systems*, 2nd revised edition, Research Stud. Press, Taunton and Wiley, New York, 1992.

[32] A. Pradera, E. Trillas and S. Cubillo, "On modus ponens generating functions," *Int. J. Uncertainty, Fuzziness and Knowledge–Based Systems*, submitted.

[33] B. Schweizer and A. Sklar, "Associative functions and abstract semigroups," *Publ. Math. Debrecen* **10** (1963) 69–81.

[34] L.A. Zadeh, "Calculus of fuzzy restrictions," *Fuzzy Sets and Their Applications to Cognitive and Decision Processes* (eds. Zadeh. L.A. et al.), Academic Press, New York, 1975, 1–39.

[35] L.A. Zadeh, "The concept of a linguistic variable and its applications to approximate reasoning, Part 1, 2 and 3," *Information Science* **8** 199–249, **8** 301–357, **9** (1975) 43–80.

Chapter 5

APPROXIMATE REASONING BASED ON LATTICE-VALUED PROPOSITIONAL LOGIC L_{vpl}

Yang Xu[1], Da Ruan[2] and Jun Liu[1]
1. Department of Applied Mathematics, Southwest Jiaotong University, Chengdu 610031, Sichuan, P.R.China
2. Belgian Nuclear Research Centre (SCK•CEN), Boeretang 200, B-2400 Mol, Belgium

Keywords: If...then... rule, lattice implication algebra, lattice-valued propositional logic L_{vpl}, approximate reasoning

Abstract: This chapter approaches the theory of approximate reasoning in the "If ... then..." type approximate-reasoning mode based on the lattice-valued proposition logic L_{vpl}. The "If...then..." rule is the rule in which both antecedent and consequence are L-type lattice-valued proposition sets. Two approximate reasoning methods, which are reasonable due to their strict semantic interpretation and strict syntactical deduction, are proposed in the lattice-valued proposition logic L_{vpl}.

1. INTRODUCTION

One focus of research in intelligent information-processing systems is to make machines more effectively model the human way of automatically processing information, especially uncertainty and imprecise information. Hence, it is indispensable to establish theories, methods and technologies on uncertain information processing. It is just the demand to develop intelligent information-processing systems that speed up the research on approximate reasoning based on uncertain information. Up to now, there have been numerous studies on this direction from various point of views, and various kinds of theories and methods on approximate reasoning have been proposed according to different uncertain information. The main stream of these discussions may be summarized approximately as follows: approximate reasoning with certainty factors [5,6], subjective Bayesian approximate reasoning [28], probability-based MYCIN-type approximate reasoning [2,31], evidence theory based approximate reasoning, [8] and approximate reasoning based on fuzzy set theory [3,9,25,26,46]. On the basis of them, some elementary and partial intelligent information-processing technologies have constantly come into being and have been applied into different intelligent information processing systems. However, these facts may be very far from the desired level of intelligent information processing and various problems still remain to be solved. Theoretically, a more reasonable, more perfect and more steady foundation of approximate reasoning theory urgently needs to be established (there have been some studies on the logical foundation of approximate reasoning based on certain logical systems [1,4,7,22,27]); In practice, it also needs to improve approximate reasoning to more naturally model human way of reasoning and form a practical technology.

It is well known that classical reasoning is only concerned with certain information. There is a strict and solid logical foundation—two-valued logical system for classical reasoning. Just due to this fact, classical reasoning has been extensively, deeply, effectively and reliably applied into many fields. However, it has shown from numerous studies that classical theories and methods for certain information cannot directly be used to uncertain information-processing systems. New theories and methods for uncertain information need to be proposed and developed based on classical theories and methods. They should be on the strict and solid logical foundation like a classical theory. Otherwise, these theories and methods would lack of well soundness, completeness, effectiveness, and universality. Certainly, this kind of logic is not two-valued logic but the development and extension of two-valued logic.

There are many types about uncertainty, such as randomicity, fuzziness, unsureness, incomparability, incompleteness, etc. This chapter mainly aims at the fuzziness and incomparability. In fact, fuzziness is a kind of the immanent uncertainty of the partition among different states in each intermediate variation of objects. However, every state would be certain to some extent. So, theories and methods for fuzziness and their logical system should characterize the corresponding dynamic variation. Incomparability is an essential attribute of the correlation between several objects, which reflects a kind of nonlinearity. Hence, the methods for incomparability should characterize the corresponding nonlinearity. Consequently, in order to establish the theories to deal with fuzziness and incomparability simultaneously, it is suitable to establish their logical foundation in the framework of the dynamic characteristic and partial-order structure.

To deal with fuzziness and incomparability in practice approximate reasoning theories using some related research works, [14,30] we successively established the lattice implication algebra [10-12,13,15-17,24,32-37,39-40,47], lattice-valued propositional logical algebra based on the lattice implication algebra [18,38], the L-type lattice-valued propositional logic and the corresponding L-type lattice-valued first-order logic [20,21,23,42], as well as the fuzzy proposition logic $FP(X)$ and fuzzy value propositional logic $FP^*(X)$[19,41]. In order to reflect the difference of degree among attributes of each intermediate-variation-state of objects and the incomparability of objects, we also established a dynamical lattice-valued logic system that can be rich enough to reflect dynamically gradational technique instead of a static one, i.e., the gradational L-type lattice-valued propositional logic system L_{vpl} [43] and the corresponding L-type first-order lattice-valued logic system L_{vfl} [44,45]. On the logical basis of L-type propositional logic system L_{vpl}, we will present two kinds of approximate reasoning theories and two approximate reasoning methods for "If…then…" rules.

The structure of this chapter is as follows. In Section 2, we briefly recall some related concepts about lattice implication algebra. In Section 3, we recall some related basic definitions and concepts in lattice-valued propositional logic system L_{vpl}. In Section 4, we propose a new interpretation for two-valued logical reasoning so that some ideas and background for approximate reasoning theories based on L_{vpl} are induced. In Section 5, we first discuss the representation of "If…then…" rules in L-type lattice-valued propositional logic L_{vpl} and the provability of approximate reasoning consequence. Then we propose two approximate reasoning methods in L_{vpl} with some properties of these methods being discussed. In Section 6, we illustrate three examples for approximate reasoning in three kinds of L-type lattice-valued logical systems. We summarize our conclusion in Section 7.

2. SOME RELATED RESULTS ON LATTICE IMPLICATION ALGEBRA

To provide a logical foundation for uncertain information-processing theory, especially for the incomparability in uncertain information as well as the transformation of uncertain information in the reasoning (when deriving new facts from the known facts), we established lattice implication algebra by combining lattice and implication algebra. In this section we briefly recall some results about lattice implication algebra related to this chapter.

Definition 2.1 Let $(L, \vee, \wedge, ^{\prime})$ be a bounded lattice with order-reversing involution " $^{\prime}$ "and the universal bounds O, I, $\rightarrow: L\times L\rightarrow L$ be a mapping. $(L, \vee, \wedge, ^{\prime}, \rightarrow)$ is called a quasi-lattice implication algebra (shortly as Q-LIA) if the following properties hold for all $x, y, z\in L$:

(I_1) $x\to(y\to z)=y\to(x\to z)$,

(I_2) $x\to x=I$,

(I_3) $x\to y=y^{-}\to x^{-}$,

(I_4) if $x\to y=y\to x=I$, then $x=y$,

(I_5) $(x\to y)\to y=(y\to x)\to x$.

$(L, \vee, \wedge, ^{-}, \to)$ is called a lattice implication algebra (shortly as LIA) if it is a Q-LIA and the following properties hold for all $x, y, z\in L$:

(l_1) $(x\vee y)\to z=(x\to z)\wedge(y\to z)$,

(l_2) $(x\wedge y)\to z=(x\to z)\vee(y\to z)$.

A LIA $(L, \vee, \wedge, ^{-}, \to)$ is called an H lattice implication algebra (shortly as H-LIA) if $x\vee y\vee[(x\wedge y)\to z]=I$ for all $x, y, z\in L$.

"$\to$" can be used to express an implication operator and a mapping, which can be distinguished from the context.

Definition 2.2 Let $(L, \vee, \wedge, ^{-}, \to)$ be a LIA. $J\subseteq L$ is said to be a filter of L if it satisfies the following conditions:

(1) $I\in L$,

(2) For all $x, y\in L$ if x and $x\to y\in J$, then $y\in L$.

$J\subseteq L$ is said to be an implicative filter of L if it satisfies the following conditions:

(3) $I\in L$,

(4) For all $x, y\in L$ if $x\to(y\to z)$ and $x\to y\in J$, then $x\to z\in J$.

Proposition 2.1 Let $(L, \vee, \wedge, ^{-}, \to)$ be a LIA. For all $x, y\in L$, $x\leq y$ iff $x\to y=I$.

Note that for all $x\in L$, $O\to x=I$, $x\to I=I$.

Proposition 2.2 Let $(L, \vee, \wedge, ^{-}, \to)$ be a LIA, and $x, y\in L$. Then $x\leq y$ iff for all $z\in L$, $x\to z\geq y\to z$ iff for all $z\in L$, $z\to x\leq z\to y$ iff for all $z\in L$, $x\to z=x\to(y\wedge z)$ iff for all $z\in L$, $z\to y=(z\vee x)\to y$.

Proposition 2.3 Let $(L, \vee, \wedge, ^{-}, \to)$ be a LIA. For all $x, y\in L$,

(1) $x\to O=x^{-}$,

(2) $(x\to y)\to y=x\vee y=(y\to x)\to x$,

(3) $[(y\to x)\to y^{-}]^{-}=x\wedge y=[(x\to y)\to x^{-}]^{-}$,

(4) $x\vee y=I$ iff $x\to y=y$,

(5) $x\wedge y=O$ iff $x\to y=x^{-}$,

(6) $((x\to y)\to y)\to y=x\to y$,

(7) If $x\leq y$, then there exists a $u\in L$, $y=u\to x$.

Theorem 2.1 $(L, \vee, \wedge, ^{-}, \to)$ is a LIA iff it satisfies the following conditions:

(1) It is a Q-LIA,

(2) $(L, \vee, \wedge)$ is a distributive lattice,

(3) For all $x, y\in L$, $(x\to y)\to y=x\vee y$,

(4) For all $x, y\in L$, $x\leq y$ iff $x\to y=I$.

Theorem 2.2 Let $(L, \vee, \wedge, ^{-}, \to)$ be a LIA, $a, b\in L$. For all $x, y\in L$,

$(a \to b)^{\sim} \leq (x \to a) \wedge (b \to y)$.

Definition 2.3 Let $(L, \vee, \wedge, {}^{\sim}, \to)$ and $(L', \vee', \wedge', {}^{\sim\prime}, \to')$ be two LIAs, where the induced partial orders are written as $\leq$ and $\leq'$ respectively, the maximal element and minimal element are written as O, I and O', I' respectively. If a mapping f: $L \to L'$ satisfies:

$$f(x \to y) = f(x) \to' f(y)$$

for all $x, y \in L$, then f is called an implication homomorphism from $(L, \vee, \wedge, {}^{\sim}, \to)$ to $(L', \vee', \wedge', {}^{\sim\prime}, \to')$ (in short, f is an implication homomorphism from L to L'). If f also satisfies:

$$f(x \vee y) = f(x) \vee' f(y),$$
$$f(x \wedge y) = f(x) \wedge' f(y),$$
$$f(x^{\sim}) = (f(x))^{\sim\prime},$$

for all $x, y \in L$, then f is called a lattice implication homomorphism from $(L, \vee, \wedge, {}^{\sim}, \to)$ to $(L', \vee', \wedge', {}^{\sim\prime}, \to')$ (in short, f is a lattice implication homomorphism from L to L').

A bijective lattice implication homomorphism is called a lattice implication isomorphism.

Theorem 2.3 Let f be an implication homomorphism from $(L, \vee, \wedge, {}^{\sim}, \to)$ to $(L', \vee', \wedge', {}^{\sim\prime}, \to')$. Then

(1) f is a $\vee$-homomorphism,

(2) If $f(O)=O$, then f is a lattice implication homomorphism.

Corollary 2.1 Let f be a mapping from $(L, \vee, \wedge, {}^{\sim}, \to)$ to $(L', \vee', \wedge', {}^{\sim\prime}, \to')$. Then

(1) f is a lattice implication homomorphism iff f is an implication homomorphism and $f(x^{\sim}) = (f(x))^{\sim\prime}$ for $x \in L$.

(2) If f is surjective, then f is a lattice implication homomorphism iff f is an implication homomorphism.

Theorem 2.4 Let $(L, \vee, \wedge, {}^{\sim}, \to)$ and $(L', \vee', \wedge', {}^{\sim\prime}, \to')$ be two LIAs, and

$$f, g: (L, \vee, \wedge, {}^{\sim}, \to) \to (L', \vee', \wedge', {}^{\sim\prime}, \to')$$

are two mappings. Then the following statements are equivalent:

(1) For all $x, y \in L$, $f(x \to y) = g(x \to y)$,

(2) For all $x, y \in L$, $f(x \vee y) = g(x \vee y)$,

(3) For all $x, y \in L$, $f(x \wedge y) = g(x \wedge y)$,

(4) For all $x \in L$, $f(x) = g(x)$.

Theorem 2.5 Let $(L, \vee, \wedge, {}^{\sim}, \to)$ be a LIA, and $x \in L$. Then

$$x \vee x^{\sim} = I \text{ iff } x \to y = x^{\sim} \vee y \text{ for every } y \in L.$$

Theorem 2.6 Let $(L, \vee, \wedge, {}^{\sim}, \to)$ be a LIA. Then $(L, \vee, \wedge, {}^{\sim}, \to)$ is an H-LIA iff $x \vee x^{\sim} = I$ for all $x \in L$.

Theorem 2.7 The collection of all LIAs is a proper class.

Definition 2.4 Let $(L, \vee, \wedge, {}^{\sim}, \to)$ be a LIA and $A \subseteq L$.

$(A]=\bigcap\{B \mid A\subseteq B\subseteq L, B \text{ is a filter of } (L, \vee, \wedge, ^{\sim}, \rightarrow)\}$

Theorem 2.8 Let $(L, \vee, \wedge, ^{\sim}, \rightarrow)$ be a LIA and $\varnothing\neq A\subseteq L$. Then

$(A]=\{x \mid x\in L$ and there exists a finite member of elements of A say $a_1, a_2, \ldots, a_n$, such that $a_1\rightarrow(a_2\rightarrow\ldots\rightarrow(a_n\rightarrow x)\ldots)=I\}$.

Theorem 2.9 (Lukasiewicz Implication Algebra on [0, 1]) $([0, 1], \vee, \wedge, ^{\sim}, \rightarrow)$ is a LIA, where, for all $x, y\in[0, 1]$,

$$x\vee y=\max(x, y),\ x\wedge y=\min(x, y),$$
$$x^{\sim}=1-x,\ x\rightarrow y=\min(1, 1-x+y).$$

It is easy to see from Theorem 2.6 that the H-LIA is the same algebraic system as a Boolean algebra. Moreover, the H-LIA is a special LIA. Hence, the LIA class contains Boolean algebra as a special case. It follows from Theorem 2.9 that there is other non-Boolean algebra besides Boolean algebra in the LIA class. The following theory gives a characterization of non-Boolean algebra in the LIA class (Called a proper-LIA).

Theorem 2.10 Let $(L, \vee, \wedge, ^{\sim}, \rightarrow)$ be a LIA and

$\Omega=\{J \mid J \text{ is a maximal implicative filter of } (L, \vee, \wedge, ^{\sim}, \rightarrow)\}$[37].

Then $(L, \vee, \wedge, ^{\sim}, \rightarrow)$ is a proper-LIA iff there exist $x, y\in L$, $x\neq y$ such that

$$J\cap\{x, y\}=\varnothing \text{ or } J\cap\{x, y\}=\{x, y\}$$

for every $J\in\Omega$.

In a Q-LIA $(L, \vee, \wedge, ^{\sim}, \rightarrow)$, we introduce an operator as follows:

$$x\otimes y=(x\rightarrow y^{\sim})^{\sim},$$

for every $x, y\in L$. It is easy to see that $x\otimes y=\mathrm{Inf}\{z \mid x\leq y\rightarrow z\}$.

Note that the $\otimes$ operator is associative. So for any $a\in L$ and any $n\in N$, we define recursively:

$$a^0=I,\ a^{n+1}=a^n\otimes a,\ a^{(n+m)}=a^n\otimes a^m.$$

Theorem 2.11 Suppose the $\otimes$ is introduced in a LIA $(L, \vee, \wedge, ^{\sim}, \rightarrow)$, denoted by $(L, \vee, \wedge, ^{\sim}, \rightarrow, \otimes)$. Then $(L, \vee, \wedge, ^{\sim}, \rightarrow, \otimes)$ a residuated lattice[29].

Theorem 2.12 In $(L, \vee, \wedge, ^{\sim}, \rightarrow, \otimes)$, the following properties hold for all $a, b, c\in L$,

(1) $a\otimes b\leq a\wedge b\leq a$,

(2) $a\leq b$ iff $a\otimes b^{\sim}=O$,

(3) $a\otimes b=a$ iff $a\wedge b^{\sim}=O$,

(4) $b\leq a\rightarrow(a\otimes b)$,

(5) $(a\rightarrow b)\otimes a\leq b$,

(6) $a\otimes a=a$ iff $a\wedge a^{\sim}=O$,

(7) $(a\otimes b)\rightarrow c=b\rightarrow(a\rightarrow c)$,

(8) $(a\rightarrow b)\otimes(b\rightarrow c)\leq a\rightarrow c$,

(9) $a\rightarrow(b\otimes c)\geq b\otimes(a\rightarrow c)$,

(10) $a\otimes(b\vee c)=(a\otimes b)\vee(a\otimes c)$,

(11) $a\otimes(b\wedge c)=(a\otimes b)\wedge(a\otimes c)$,

(12) $a\otimes b=(a\vee b)\otimes(a\wedge b)$,
(13) $a\vee b=I$ implies $a\rightarrow(b\otimes c)=b\otimes(a\rightarrow c)$,
(14) $a\vee b=I$ implies $(a\otimes a)\vee(b\otimes b)=I$.
(15) $a\vee b=I$ implies $a^n\vee b^n=I$, for every $n\in N$,
(16) $(a\vee b)^n=a^n\vee b^n$, $(a\wedge b)^n=a^n\wedge b^n$, for every $n\in N$.

Let $(L, \vee, \wedge, ', \rightarrow)$ be a LIA. We introduce the following notation:

$$[a_1, a_2, \ldots, a_n, x]=a_1\rightarrow(a_2\rightarrow(\ldots\rightarrow(a_n\rightarrow x)\ldots))$$

for every $a_1, a_2, \ldots, a_n, a, x\in L$ and for every $n\in N$.

If $a_1=a_2=\ldots=a_n=a$, then

$$[a, x]^0=x,\ [a, x]^1=[a, x],\ [a, x]^n=[a, a, \ldots, a, x].$$

Proposition 2.4 Let $(L, \vee, \wedge, ', \rightarrow)$ be a LIA. Then

(1) For every $n\in N$, $[a_1, a_2, \ldots, a_n, x]=I$ iff $a_1\otimes a_2\otimes\ldots\otimes a_n\leq x$.
(2) For every $n\in N$ and $a\in L$, $[a, O]^n\rightarrow O=a^n$.
(3) If $a, b, x\in L$, $a\geq b$, then for every $n\in N$, $[a, x]^n\leq[b, x]^n$.
(4) Let $a, x\in L$ for every $n, m\in N$, if $n\geq m$, then $[a, x]^n\geq[a, x]^m$.

3. SOME RELATED RESULTS ABOUT LATTICE-VALUED PROPOSITIONAL LOGIC SYSTEM BASED ON LIA—L_{VPL} [43]

In order to establish an approximate reasoning theory to deal with fuzziness and incomparability, we continually investigate the lattice-valued logical system based on LIA, mainly including lattice-valued propositional logic $LP(X)$, the L-type lattice-valued propositional logic L-$LP(X)$, the L-type first-order lattice-valued logic FM, the fuzzy propositional logic $FP(X)$, the fuzzy-valued propositional logic $FP^*(X)$, the gradational L-type propositional lattice-valued logic L_{vpl}, and the gradational L-type first-order lattice-valued logic L_{vfl}. Based on lattice-valued propositional logical system L_{vpl}, this chapter presents a kind of approximate reasoning theories with semantic interpretation and syntactical deduction. This section recalls some concepts and results about L_{vpl} related to the chapter.

In the following, we suppose $(L, \vee, \wedge, ', \rightarrow, O, I)$ (shortly as L) be a complete lattice implication algebra.

3.1 The Language of L_{vpl}

The alphabet of L_{vpl} consists of the following symbols:
(1) Variables $x, y, \ldots$
(2) Symbols for truth value a, $a\in L$;
(3) Logical connectives: $\vee, \wedge, ', \rightarrow$;
(4) Auxiliary symbols (,) etc.

Remark: the algebraic operator in L and the logical connectives in L_{vpl} are expressed by the same symbols, which are consistent but different in the concept.

The formulae of L_{vpl} are recursively defined as follows:

(1) A variable is a formula;

(2) Every truth-value $a \in L$ is a formula;

(3) If p, q are formulae, then $p \vee q, p \wedge q, p^{\sim}, p \rightarrow q$ are formulae;

(4) There are no formulae other than those constructed from the ones described in (1) and (2), by means of successive applications of (3).

For the convenience, we use the following notations:

$$p \leftrightarrow q = (p \rightarrow q) \wedge (q \rightarrow p).$$

$$p \otimes q = (p \rightarrow q^{\sim})^{\sim}.$$

$$F_p = \{q \mid q \text{ is a formula of } L_{vpl}\}.$$

$$\mathfrak{I} \subseteq L^{F_p} = \{X \mid X\colon F_p \rightarrow L\}.$$

$$\mathfrak{I}_h \subseteq \mathfrak{I}_H = \{T \mid T\colon F_p \rightarrow L \text{ is a homomorphic mapping}\}.$$

3.2 The Rule of Inference in L_{vpl}

For every $n \in \{0, 1, 2, \ldots\}$, a mapping

$$r_n\colon D_n(r_n) \rightarrow F_p$$

is called a partial n-ary operation on F_p, where $D_n(r_n) \subseteq F_p^{\,n}$.

A mapping

$$t_n\colon L^n \rightarrow L$$

is said to be an n-ary truth-valued operation on L, if

(1) For arbitrary $\alpha \in L$, and $(\alpha_1, \alpha_2, \ldots, \alpha_n) \in L^n$,

$$\alpha \rightarrow t_n(\alpha_1, \alpha_2, \ldots, \alpha_n) \geq t_n(\alpha \rightarrow \alpha_1, \alpha \rightarrow \alpha_2, \ldots, \alpha \rightarrow \alpha_n).$$

(2) For arbitrary $(\alpha_1, \alpha_2, \ldots, \alpha_n) \in L^n$,

if $(\alpha_1, \alpha_2, \ldots, \alpha_n) \leq (\beta_1, \beta_2, \ldots, \beta_n)$, then $t_n(\alpha_1, \alpha_2, \ldots, \alpha_n) \leq t_n(\beta_1, \beta_2, \ldots, \beta_n)$.

We write

$$R_n = \{r_n \mid r_n\colon D_n(r_n) \rightarrow F_p\}.$$

$$T_n = \{t_n \mid t_n\colon L^n \rightarrow L \text{ is an } n\text{-ary truth-valued operation on } L\}.$$

$$\mathfrak{R} = \bigcup_{n=0}^{\infty} \mathfrak{R}_n, \text{ where, } \mathfrak{R}_n \subseteq R_n \times T_n.$$

Specially, we give some partial n-ary operations and some truth-valued operations needed in the chapter.

Set

$$D^{\circ}_2 = \{(p, p \rightarrow q) \mid p, q \in F_p\}.$$

$$D^{*}_2 = \{(p \rightarrow g, p \rightarrow q) \mid p, g, q \in F_p\}.$$

$$D'_2 = \{(p \rightarrow q, q \rightarrow g) \mid p, g, q \in F_p\}.$$

$$r^{\circ}_2\colon D^{\circ}_2 \rightarrow F_p,\ r^{\circ}_2(p, p \rightarrow q) = q.$$

$$r^{*}_2\colon D^{*}_2 \rightarrow F_p,\ r^{*}_2(p \rightarrow g, p \rightarrow q) = p \rightarrow (g \wedge q).$$

$r'_2: D'_2 \to F_p$, $r'_2(p\to q, q\to g)=p\to g$.
$r_1(\theta_0): F_p\to F_p$, $r_1(\theta_0)(p)= \theta_0\to p$.
$t^{o}_2: L\times L\to L$, $t^{o}_2(\alpha, \beta)=\alpha\otimes\beta$.
$t^{*}_2: L\times L\to L$, $t^{*}_2(\alpha, \beta)=\alpha\wedge\beta$.
$t_1(\theta_0): L\times L\to L$, $t_1(\theta_0)(\alpha)=\theta_0\to\alpha$.

Definition 3.1 Let $X\in L^{F_p}$, $(r, t)\in \mathfrak{R}_n$, $\alpha\in L$. If

$$X\circ r\supseteq \alpha\otimes(t\circ \prod_1^n X)$$

in $D_n(r)$, then X is called to be α -I type closed w.r.t. (r, t). If

$$X\circ r\supseteq t\circ(\prod_1^n \alpha\otimes X)$$

in $D_n(r)$, then X is called to be α -II type closed w.r.t. (r, t).

If X is α -I (α -II) type closed w.r.t. (r, t) for arbitrary $(r, t)\in\mathfrak{R}$, then X is called to be α -I (α -II) type closed w.r.t. $\mathfrak{R}$.

Proposition 3.1

(1) If L is an H-LIA, then for every $T\in\mathfrak{I}_H$, T is α -i(i=I, II) type closed w.r.t. (r^{o}_2, t^{o}_2).

(2) For every $T\in\mathfrak{I}_H$, T is α -i(i=I, II) type closed w.r.t. (r^{*}_2, t^{*}_2) and $(r_1(\theta_0), t_1(\theta_0))$.

(3) If $\alpha\le\bigwedge_{\theta\in L}(\theta\vee\widetilde{\theta})$, then for every $T\in\mathfrak{I}_H$, T is α -i (i=I, II) type closed w.r.t. (r^{o}_2, t^{*}_2) and (r'_2, t^{*}_2).

Proposition 3.2 Let $X\in L^{F_p}$, $\alpha\in L$.

(1) If X is α -i (i=I, II) type closed w.r.t. $\mathfrak{R}$, then for arbitrary $\beta\in L$, $\beta\to X$ is α -i (i=I, II) type closed w.r.t. $\mathfrak{R}$.

(2) If $\aleph\subseteq\{X\mid X$ is α -i(i=I, II) type closed w.r.t. $\mathfrak{R}\}$, then $\bigcap_{X\in\aleph} X$ is α -i (i=I, II) type closed w.r.t. $\mathfrak{R}$.

Definition 3.2 Let $\mathfrak{I}\subseteq L^{F_p}$, $\alpha\in L$. $\mathfrak{R}$ is said α-i type sound w.r.t. $\mathfrak{I}$, if for every $T\in\mathfrak{I}$, T is α-i type closed w.r.t. $\mathfrak{R}$ (i=I, II).

3.3 Semantics of L_{vpl}

Definition 3.3 Let $\alpha, \beta\in L$. Two mappings are defined as follows:

(1) $C_{\mathfrak{I}}: L^{F_p}\to L^{F_p}$,

$X\mapsto C_{\mathfrak{I}}^X$,

$$C_{\mathfrak{I}}^X(p)=\bigwedge_{T\in\mathfrak{I}}[\prod(X\subseteq T)\to T(p)],$$

where, $\prod(X \subseteq T) = \underset{q \in F_p}{\wedge} [X(q) \to T(q)]$.

(2) $C^{\beta}_{(C^{\phi}_{\mathfrak{J}}, \Re(\alpha-i))}: L^{F_p} \to L^{F_p}$,

$X \mapsto C^{\beta, X}_{(C^{\phi}_{\mathfrak{J}}, \Re(\alpha-i))}$,

$C^{\beta, X}_{(C^{\phi}_{\mathfrak{J}}, \Re(\alpha-i))}(p)$

$= \wedge \{Y(p) \mid Y \supseteq \beta \otimes (C^{\phi}_{\mathfrak{J}} \cup X),\ Y$ is α-i type closed w.r.t. $\mathfrak{R}\}$, i=I, II.

Definition 3.4 Let α, β, $\tau \in L$, $X \in L^{F_p}$ and $\mathfrak{J} \subseteq L^{F_p}$. If

$$\vee \{C^{\beta \otimes X}_{\mathfrak{J}}(p) \otimes C^{\beta \otimes X}_{\mathfrak{J}}(\widetilde{p}) \mid p \in F_p\} \leq \tau,$$

then X is said to be $\tau^{\sim}$-i type consistent w.r.t. (α, β, $\mathfrak{J}$)(i=I, II).

Proposition 3.3 The mapping $C_{\mathfrak{J}}$ satisfies the following properties:

(1) Let $X_1, X_2 \in L^{F_p}$. Then

$$C^{X_1}_{\mathfrak{J}} \cup C^{X_2}_{\mathfrak{J}} \subseteq C^{X_1 \cup X_2}_{\mathfrak{J}},$$

$$C^{X_1}_{\mathfrak{J}} \cap C^{X_2}_{\mathfrak{J}} \supseteq C^{X_1 \cap X_2}_{\mathfrak{J}}.$$

(2) If $\mathfrak{J}_1 \subseteq \mathfrak{J}_2 \subseteq L^{F_p}$, then

$$C^{X}_{\mathfrak{J}_2} \subseteq C^{X}_{\mathfrak{J}_1}$$

holds for arbitrary $X \in L^{F_p}$.

(3) Let $p_1, p_2 \in F_p$, $X \in L^{F_p}$. Then

$$C^{X}_{\mathfrak{J}_h}(p_1) \vee C^{X}_{\mathfrak{J}_h}(p_2) \leq C^{X}_{\mathfrak{J}_h}(p_1 \vee p_2),$$

$$C^{X}_{\mathfrak{J}_h}(p_1) \wedge C^{X}_{\mathfrak{J}_h}(p_2) \geq C^{X}_{\mathfrak{J}_h}(p_1 \vee p_2).$$

(4) $C^{X}_{\mathfrak{J}_h}(O) \leq \underset{p \in F_p}{\vee} X(p)$, $C^{X}_{\mathfrak{J}_h}(I) = I$.

(5) $C_{\mathfrak{J}}$ is a closure operation, i.e., for arbitrary X, $Y \in L^{F_p}$,

$X \subseteq C^{X}_{\mathfrak{J}}$,

$X \subseteq Y$ implies $C^{X}_{\mathfrak{J}} \subseteq C^{Y}_{\mathfrak{J}}$,

$C^{X}_{\mathfrak{J}} = C^{C^{X}_{\mathfrak{J}}}_{\mathfrak{J}}$.

(6) If $T \in \mathfrak{J}$, then $C^{T}_{\mathfrak{J}} = T$.

Proposition 3.4 Let β, β_1, β_2, α, α_1, $\alpha_2 \in L$, $X, X_1, X_2 \in L^{F_p}$, $\mathfrak{J} \subseteq L^{F_p}$.

(1) If $\beta_1 \leq \beta_2$, $\alpha_1 \leq \alpha_2$, and $X_1 \subseteq X_2$, then

$$C^{\beta_1, X_1}_{(C^{\phi}_{\mathfrak{J}}, \Re(\alpha_1-i))} \subseteq C^{\beta_2, X_2}_{(C^{\phi}_{\mathfrak{J}}, \Re(\alpha_2-i))},\ (i\text{=I, II}).$$

(2) If $\mathfrak{R}$ is said α-i type sound w.r.t. $\mathfrak{J}$, then

$$C^{\beta,X}_{(C^{\phi}_{\mathfrak{J}},\Re(\alpha-i))} \subseteq C^{\beta\otimes X}_{\mathfrak{J}}, \ (i\text{=I, II}).$$

(3) $C^{\beta, C^{\beta,X}_{(C^{\phi}_{\mathfrak{J}},\Re(\alpha-i))}}_{(C^{\phi}_{\mathfrak{J}},\Re(\alpha-i))} = C^{\beta,\beta\otimes X}_{(C^{\phi}_{\mathfrak{J}},\Re(\alpha-i))}$, ($i$=I, II).

3.4 Syntax of L_{vpl}

Definition 3.5 Let $X\in L^{F_p}$, $\mathfrak{J}\subseteq L^{F_p}$, $p\in F_p$, θ, α, $\beta\in L$. (P^i, (n), X, (p, θ)-(α, β)) is said to be a (α, β)-i type proof with the truth degree θ from X to p (shortly, θ-(α, β)-i type proof from X to p), if the mapping

P^i : (n) $\rightarrow$ $F_p\times L$, (where (n)={1, 2, ..., n}),

$j \mapsto (p_j, \theta_j)$

satisfies

(1) $(p_n, \theta_n)=(p, \theta)$ and

(2) $\theta_j=\beta\otimes C^{\varnothing}_{\mathfrak{J}}(p_j)$ or $\theta_j=\beta\otimes X(p_j)$ or there exist $j_1, \dots, j_k<j$, and $(r, t)\in \Re_k$ such that

$(p_j, \theta_j)=(r(p_{j_1},\dots,p_{j_k}), \alpha\otimes t(\theta_{j_1},\dots,\theta_{j_k}))$ for i=I,

$(p_j, \theta_j)=(r(p_{j_1},\dots,p_{j_k}), t(\alpha\otimes\theta_{j_1},\dots,\alpha\otimes\theta_{j_k}))$ for i=II,

where n is said to be the length of θ-(α, β)-i type proof from X to p under P^i, denoted by $l(P^i)$. i=I, II.

Theorem 3.1 Let $X\in L^{F_p}$, $\mathfrak{J}\subseteq L^{F_p}$, α, $\beta\in L$, the truth-valued operations in $\Re$ satisfy semicontinuity[14]. Then for arbitrary $p\in F_p$,

$C^{\beta,X}_{(C^{\phi}_{\mathfrak{J}},\Re(\alpha-i))}(p)=\vee\{\theta \mid$ there exists (P^i, (n), X, (p, θ)-(α, β))}, (i=I, II).

Theorem 3.2 Let $\mathfrak{J}\subseteq L^{F_p}$. If for arbitrary α, $\beta\in L$, $\Re$ is α-i type sound w.r.t. $\mathfrak{J}$, and for arbitrary $X\in L^{F_p}$, $C^{\beta,X}_{(C^{\phi}_{\mathfrak{J}},\Re(\alpha-i))}\in\mathfrak{J}$, then

$$C^{\beta,X}_{(C^{\phi}_{\mathfrak{J}},\Re(\alpha-i))} = C^{\beta\otimes X}_{\mathfrak{J}}, \ (i\text{=I, II}).$$

4. A KIND OF INTERPRETATION ABOUT $X \| == Y$

In two-valued logic C_2, we write F_2 as the set of all well-formed formulae (shortly wff) in C_2, $\mathfrak{J}^*$ as the set of all valuations in C_2. Let X, $Y\subseteq F_2$. A set Y of wffs is said to be logically inferred from the set X of wffs, written $X\|== Y$, if for any $q\in Y$, $T(X)\subseteq\{I\}$ for any $T\in\mathfrak{J}^*$ implies $T(q)=I$.

Hence, it is easy to prove that

$$X\|\!\!==Y \text{ iff } X\|\!\!==q \text{ for any } q\in Y$$

$$\text{iff } \bigwedge_{T\in\mathfrak{I}^*}[(\bigwedge_{p\in X}T(p))\rightarrow T(q)]\geq Y(q)$$

$$\text{iff } \bigwedge_{T\in\mathfrak{I}^*}\{[\bigwedge_{p\in F_2}(X(p)\rightarrow T(p))]\rightarrow T(q)\}\geq Y(q), \qquad (4.\ 1)$$

where, $\rightarrow$ is the Kleene implication operator and

$$X(p)=\begin{cases} I, & p\in X, \\ O, & p\notin X. \end{cases} \quad Y(p)=\begin{cases} I, & p\in Y, \\ O, & p\notin Y. \end{cases}$$

According to the notation in L_{vpl}, (4. 1) means that

$$C_{\mathfrak{I}^*}^{X}\supseteq Y \qquad (4.\ 2)$$

As for $X\|\!\!==Y$, X and Y are requested to be consistent. Otherwise, the corrected inference consequence may not be obtained even using corrected inference.

From the equivalent description about $X\|\!\!==Y$ above, we know that

(1) In the two-valued logic, upon X and Y, which are consistent, are given, whether or not $X\|\!\!==Y$ holds completely depends on whether or not (4.1) which is constructed by $\mathfrak{I}^*$ and $\rightarrow$ holds. Therefore, whether or not $X\|\!\!==Y$ holds completely depends on the choice of $\mathfrak{I}^*$ and $\rightarrow$. It is also one of reasons that implication operators attract considerable attention.

(2) In fact, the propositional symbol is a formal notation for proposition. It would be no meaning without being given a truth-value. Hence, the reasoning in the two-valued logic is a transformation among truth-values of propositions in substance. Moreover, the reasoning among uncertain propositions is also a transformation of the corresponding truth-value in substance. Establishing a logical system as a foundation and investigating the law of truth-value transformation for the modelling of uncertain reasoning is in fact reasonable and feasible.

(3) Note that (4.1) is only a semantic form. It is easy to see from the soundness and completeness in the two-valued logic that there exists a corresponding syntax form for (4.1). Moreover, note that (4.1) can be extended expediently to approximate reasoning, so, it is feasible to extend the corresponding syntax system to approximate reasoning. Therefore, for the approximate reasoning, we can establish a kind of formal logical system in which semantics is compatible with syntax. Therefore, approximate reasoning may be implemented according to semantics in this logical system and using the syntax in this logical system can strictly prove the approximate reasoning consequence. Consequently, these kind of approximate reasoning theories can be established on a strict and solid logical foundation.

In L_{vpl} [43], we have established a logical system with a semantic system and the corresponding syntactical system compatible with each other constructed by more general semantic interpretation and more general implication operator "→". Based on L_{vpl} and the academic idea reflected by (4. 1), it is certainly reasonable and feasible to construct a class (two kinds) of approximate reasoning theories and methods.

5. APPROXIMATE REASONING BASED ON L_{VPL}

Due to the fact that the single-input-single-output uncertain reasoning theory can be extended to multiple, multi-dimension as well as multiple and multi-dimension cases by using the similar method as in [46], this chapter will only be concerned with the single-input-single-output uncertain reasoning models.

In the following, i=I, II, and if there are no special explanations, then $\alpha, \beta, \tau \in L$, $X, X', Y \in L^{F_p}$ and $\mathfrak{J} \subseteq L^{F_p}$.

5.1 Representation of "If...then..." Rule in L_{vpl}

Single-input-single-output approximate reasoning model:

Rule: If X, then Y.
Premise: X',

Conclusion: Y'. (5. 1)

where $X, X', Y, Y' \in L^{F_p}$.

Upon L_{vpl} is given, its truth-value field $(L, \vee, \wedge, ^{-}, \rightarrow, O, I)$ is fixed. So, the consistency of semantics and syntax in L_{vpl} mainly adjusted by $\mathfrak{J}$. Therefore, we suppose in the following that → is the implication operator in $(L, \vee, \wedge, ^{-}, \rightarrow, O, I)$, and the study will focus on the $\mathfrak{J}$.

Definition 5.1 The rule in (5.1) is said to be $(\alpha, \beta, \tau, \mathfrak{J})$-$i$ type representable in L_{vpl}, if there exist $\alpha, \beta, \tau \in L, \mathfrak{J} \subseteq L^{F_p}$ such that

(1) X, Y are τ^{-}-i type consistent w.r.t. $(\alpha, \beta, \mathfrak{J})$.

(2) $C_{\mathfrak{J}}^{X} \supseteq \tau^{-} \otimes Y$ (5. 2)

Proposition 5.1 Let $X, Y \in L^{F_p}$. If Y is τ^{-}-i type consistent w.r.t. $(\alpha, \beta, \mathfrak{J})$ and $X \subseteq Y$, then so is X.

Corollary 5.1 Let $X, Y \in L^{F_p}$. If X, Y are τ^{-}-i type consistent w.r.t. $(\alpha, \beta, \mathfrak{J})$, then $X \cap Y$, $X \otimes Y$, $\theta \otimes X (\theta \in L)$ are all τ^{-}-i type consistent w.r.t. $(\alpha, \beta, \mathfrak{J})$.

It follows from Proposition 5.1 and (5) in Proposition 3.3 that

Proposition 5.2 In (5.1), if $Y \subseteq X$ and X is τ^{-}–i type consistent w.r.t.

$(\alpha, \beta, \mathfrak{I})$, then the rule in (5.1) is $(\alpha, \beta, \tau, \mathfrak{I})$-$i$ type representable in L_{vpl}.

Proposition 5.3 In (5.1), if $X \subseteq Y$, and there exists a $T_0 \in \mathfrak{I}_H$, $Y \subseteq T_0$, then for all $\alpha, \beta, \tau \in L$, the rule in (5.1) is $(\alpha, \beta, \tau, \mathfrak{I})$-$i$ type representable in L_{vpl}, where $\mathfrak{I}=\{T \mid T \supseteq Y\}$.

Proof. Set $\mathfrak{I}=\{T \mid T \supseteq Y\}$. Then for all $\alpha, \beta, \tau \in L$,

(1) $C_{\mathfrak{I}}^{\beta \otimes Y} = C_{\mathfrak{I}}^{\beta \otimes \bigcap_{T \in \mathfrak{I}} T} \subseteq C_{\mathfrak{I}}^{\bigcap_{T \in \mathfrak{I}} (\beta \otimes T)} \subseteq \bigcap_{T \in \mathfrak{I}} C_{\mathfrak{I}}^{\beta \otimes T} \subseteq \bigcap_{T \in \mathfrak{I}} C_{\mathfrak{I}}^{T} = \bigcap_{T \in \mathfrak{I}} T \subseteq T_0$.

Hence, for any $p \in F_p$,

$$C_{\mathfrak{I}}^{\beta \otimes Y}(p) \otimes C_{\mathfrak{I}}^{\beta \otimes Y}(\widetilde{p}) \le T_0(p) \otimes T_0(\widetilde{p}),$$

$$\tau^{\sim} \otimes T_0(p) \le T_0(p) = (T_0(\widetilde{p}))^{\sim},$$

$$\tau^{\sim} \le T_0(p) \to (T_0(\widetilde{p}))^{\sim} = (T_0(p) \otimes T_0(\widetilde{p}))^{\sim},$$

$$T_0(p) \otimes T_0(\widetilde{p}) \le \tau .$$

Consequently, X and Y are $\tau^{\sim}$-i type consistent w.r.t. $(\alpha, \beta, \mathfrak{I})$.

(2) Note that $X \subseteq Y$ and $\mathfrak{I}=\{T \mid T \supseteq Y\}$ imply $C_{\mathfrak{I}}^{X} = C_{\mathfrak{I}}^{\phi} = Y \supseteq \tau^{\sim} \otimes Y$.

Corollary 5.2 In (5.1), if $Y \in \mathfrak{I}_H$, $X \subseteq Y$, then for all $\alpha, \beta, \tau \in L$, the rule in (5.1) is $(\alpha, \beta, \tau, \mathfrak{I})$-$i$ type representable in L_{vpl}, where $\mathfrak{I}=\{T \mid T \supseteq Y\}$.

Corollary 5.3 In (5.1), if $X \subseteq Y$ and Y is $\tau^{\sim}$–i type consistent w.r.t. $(\alpha, \beta, \mathfrak{I})$, then the rule in (5.1) is $(\alpha, \beta, \tau, \mathfrak{I})$-$i$ type representable in L_{vpl}, where $\mathfrak{I}=\{T \mid T \supseteq Y\}$.

Proposition 5.4 In (5.1), set

$$\mathfrak{I}=\{T \mid T \supseteq X \to (\tau^{\sim} \otimes Y \otimes T)\}.$$

Let X and Y be $\tau^{\sim}$-i type consistent w.r.t. $(\alpha, \beta, \mathfrak{I})$, then the rule in (5.1) is $(\alpha, \beta, \tau, \mathfrak{I})$-$i$ type representable in L_{vpl}.

Proof. We only need to prove that $C_{\mathfrak{I}}^{X} \supseteq \tau^{\sim} \otimes Y$.

For any $q \in F_p$,

$$C_{\mathfrak{I}}^{X}(q) = \bigwedge_{T \in \mathfrak{I}} [\Pi(X \subseteq T) \to T(q)]$$

$$\ge \bigwedge_{T \in \mathfrak{I}} [(X(q) \to T(q)) \to T(q)]$$

$$\ge [X(q) \to T(q)] \to [X(q) \to (\tau^{\sim} \otimes Y(q) \otimes T(q))]$$

$$\ge [X(q) \to T(q)] \to [(\tau^{\sim} \otimes Y(q)) \otimes (X(q) \to T(q))]$$

$$\ge (\tau^{\sim} \otimes Y(q)) \otimes [(X(q) \to T(q)) \to (X(q) \to T(q))]$$

$$= \tau^{\sim} \otimes Y(q).$$

Definition 5.2 Let $\mathfrak{I} \subseteq L^{F_p}$, $X \in L^{F_p}$, $\beta \in L$. We define recursively:

$$(f_{\mathfrak{I}}^{\beta})^{(n)}(X),\ n=0, 1, \dots,$$

$$(f_{\mathfrak{I}}^{\beta})^{(0)}(X) = X,\ (f_{\mathfrak{I}}^{\beta})^{(1)}(X) = C_{\mathfrak{I}}^{\beta \otimes X};$$

$$(f_{\mathfrak{I}}^{\beta})^{(n)}(X)=(f_{\mathfrak{I}}^{\beta})^{(1)}((f_{\mathfrak{I}}^{\beta})^{(n-1)}(X)),\ n=2,3,\ \ldots;$$

Proposition 5.5 Let $\mathfrak{I}\subseteq L^{F_p}$, $X\in L^{F_p}$, $\beta\in L$.

(1) $(f_{\mathfrak{I}}^{\beta})^{(n)}(X)\subseteq(f_{\mathfrak{I}}^{\beta})^{(n-1)}(X)$, n=2, 3, … .

(2) If X is $\tau^{\sim}$-i type consistent w.r.t. (α, β, $\mathfrak{I}$), then so are $(f_{\mathfrak{I}}^{\beta})^{(n)}(X)$, n=0, 1, … .

(3) If $\mathfrak{I}=\{T\mid T\supseteq X\}$, then $(f_{\mathfrak{I}}^{\beta})^{(n)}(X)=X$, n=0, 1, 2, … .

Proof. (1) If n=2,

$$(f_{\mathfrak{I}}^{\beta})^{(2)}(X)=C_{\mathfrak{I}}^{\beta\otimes C_{\mathfrak{I}}^{\beta\otimes X}}\subseteq C_{\mathfrak{I}}^{C_{\mathfrak{I}}^{\beta\otimes X}}=C_{\mathfrak{I}}^{\beta\otimes X}=(f_{\mathfrak{I}}^{\beta})^{(1)}(X).$$

Suppose for $n=k$,

$$(f_{\mathfrak{I}}^{\beta})^{(k)}(X)\subseteq(f_{\mathfrak{I}}^{\beta})^{(k-1)}(X),\ k>2.$$

Then for $n=k+1$,

$$(f_{\mathfrak{I}}^{\beta})^{(k+1)}(X)=C_{\mathfrak{I}}^{\beta\otimes(f_{\mathfrak{I}}^{\beta})^{(k)}(X)}\subseteq C_{\mathfrak{I}}^{(f_{\mathfrak{I}}^{\beta})^{(k)}(X)}=(f_{\mathfrak{I}}^{\beta})^{(k)}(X).$$

(2) It is easy to be seen by Proposition 5.1, Definition 5.2 and (1) here.

(3) If n=1. $(f_{\mathfrak{I}}^{\beta})^{(1)}(X)=C_{\mathfrak{I}}^{\beta\otimes X}\subseteq C_{\mathfrak{I}}^{X}=X$;

For arbitrary $q\in F_p$,

$$\begin{aligned}C_{\mathfrak{I}}^{\beta\otimes X}(q)&=\bigwedge_{T\in\mathfrak{I}}[\Pi(\beta\otimes X\subseteq T)\to T(q)]\\ &\geq\bigwedge_{T\in\mathfrak{I}}[(\beta\otimes X(q)\to T(q))\to T(q)]\\ &=\bigwedge_{T\in\mathfrak{I}}[(\beta\otimes X(q))\vee T(q)]\\ &\geq\bigwedge_{T\in\mathfrak{I}}T(q)=X(q).\end{aligned}$$

Hence, $(f_{\mathfrak{I}}^{\beta})^{(1)}(X)=X$.

Set $(f_{\mathfrak{I}}^{\beta})^{(n)}(X)=X$, then

$$(f_{\mathfrak{I}}^{\beta})^{(n+1)}(X)=(f_{\mathfrak{I}}^{\beta})^{(1)}(X)=X.$$

If $\mathfrak{I}$ is given, then for any $X\in L^{F_p}$, and for any $\alpha,\beta,\tau\in L$, we can obtain the following rule chain:

Rule 1: if $(f_{\mathfrak{I}}^{\beta})^{(0)}(X)$, then $(f_{\mathfrak{I}}^{\beta})^{(1)}(X)$.

Rule n: if $(f_{\mathfrak{I}}^{\beta})^{(n-1)}(X)$, then $(f_{\mathfrak{I}}^{\beta})^{(n)}(X)$. n=2, 3, ….

which is called a rule chain generated by X. The following Proposition 5.6 shows every rule in the rule chain generated by X is (α, β, τ, $\mathfrak{I}$)-i type representable in L_{vpl}.

Proposition 5.6 In the rule chain above, if X is $\tau^{\sim}$-i type consistent w.r.t. (α, β, $\mathfrak{I}$), then every rule is (α, β, τ, $\mathfrak{I}$)-i type representable in L_{vpl}.

Proof. Since X is $\tau^{\sim}$-i type consistent w.r.t. (α, β, $\mathfrak{I}$), then it follows from (1) in Proposition 5.5 and Proposition 5.1 that$(f_{\mathfrak{I}}^{\beta})^{(n)}(X)$ is $\tau^{\sim}$-i type consistent w.r.t. (α, β, $\mathfrak{I}$), n=1, 2, … .

Also it follows from (1) in Proposition 5.5 that:

$$(f_{\mathfrak{I}}^{\beta})^{(n-1)}(X) \supseteq \tau^{\sim} \otimes (f_{\mathfrak{I}}^{\beta})^{(n)}(X), n=2, 3, \ldots .$$

Moreover, note that: $C_{\mathfrak{I}}^{X} \supseteq (f_{\mathfrak{I}}^{\beta})^{(1)}(X) \supseteq \tau^{\sim} \otimes (f_{\mathfrak{I}}^{\beta})^{(1)}(X)$.

Hence, every rule in the above rule chain is (α, β, τ, $\mathfrak{I}$)-i type representable in L_{vpl}.

Remark 1: if β=I, then there exists only one rule in the above rule chain, i.e.,

If X, then $C_{\mathfrak{I}}^{X}$.

Remark 2: if β= I, and $X \in \mathfrak{I}$, then the only rule existing in the above rule chain will be changed back to the rule:

If X, then X.

Proposition 5.7 If $\mathfrak{I} \subseteq \{T \mid T$ is $\tau^{\sim}$-i type consistent w.r.t. (α, β, $\mathfrak{I}^{*}$)$\}$, then

Rule: if $\bigcap_{T\in\mathfrak{I}} T$, then $\bigcap_{T\in\mathfrak{I}} (\theta \otimes T)$, for arbitrary $\theta \in L$,

Rule: if $\bigcap_{T\in\mathfrak{I}} T$, then $\theta \otimes \bigcap_{T\in\mathfrak{I}} T$, for arbitrary $\theta \in L$,

are (α, β, τ, $\mathfrak{I}^{*}$)-i type representable in L_{vpl}.

5.2 Provability about Single-Input-Single-Output Approximate Reasoning

In L_{vpl}, we propose approximate reasoning methods in accordance with semantics and show the provability of approximate reasoning consequence in accordance with syntax.

Just like the reasoning in the two-valued logic, if approximate reasoning is based on the rule in (5. 1), then generally it is required that not only the rule should be representable to certain extent, but also new input information should be consistent to certain extent. Otherwise, whatever the reasoning method, the reasoning consequence cannot be reasonable. Therefore, we propose

Definition 5.3 The approximate reasoning model (5.1) is said a (α, β, τ, $\mathfrak{I}$)-i type regular approximate reasoning model, if

(1) the rule in (5.1) is (α, β, τ, $\mathfrak{I}$)-i type representable;

(2) new input information X' is $\tau^{\sim}$–i type consistent w.r.t. (α, β, $\mathfrak{I}$).

Approximate reasoning consequence can be given by the following equation:

$$Y' = C_{\mathfrak{I}}^{\beta \otimes X'}. \tag{5. 3}$$

Where β, $\mathfrak{I}$ are obtained from the representation of the rule in (5.1).

It is noticeable that in general α, β, τ and $\mathfrak{I}$ are not obtained uniquely by the representation of uncertainty rule. So the reasoning consequence depends on the selection of α, β, τ and $\mathfrak{I}$ despite the uncertainty rule is representable and the uncertain reasoning model is regular. This is just because L_{vpl} is a class of lattice-valued logical systems that are different due to the different α, β, τ and $\mathfrak{I}$. Hence, the different selection of α, β, τ and $\mathfrak{I}$ in approximate reasoning means the different approximate reasoning in different logical systems. Consequently, the reasoning consequence depends on the selection of α, β, τ and $\mathfrak{I}$ (i.e., the corresponding logical system).

It follows from L_{vpl} that if there exist α, β, $\mathfrak{I}$ and $\mathfrak{R}$ such that

$$C_{(C_{\mathfrak{I}}^{\phi}, \mathfrak{R}(\alpha-i))}^{\beta, X'} \in \mathfrak{I},$$

then the reasoning consequence obtained from (5.3) can also be obtained by a strict formal deduction in system L_{vpl}.

It follows from the definition of $C_{(C_{\mathfrak{I}}^{\phi}, \mathfrak{R}(\alpha-i))}^{\beta, X'}$ and Proposition 3. 2 that

Proposition 5.8 Given $\mathfrak{R}$ and α. If

$$\mathfrak{I} = \{ T \mid T \text{ is } \alpha\text{-}i \text{ type closed w.r.t. } \mathfrak{R}\},$$

then $C_{(C_{\mathfrak{I}}^{\phi}, \mathfrak{R}(\alpha-i))}^{\beta, X'} \in \mathfrak{I}$.

Proposition 5.9 Given $\mathfrak{R}$ and α. If $\mathfrak{I} = \{T \mid T \supseteq X\}$, then when

(1) $\beta \otimes X' \supseteq X$, or

(2) $\beta \otimes \tau^{\sim} \otimes Y \supseteq X$,

we have

$$C_{(C_{\mathfrak{I}}^{\phi}, \mathfrak{R}(\alpha-i))}^{\beta, X'} \in \mathfrak{I}.$$

Proof. Note that for any $Y^* \in \{Y \mid Y \supseteq \beta \otimes (C_{\mathfrak{I}}^{\phi} \cup X'),\ Y \text{ is } \alpha\text{-}i \text{ type closed w.r.t. } \mathfrak{R}\}$

$$Y^* \supseteq \beta \otimes (C_{\mathfrak{I}}^{\phi} \cup X') \supseteq \beta \otimes X' \supseteq X,$$

$$Y^* \supseteq \beta \otimes (C_{\mathfrak{I}}^{\phi} \cup X') \supseteq \beta \otimes C_{\mathfrak{I}}^{\phi} = \beta \otimes C_{\mathfrak{I}}^{X} \supseteq \beta \otimes \tau^{\sim} \otimes Y \supseteq X.$$

Hence, $\bigcap \{Y \mid Y \supseteq \beta \otimes (C_{\mathfrak{I}}^{\phi} \cup X'),\ Y \text{ is } \alpha\text{-}i \text{ type closed w.r.t. } \mathfrak{R}\} \supseteq X$, i.e.,

$$C_{(C_{\mathfrak{I}}^{\phi}, \mathfrak{R}(\alpha-i))}^{\beta, X'} \in \mathfrak{I}.$$

Proposition 5.10 Given $\mathfrak{R}$ and α. In (5. 1), if $\mathfrak{I} = \{T \mid T \supseteq \beta \otimes X'\}$, then

$$C_{(C_{\mathfrak{I}}^{\phi}, \mathfrak{R}(\alpha-i))}^{\beta, X'} \in \mathfrak{I}.$$

Proof. Follows from $Y' \supseteq \beta \otimes (C_{\mathfrak{J}}^{\phi} \cup X') \supseteq \beta \otimes X'$.

5.3 Some Properties on Approximate Reasoning

Unless a special statement, we always suppose in the following that the approximate reasoning model (5.1) is a $(\alpha, \beta, \tau, \mathfrak{J})$-$i$ type regular approximate reasoning model. It follows from Proposition 3.3 that

Proposition 5.11 The following properties hold in (5.3):

(1) $Y'=C_{\mathfrak{J}}^{\beta\otimes X'}$ is $\tau^{\sim}$-i type consistent w.r.t. $(\alpha, \beta, \mathfrak{J})$.

(2) If $X'=X$, then $Y'=C_{\mathfrak{J}}^{\beta\otimes X}$.

(3) Let $X_1, X_2 \in L^{F_p}$. Then

$$C_{\mathfrak{J}}^{\beta\otimes X_1} \cup C_{\mathfrak{J}}^{\beta\otimes X_2} \subseteq C_{\mathfrak{J}}^{\beta\otimes (X_1\cup X_2)},$$

$$C_{\mathfrak{J}}^{\beta\otimes X_1} \cap C_{\mathfrak{J}}^{\beta\otimes X_2} \supseteq C_{\mathfrak{J}}^{\beta\otimes (X_1\cap X_2)}.$$

If L is a completely distributive lattice, then "=" holds in the above two expressions.

(4) If $\mathfrak{J}_2 \subseteq \mathfrak{J}_1 \subseteq L^{F_p}$, $\alpha_1 \le \alpha_2, \beta_1 \le \beta_2$, and the rule in (5.1) is not only $(\alpha_1, \beta_1, \tau, \mathfrak{J}_1)$-$i$ type representable in L_{vpl}, but also $(\alpha_2, \beta_2, \tau, \mathfrak{J}_2)$-$i$ type representable in L_{vpl}, then for arbitrary $X' \in L^{F_p}$,

$$C_{\mathfrak{J}_1}^{\beta_1\otimes X'} \subseteq C_{\mathfrak{J}_2}^{\beta_2\otimes X'}.$$

(5) Let $p_1, p_2 \in F_p$, $X' \in L^{F_p}$. If $\mathfrak{J}=\mathfrak{J}_h$, then

$$C_{\mathfrak{J}_h}^{\beta\otimes X'}(p_1) \vee C_{\mathfrak{J}_h}^{\beta\otimes X'}(p_2) \le C_{\mathfrak{J}_h}^{\beta\otimes X'}(p_1 \vee p_2),$$

$$C_{\mathfrak{J}_h}^{\beta\otimes X'}(p_1) \wedge C_{\mathfrak{J}_h}^{\beta\otimes X'}(p_2) \ge C_{\mathfrak{J}_h}^{\beta\otimes X'}(p_1 \vee p_2).$$

$$C_{\mathfrak{J}_h}^{\beta\otimes X'}(O) \le \bigvee_{p\in F_p} (\beta \otimes X'(p)) \le \beta \otimes \bigvee_{p\in F_p} X'(p),$$

$$C_{\mathfrak{J}_h}^{\beta\otimes X'}(I)=I.$$

(6) For arbitrary $X', X_1, X_2 \in L^{F_p}$,

$$\beta \otimes X' \subseteq C_{\mathfrak{J}}^{\beta\otimes X'},$$

$$X_1 \subseteq X_2 \text{ implies } C_{\mathfrak{J}}^{\beta\otimes X_1} \subseteq C_{\mathfrak{J}}^{\beta\otimes X_2},$$

$$C_{\mathfrak{J}}^{\beta\otimes X'} = C_{\mathfrak{J}}^{C_{\mathfrak{J}}^{\beta\otimes X'}}.$$

(7) If $T \in \mathfrak{J}$, then $\beta \otimes T \subseteq C_{\mathfrak{J}}^{\beta\otimes T} \subseteq T$. If $\beta \otimes T \in \mathfrak{J}$, then $\beta \otimes T = C_{\mathfrak{J}}^{\beta\otimes T}$.

It follows from Proposition 5.6 and Definition 5.2 that

Proposition 5.12 Every rule in the rule chain generated by X satisfies Modus Ponens.

Proposition 5.13 Given $\mathfrak{J}$. If $\mathfrak{R}$ is α-i type sound w.r.t. $\mathfrak{J}$, then

(1) $C_{\mathfrak{J}}^{\beta\otimes X'}$ is α-i type closed w.r.t. $\mathfrak{R}$.

(2) If $\mathfrak{J}=\{T \mid T$ is α-i type closed w.r.t. $\mathfrak{R}\}$, $C_{\mathfrak{J}}^{\beta\otimes X'}$ is provable in L_{vpl}.

Proof. (1) Since $\mathfrak{R}$ is α-i type sound w.r.t. $\mathfrak{J}$, then for any $T\in\mathfrak{J}$, T is α-i type closed w.r.t. $\mathfrak{R}$. It follows from Proposition 3.2 that $\Pi(\beta\otimes X'\subseteq T)\to T$ is α-i type closed w.r.t. $\mathfrak{R}$ for any $T\in\mathfrak{J}$. Moreover, from Proposition 3.2, we know that:

$$C_{\mathfrak{J}}^{\beta\otimes X'}=\bigcap_{T\in\mathfrak{J}}[\Pi(\beta\otimes X'\subseteq T)\to T] \text{ is } \alpha\text{-}i \text{ type closed w.r.t. } \mathfrak{R}.$$

(3) It follows from Theorem 3.2 and Proposition 5.8 that

$$C_{\mathfrak{J}}^{\beta\otimes X'}=C_{(C_{\mathfrak{J}}^{\phi},\mathfrak{R}(\alpha-i))}^{\beta,X'}\in\mathfrak{J}.$$

Hence, $C_{\mathfrak{J}}^{\beta\otimes X'}$ is provable in L_{vpl}.

Corollary 5.4 Given $\mathfrak{J}$. Suppose that $\mathfrak{R}$ is α-i type sound w.r.t. $\mathfrak{J}$.

(1) If $(r^o{}_2, t^o{}_2)\in\mathfrak{R}$, then

For i=I, $C_{\mathfrak{J}}^{\beta\otimes X'}(q)\geq\alpha\otimes C_{\mathfrak{J}}^{\beta\otimes X'}(p)\otimes C_{\mathfrak{J}}^{\beta\otimes X'}(p\to q)$ for arbitrary p, $q\in F_p$;

For i=II, $C_{\mathfrak{J}}^{\beta\otimes X'}(q)\geq\alpha\otimes\alpha\otimes C_{\mathfrak{J}}^{\beta\otimes X'}(p)\otimes C_{\mathfrak{J}}^{\beta\otimes X'}(p\to q)$ for arbitrary p, $q\in F_p$.

(2) If $(r^0{}_2, t^*{}_2)\in\mathfrak{R}$, then $C_{\mathfrak{J}}^{\beta\otimes X'}(q)\geq\alpha\otimes(C_{\mathfrak{J}}^{\beta\otimes X'}(p)\wedge C_{\mathfrak{J}}^{\beta\otimes X'}(p\to q))$, for arbitrary p, $q\in F_p$.

(3) If $(r^*{}_2, t^*{}_2)\in\mathfrak{R}$, then

$$C_{\mathfrak{J}}^{\beta\otimes X'}(p\to(g\wedge q))\geq\alpha\otimes(C_{\mathfrak{J}}^{\beta\otimes X'}(p\to g)\wedge C_{\mathfrak{J}}^{\beta\otimes X'}(p\to q))$$

for arbitrary g, p, $q\in F_p$.

(4) If $(r'_2, t^*{}_2)\in\mathfrak{R}$, then

$$C_{\mathfrak{J}}^{\beta\otimes X'}(p\to g)\geq\alpha\otimes(C_{\mathfrak{J}}^{\beta\otimes X'}(p\to q)\wedge C_{\mathfrak{J}}^{\beta\otimes X'}(q\to g))$$

for arbitrary g, p, $q\in F_p$.

(5) If $(r_1^{\theta_0}, t_1^{\theta_0})\in\mathfrak{R}$, then

For i=I, $C_{\mathfrak{J}}^{\beta\otimes X'}(\theta_0\to p)\geq\alpha\otimes(\theta_0\to C_{\mathfrak{J}}^{\beta\otimes X'}(p))$ for arbitrary $p\in F_p$;

For i=II, $C_{\mathfrak{J}}^{\beta\otimes X'}(\theta_0\to p)\geq\theta_0\to(\alpha\otimes C_{\mathfrak{J}}^{\beta\otimes X'}(p))$ for arbitrary $p\in F_p$.

(6) In general, if $(r_n, t_n)\in\mathfrak{R}$, then

For i=I, $C_{\mathfrak{J}}^{\beta\otimes X'}(r_n(p_1, \ldots, p_n))\geq\alpha\otimes t_n(C_{\mathfrak{J}}^{\beta\otimes X'}(p_1),\ldots,C_{\mathfrak{J}}^{\beta\otimes X'}(p_n))$ for arbitrary $p_1, \ldots, p_n\in F_p$;

For i=II, $C_{\mathfrak{J}}^{\beta\otimes X'}(r_n(p_1, \ldots, p_n))\geq t_n(\alpha\otimes C_{\mathfrak{J}}^{\beta\otimes X'}(p_1),\ldots,\alpha\otimes C_{\mathfrak{J}}^{\beta\otimes X'}(p_n))$, for arbitrary $p_1, \ldots, p_n\in F_p$.

6. EXAMPLES

In this section, we take

$\mathscr{R}$=$\{(r_1^{\theta_0}, t_1^{\theta_0}) \mid \theta_0 \in L\} \cup \{(r^0{}_2, t^*{}_2), (r^*{}_2, t^*{}_2), (r'{}_2, t^*{}_2)\}$, (6. 1)

$\mathfrak{I}_i$=$\{T \mid T\in L^{F_p}$, where T is α -i type closed w.r.t. $\mathscr{R}\}$, (i=I, II). (6. 2)

From [43], we know that if $\alpha\ (\in L) \le \bigwedge_{\theta\in L}(\theta\vee\widetilde{\theta})$, then

(1) $\mathfrak{I}_H \subseteq \mathfrak{I}_i$, i=I, II;

(2) $C^{\beta,X}_{(C^{\phi}_{\mathfrak{I}},\mathfrak{R}(\alpha-i))} \in \mathfrak{I}_i$ for arbitrary $X\in L^{F_p}$ and for arbitrary $\beta\in L$;

(3) $\mathscr{R}$ is α -i type sound w.r.t. $\mathfrak{I}_i$, i=I, II;

(4) $C^{\beta,X}_{(C^{\phi}_{\mathfrak{I}},\mathfrak{R}(\alpha-i))} = C^{\beta\otimes X}_{\mathfrak{I}_i}$ (i=I, II) for arbitrary $X\in L^{F_p}$ and for arbitrary $\beta\in L$;

(5) $C^{\beta\otimes X}_{\mathfrak{I}_i}(p) = \vee\{\theta \mid$ there exists $(P^i, (n), X, (p, \theta)\text{-}(\alpha,\beta))\}$

holds for arbitrary $p\in F_p$, for arbitrary $X\in L^{F_p}$ and for arbitrary $\beta\in L$ (i=I, II).

Example 6.1 In L_{vpl}, we obtained the Lukasiewicz Proposition Logic Systems L_u, where the LIA is ([0, 1], $\vee$, $\wedge$, $\tilde{}$, $\to$). Concretely, for any $\sigma, \theta\in[0, 1]$,

$$\sigma\vee\theta = \max(\sigma, \theta),\ \sigma\wedge\theta = \min(\sigma, \theta),$$
$$\sigma^{\sim} = 1-\sigma,\ \theta\to\sigma = \min(1, 1-\theta+\sigma).$$

$\mathscr{R}$ is (6. 1) and $\mathfrak{I}$ is (6. 2).

Let $X, X', Y\in[0, 1]^{F_p}$ be given respectively as follows:

$$X(p)=\begin{cases} a, & p=a\in[0, 1],\\ 0.2, & \text{otherwise.}\end{cases}\quad Y(p)=\begin{cases} a^2, & p=a\in[0, 1],\\ 0.2, & \text{otherwise.}\end{cases},$$

$X'(p)=0.5\otimes X$.

Take α=0.4, β=0.8. Note that:

(1) there exists $T_0\in\mathfrak{I}_H$, such that

$$C^{0.8\otimes X}_{\mathfrak{I}}(q) = \bigwedge_{T\in\mathfrak{I}}[(\bigwedge_{p\in F_p}(0.8\otimes X(p)\to T(p)))\to T(q)]$$
$$= \bigwedge_{T\in\mathfrak{I}}[(\bigwedge_{\theta>0.2}(0.8\otimes\theta\to T(\theta)))\to T(q)] \le T_0(q).$$

Hence, X is 1-i type consistent w.r.t. (0.4, 0.8, $\mathfrak{I}$), i=I, II.

(2) Since X', $Y\subseteq X$, then X', Y are 1-i type consistent w.r.t. (0.4, 0.8, $\mathfrak{I}$), i=I, II.

(3) $C^X_{\mathfrak{I}} \supseteq X \supseteq Y = 1\otimes Y$.

Consequently, we can obtain the following (0.4, 0.8, 0, $\mathfrak{I}$)-i type regular approximate reasoning model:

Rule: if X, then Y,

Input: X'.

So, an approximate reasoning consequence can be obtained as follows:

$$Y'(q)=C_{\mathfrak{I}}^{0.8\otimes X'}(q)=\bigwedge_{T\in\mathfrak{I}}[(\bigwedge_{\theta>0.5}(0.3\otimes\theta\to T(\theta)))\to T(q)].$$

The approximate reasoning consequence is 1-i type consistent w.r.t. (0.4, 0.8, $\mathfrak{I}$) (i=I, II) in L_u, and it can be proved formally.

Example 6.2 In L_{vpl}, we obtained F-valued Lukasiewicz Proposition Logic Systems FL_u, where the LIA is ($[0, 1]^{[0, 1]}$, $\vee$, $\wedge$, $\bar{}$, $\to$). Concretely, for any $f, g\in[0, 1]^{[0, 1]}$,

$(f\vee g)(x)=\max(f(x), g(x))$, $(f\wedge g)(x)=\min(f(x), g(x))$,

$\bar{f}(x)=1-f(x)$, $(f\to g)(x)=\min(1, 1-f(x)+g(x))$.

The maximal element and the minimal element are respectively:

$\underline{0}: [0, 1]\to[0, 1]$, $\underline{0}(a)=0$, for all $a\in[0, 1]$,

$\underline{1}: [0, 1]\to[0, 1]$, $\underline{1}(a)=1$, for all $a\in[0, 1]$.

$\mathfrak{R}$ is (6.1) and $\mathfrak{I}$ is (6.2).

Let $X, X', Y\in([0, 1]^{[0, 1]})^{F_p}$ are defined respectively as follows:

$$X(p)=\begin{cases} f, & p=f\in[0,1]^{[0,1]},\\ x\wedge 0.4, & \text{otherwise.}\end{cases}\quad Y=x^2\otimes X,\ X'=x^3\wedge X.$$

Take $\alpha=0.4\wedge x$, $\beta=0.6\wedge x$. Where, x: $[0, 1]\to[0, 1]$, $x(a)=a$.

Note that:

(1) There exists $T_0\in\mathfrak{I}_H$ such that

$$C_{\mathfrak{I}}^{(0.6\wedge x)\otimes X}(q)=\bigwedge_{T\in\mathfrak{I}}[(\bigwedge_{p\in F_p}((0.6\wedge x)\otimes X(p)\to T(p)))\to T(q)]$$

$$=\bigwedge_{T\in\mathfrak{I}}[(\bigwedge_{f\in[0,1]^{[0,1]}}((0.6\wedge x)\otimes f\to T(f)))\to T(q)]\le T_0(q).$$

Hence, X is $\underline{1}$-i type consistent w.r.t. ($0.4\wedge x$, $0.6\wedge x$, $\mathfrak{I}$), i=I, II.

(2) Since X', $Y\subseteq X$, then X', Y are $\underline{1}$-i type consistent w.r.t. ($0.4\wedge x$, $0.6\wedge x$, $\mathfrak{I}$), i=I, II.

(3) $C_{\mathfrak{I}}^{X}\supseteq X\supseteq Y=\underline{1}\otimes Y$.

Consequently, a ($0.4\wedge x$, $0.6\wedge x$, 0, $\mathfrak{I}$)-i type regular approximate reasoning model can be obtained:

Rule: If X, then Y,

Input: X'.

So an approximate reasoning consequence can be obtained as follows:

$$Y'(q)=C_{\mathfrak{I}}^{(0.6\wedge x)\otimes X'}(q)$$

$$= \underset{T\in\Im}{\wedge}[(\underset{p\in F_p}{\wedge}((0.6\wedge x)\otimes(X(p)\wedge x^2))\to T(p)))\to T(q)]$$

$$= \underset{T\in\Im}{\wedge}[(\underset{f\in[0,1]^{[0,1]}}{\wedge}((0.6\wedge x)\otimes(f\wedge x^2))\to T(f)))\to T(q)].$$

The approximate reasoning consequence is $\underline{1}$-i type consistent w.r.t. ($0.4\wedge x$, $0.6\wedge x$, $\Im$)(i=I, II) in FL_u, and can be proved formally in FL_u.

Example 6.3 In L_{vpl}, we obtained a logic system C_5, where the LIA is ($\{O, a, b, c, I\}$, $\vee$, $\wedge$, $^{\sim}$, $\to$). Concretely, $O<a<b<c<I$, for any σ, $\theta\in\{O, a, b, c, I\}$,

$$\sigma\vee\theta=\max(\sigma, \theta),\ \sigma\wedge\theta=\min(\sigma, \theta),$$

"$\sim$" and "$\to$" are given by the following Table 1 and Table 2 respectively

Table 1

θ	$\theta^{\sim}$
O	I
a	c
b	b
c	a
I	O

Table 2

$\to$	O	a	b	c	I
O	I	I	I	I	I
a	c	I	I	I	I
b	b	c	I	I	I
c	a	b	c	I	I
I	O	a	b	c	I

$\Re$ is (6. 1) and $\Im$ is (6. 2).

Let X, X', $Y\in\{O,a,b,c,I\}^{F_p}$ are defined respectively as follows:

$$X(p)=\begin{cases} p, & p\in\{O,a,b,c,I\}, \\ O, & \text{otherwise.}\end{cases},\ Y=C_{\Im}^{X},$$

$X'(O)=O$; $X'(a)=a$; $X'(b)=a$; $X'(c)=b$;

$X'(I)=I$; $X'(p)=O$, $p\in F_p-\{O, a, b, c, I\}$.

Take $\alpha=b$, $\beta=a$. Note that:

(1) There exists $T_0\in\Im_H$ such that

$$C_{\Im}^{\beta\otimes X}(q) = \underset{T\in\Im}{\wedge}[(\underset{p\in F_p}{\wedge}(\beta\otimes X(p)\to T(p)))\to T(q)]$$

$$= \underset{T\in\Im}{\wedge}[(a\to T(I))\to T(q)]\le T_0(q).$$

Hence, X is I-i type consistent w.r.t. (b, a, $\Im$), i=I, II.

(2) Note that $C_{\Im}^{\beta\otimes Y}=C_{\Im}^{\beta\otimes C_{\Im}^{X}}\subseteq C_{\Im}^{C_{\Im}^{X}}=C_{\Im}^{X}$, and there exists $T_1\in\Im_H$ such that

$$C_{\Im}^{X}(q) = \underset{T\in\Im}{\wedge}[(\underset{p\in F_p}{\wedge}(X(p)\to T(p)))\to T(q)]$$

$$= \underset{T\in\Im}{\wedge}[((a\to T(a))\wedge(b\to T(b))\wedge(c\to T(c))\wedge T(I))\to T(q)]$$

$$\le T_1(q).$$

Hence, $C_{\Im}^{\beta\otimes Y}(q)\le T_1(q)$. Consequently, Y is I-i type consistent w.r.t. (b, a, $\Im$), i=I, II.

(3) Since $X' \subseteq X$, so X' is I-i type consistent w.r.t. $(b, a, \mathfrak{I})$, i=I, II.

(4) $C_{\mathfrak{I}}^{X} = Y = I \otimes Y$.

Consequently, a $(b, a, O, \mathfrak{I})$-i type regular approximate reasoning model as follows can be obtained:

Rule: if X, then Y,

Input: X'.

So, the approximate reasoning consequence can be obtained as follows:

$$Y'(q) = C_{\mathfrak{I}}^{a \otimes X'}(q) = \bigwedge_{T \in \mathfrak{I}} [(\bigwedge_{p \in F_p} (a \otimes X'(p) \to T(p))) \to T(q)]$$
$$= \bigwedge_{T \in \mathfrak{I}} [(a \to T(I)) \to T(q)].$$

The approximate reasoning is I-i type consistent w.r.t. $(b, a, \mathfrak{I})$(i=I, II) in C_5, and can also be proved formally in C_5.

7. CONCLUSION

In this chapter we established two kinds of approximate reasoning theories and approximate reasoning methods in "If…then…" type approximate reasoning mode based on the strict lattice-valued proposition logic system L_{vpl}. We discussed some important properties about these approximate reasoning methods. We concluded that these two approximate reasoning methods would be sound and complete from the theoretical point of view due to their strict semantic interpretation and strict syntactical deduction for the approximate reasoning consequence in lattice-valued proposition logic L_{vpl}.

Acknowledgements

We gratefully acknowledge Professor Etienne E. Kerre for his kind help and valuable suggestions about this paper.
This work was supported by the Flanders-China Cooperation and the National Natural Science Foundation of China (Grant No. 69674015, 69774016).

REFERENCES

[1] C. Boutilier, Model logics for qualitative possibility theory, *International Journal of Approximate Reasoning*, 10(2)(1984) 173-201.

[2] D.A. Waterman, *Expert Systems*, Addison-Wesley, Reading, MA, 1986.

[3] D. Dubois, J. Lang, H. Prade, Fuzzy sets in approximate reasoning, Part I: Inference with possibility distributions, Part II: Logical approaches, *Fuzzy Sets and Systems*, 40(1991)143-202, 203-244.

[4] D. Dubois, H. Prade, Combining hypothetical reasoning and plausible inference in possibility logic, *Multiple-Valued Logic*, 1(3)(1996) 219-239.

[5] E.H. Shortliffe, G.B. Buhanan, A model of inexact reasoning in medicine, *Mathematical Biosciences*, 23(1975) 351-379.

[6] E.H. Shortliffe, *Computer-Based Medical Consultion*, MYCIN, Elsevier, North-Holland, New York, 1976.

[7] E. Orlowska, Many-valuedness and uncertainty, *Multiple-Valued Logic*, 4(3)(1999) 207-227.

[8] G. Shafer, *A Mathematical Theory of Evidence*, Princeton University Press, Princeton, New Jersey, 1976.

[9] J.F. Baldwin, A new approach to approximate reasoning using a fuzzy logic, *Fuzzy Sets and Systems*, (2)(1979) 309-325.

[10] J. Liu, Y. Xu, On certain filters in lattice implication algebras, *Chinese Quarterly Journal of Mathematics*, 11(4)(1996) 106-110.

[11] J. Liu, Y. Xu, On filters and structure of lattice implication algebras, *Chinese Science Bulletin*, 42(18)(1997) 1517-1520.

[12] J. Liu, Y. Xu, On prime filters and decomposition theorem of lattice implication algebras, *Journal of Fuzzy Mathematics*, 6(4)(1998) 1001-1008.

[13] J. Liu , Y. Xu, On the representation of lattice implication algebras, *Journal of Fuzzy Mathematics*, 7(1)(1999) 251-258.

[14] J. Pavelka, On fuzzy logic I: Many-valued rules of inference, II: Enriched residuated lattices and semantics of propositional calculi, III: Semantical completeness of some many-valued propositional calculi, *Zeitschr. F. Math. Logik und Grundlagend. Math.* , 25(1979)45-52, 119-134, 447-464.

[15] K.Y. Qin, Y. Xu, Fuzzy power sets based on lattice implication algebras, *Fuzzy Systems and Mathematics* (in Chinese), 7(2)(1993) 17-24.

[16] K.Y. Qin, Z.M. Song, Y. Xu , On product and factorization of Lattice implication algebras, *Chinese Quarterly Journal of Mathematics*, 8(4)(1993)53-57.

[17] K.Y. Qin, Y. Xu, Lukasiewicz implication algebra on [0, 1], *Proc. of Fifth IFSA*, Seoul, Vol. 7, 1993, pp. 410-412.

[18] K.Y. Qin, Y. Xu, Lattice-valued proposition logic(□) , *Journal of Southwest Jiaotong University*, 2(1)(1994) 22-27.

[19] K.Y. Qin, Y. Xu, Fuzzy propositional logic system *FP*(*X*)(I), *Journal of Fuzzy Mathematics*, 2(2)(1994) 351-357.

[20] K.Y. Qin, Y. Xu, Lattice valued propositional logic system *L*(*X*)(I), *Fuzzy Systems and Mathematics* (in Chinese), 11(4)(1997) 5-11.

[21] K.Y. Qin, Y. Xu, Z.M. Song, Lattice valued propositional logic system *L*(*X*)(II), *Fuzzy Systems and Mathematics* (in Chinese), 12(1)(1998) 10-19.

[22] K.Y. Qin, Y. Xu, Z.M. Song, Some kinds of approximate reasoning based on system *L*(*X*), *Fuzzy Systems and Mathematics* (in Chinese), 12(2)(1998) 55-60.

[23] K.Y. Qin, Y. Xu, Z.M. Song, Compactness and Lowenheim-Skolem property of logic system FM, *Chinese Science Bulletin* (in Chinese), 43(4)(1998) 371-375.

[24] K.Y. Qin, Y. Xu, The category of fuzzy lattice implication algebras, *Mathematica Applicata* (in Chinese), 11(3)(1998) 13-16.

[25] L.A. Zadeh, A fuzzy sets as a basis for a theory of possibility, *Fuzzy Sets and Systems*, (1)(1978) 3-28.

[26] L.A. Zadeh, A theory of approximate reasoning, in J.E. Hayes, D. Michie, L.I. Kulick (eds.), *Machine Intelligence*, New York, (9)(1979) 149-194.
[27] M.R. Genesereth, N.J. Nilsson, *Logical Foundations of Artificial Intelligence*, Morgan-Kaufmann, Los Altos, Calif. , 1987.
[28] R.O. Duda, P.E. Haet, N.J. Nileson, Subjective Bayesian methods for rule-based inference systems, *AFIPS Conference Proc.* , Vol. 45, AFIPS Press, 1976, 1075-1082.
[29] R.P. Dilworth, M. Ward, Residuated lattices, *Trans. Amer. Math. Soc.* , 45(1939) 335-354.
[30] V. Novak, First-order fuzzy logic, *Studia Logica* 46 (1) (1982) 87-109.
[31] X. Luo, C. Zhang, LSLNCF: A hybrid uncertain reasoning model based on probability, *International Journal of Uncertainty, Fuzziness and Knowledge-Based Systems*, 6(4)(1998) 401-422.
[32] Y.B. Jun, Y. Xu, On right regular maps of lattice implication algebras, *J. of Inst. of Math. Comp. Sci.* (Math. Ser.), 10(1)(1997) 51-58.
[33] Y.B. Jun, E. H. Roh, Y. Xu, LI-ideals in lattice implication algebras, *Bull. Korean Math. Soc.*, 35(1)(1998) 13-24.
[34] Y.B. Jun, Y. Xu, K.Y. Qin, Positive implicative and associative filters of lattice implication algebras, *Bull. Korean Math. Soc.* , 35(1)(1998) 53-61.
[35] Y.B. Jun, Y. Xu, Quasi-uniformity on lattice implication algebras, *Chinese Quarterly Journal of Mathematics*, 13(1)(1998) 84-86.
[36] Y. Xu, Lattice implication algebras, *Journal of Southwest Jiaotong University* (in Chinese), 89(1)(1993) 20-27.
[37] Y. Xu, K.Y. Qin, On filters of lattice implication algebras, *Journal of Fuzzy Mathematics*, 1(2)(1993)251-260.
[38] Y. Xu, K.Y. Qin, Lattice-valued propositional logic (I), *Journal of Southwest Jiaotong University*, 1(2)(1993) 123-128.
[39] Y. Xu, K.Y. Qin, On the category of lattice implication algebras, *Proceedings. of the First Asian Fuzzy Systems Symposium*, Singapore, Vol. 11, 1993, pp. 1098-1101.
[40] Y. Xu, K.Y. Qin, Characteristic of proper lattice implication algebras, *Chinese Quarterly Journal of Mathematics*, 9(4)(1994) 14-17.
[41] Y. Xu, K.Y. Qin, Fuzzy valued propositional logic system *FP*(X)*, *Proc. of China Association for Science and Technology Second Academic Annual Meeting of Youths*, (Sichuan), Press of Southwest Jiaotong University, 8(1995) 253-257.
[42] Y. Xu, K.Y. Qin, Z.M. Song, Syntax of first order lattice valued logic system FM, *Chinese Science Bulletin*, 42(16)(1997) 1337-1340.
[43] Y. Xu, K.Y. Qin, J. Liu, Z.M. Song, *L*-valued proposition logic L_{vpl}, *Information Sciences*, 114(1999) 205-235.
[44] Y. Xu, J. Liu, Z.M. Song, K.Y. Qin, *L*-valued first-order logic L_{vfl} (I): Semantics of L_{vfl}, *International Journal of General Systems* (accepted)
[45] Y. Xu, Z.M. Song, K.Y. Qin, J. Liu, *L*-valued first-order logic L_{vfl} (II): Syntax of L_{vfl}, *Multiple-Valued Logic* (accepted)
[46] Y. Xu, E.E. Kerre, D. Ruan, Z.M. Song, Fuzzy reasoning based on extension principle (submitted)
[47] Z.M. Song, Y. Xu, On congruence relations of lattice implication algebras, *Mathematica Applicata* (in Chinese), 10(3)(1997) 121-124.

Chapter 6

MINING INTERESTING POSSIBILISTIC SET-VALUED RULES

Alexandr A. Savinov
GMD — German National Research Center for Information Technology
AiS.KD — Autonomous Intelligent Systems Institute, Knowledge Discovery Team
Schloss Birlinghoven, Sankt-Augustin, D-53754 Germany
savinov@gmd.de

Key words: Data mining, Rule induction, Set-valued possibilistic rule, Prime disjunction, Dual transformation

Abstract: We describe the problem of mining set valued rules in large relational tables containing categorical attributes taking a finite number of values. Such rules allow for an interval of possible values to be selected for each attribute in condition instead of a single value for association rules, while conclusion contains a projection of the data restricted by the condition onto a target attribute. An example of such a rule might be "IF HOUSEHOLDSIZE = {Two OR Tree} AND OCCUPATION={Professional OR Clerical} THEN PAYMENT_METHOD = {CashCheck (Max=249, Sum=4952) OR DebitCard (Max=175, Sum=3021)} WHERE Confidence=85%, Support=10%." We use an original conceptional and formal framework for representing multidimensional distribution induced from data by a number of so-called prime disjunctions upper bounding its surface. Each prime disjunction represents a wide multidimensional interval of impossible combinations of attribute values. This original formalism generalises the conventional boolean approach in two directions: (i) finite-valued attributes (instead of only 0 and 1), and (ii) continuous-valued semantics instead of (true and false). In addition, we describe an efficient algorithm, which carries out the generalised dual transformation from possibilistic disjunctive normal form (DNF) representing data into conjunctive normal form (CNF) representing knowledge and thus generates all the most interesting prime disjunctions. Once obtained they can be used to build different forms of rules or for other purposes (prediction, clustering etc.).

1. INTRODUCTION

The problem of rule induction can be stated as follows. Given a database consisting of a number of records, where each record is a sequence of attribute values. Find rules which by their conditions select wide intervals in a multidimensional space, where the distribution of values in conclusion is highly inhomogeneous, i.e., contains large quantity of information. In the case where variables in condition and conclusion may take only one value we obtain so-called association rules [1,2,13] looking like the following:

IF $x_1 = a_{13}$ AND $x_2 = a_{25}$ THEN $x_3 = a_{32}$
WHERE Support=s% AND Confidence=c%

where s and c are rule statistic parameters. If variables may take as a value any subset of the domain then we obtain so-called set-valued rules like the following:

IF $x_1 = \{a_{12}, a_{15}\}$ AND $x_2 = \{a_{21}, a_{27}\}$ THEN $x_3 = \{a_{33} : p_{33},\ a_{36} : p_{36}\}$

where x_i is a variable, a_{ij} are their possible values, and p_{ij} are semantic parameters. Each variable in such a rule may take any value from the corresponding subset, e.g., in the above rule x_1 has two possible values — a_{12} or a_{15}, while x_3 may take only value a_{33} or a_{36}. The rule then guarantees that variable(s) in conclusion obeys to constraints represented by parameters p_{ij}. At the very beginning we would like to point out a significant difference in computational complexity of these two approaches. If variables $x_1, x_2, \ldots, x_n$ have $n_1, n_2, \ldots, n_n$ values, respectively, then for association rules the number of conditions is $n_1 \times n_2 \times \ldots \times n_n$ while for set-valued rules the potential search space size is $2^{n_1} \times 2^{n_2} \times \ldots \times 2^{n_n} = 2^{n_1 + n_2 + \ldots + n_n}$.

In the chapter we consider the problem of mining set-valued rules and suppose that each variable, i.e., each table column, takes only a finite number of values. The semantics induced from data is represented by a frequency distribution over the universe of discourse equal to the Cartesian product of all sets of the values. Theoretically, all possible set-valued conditions produce rules, i.e., we have as many potential rules as the size of the space of conditions. The rule conclusion is calculated as a formal projection of the distribution restricted by the condition onto a target variable (Fig. 1). (For simplicity we suppose that there is only one target variable.) Then the problem, which has been paid a lot of attention in recent years, is whether one or another rule is really interesting or not. For example, for association rules interestingness is based mainly on confidence and support.

For logical approaches it may be, e.g., a classification power or the number of covered examples. In this chapter we suppose that rule interestingness is equivalent to its surprisingness, i.e., informally, to what extent information in the rule is unusual and differs from what has been expected. This notion is traditionally formalised by means of one or another measure of the quantity of information. For example, the crisp rule

$$\text{IF } x_1 = \{a_{12}, a_{15}\} \text{ THEN } x_3 = \{a_{31}, a_{32}, a_{33}, a_{34}, a_{35}\}$$

is obviously not interesting since it says exactly what we have expected — the target variable may take any value, while the rule

$$\text{IF } x_1 = \{a_{11}, a_{14}\} \text{ THEN } x_3 = \{a_{31}, a_{33}, a_{35}\}$$

is more interesting since, contrary to our expectations, it has turned out that several values are not possible. The main problem, however, is that information in the rule and its surprisingness should be compared not only with our default expectations expressed as a projection of the whole unconditioned distribution but with other rules as well. For example, the rule

$$\text{IF } x_1 = \{a_{11}, a_{14}\} \text{ AND } x_2 = \{a_{22}, a_{23}\} \text{ THEN } x_3 = \{a_{31}, a_{33}, a_{35}\}$$

may be found interesting in relation to our default expectations but it contains nothing new in relation to the previous rule, which contains more information due to a simpler condition. Thus informally, the more general the rule condition (the wider the interval selected by the condition) and the more specific the rule conclusion (the closer the conclusion distribution to the singular form), the more interesting and informative it is. Note that for each concrete rule there is some limit to which we can widen (generalise) its condition and restrict the conclusion, i.e., there is the notion of the most strong rule.

To find such maximally general in condition and specific in conclusion rules we use an approach according to which any finite multidimensional possibility distribution can be formally represented by a set of special logical constructions called possibilistic prime disjunctions. Using them we avoid many problems connected with the separate treatment of conditions and conclusions. Prime disjunction has an important property of optimality — it is the most informative among all constructions of such a form and an attempt to generalise it results in that it becomes wrong, i.e., does not follow from the original distribution and cannot be used to represent it. It is precisely the property that later on guarantees an absence of a rule, which is informative however is not surprising enough in relation to some other rule. Prime disjunctions are helpful of themselves since they can be interpreted as negative associations, i.e., they intensionally represent the widest intervals of combinations of values where the distribution induced from data has the lowest values. In this chapter, however, it is important that prime disjunctions can be easily transformed into the form of rules representing the most interesting dependencies among attributes.

In addition to the original conceptual and formal frameworks based on the notion of possibilistic prime disjunction and knowledge representation language (Sections 2 and 3) we propose a concrete algorithm which efficiently generates all the most interesting prime disjunctions. There are two general approaches to the problem of generation of prime disjunctions. The first approach (Section 4), called the covering method, builds one prime disjunction at every step. The algorithm starts from the strongest disjunction, which does not cover any record and then searches through their space by weakening the current disjunction unless all the distribution points are covered. For our purposes this approach has one big disadvantage that the whole data set has to be in memory to store the cover state.

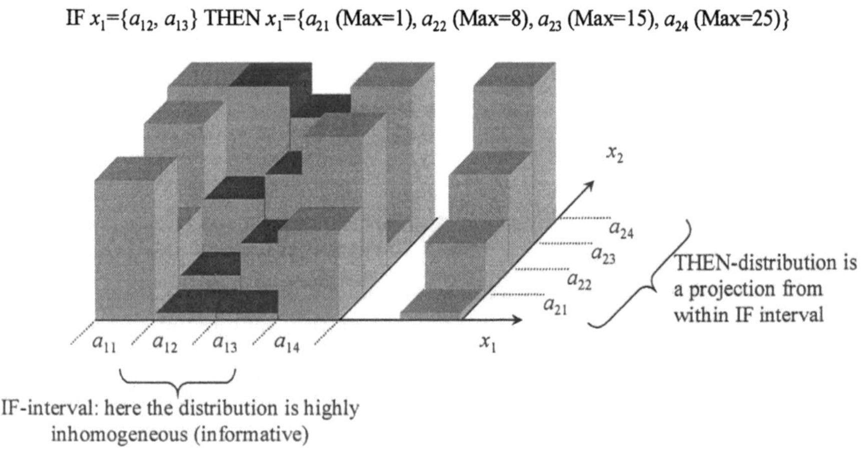

Figure 1. An example of possibilistic set-valued rule.

Another approach (Section 5) builds all prime disjunctions in parallel while processing all data elements in succession. The algorithm assumes that at each moment the semantics is equal to the number of processed records and the current set of prime disjunctions is updated each time new data element is processed. Formally this algorithm is based on the generalised formula for transforming DNF representing data into CNF consisting of prime disjunctions. An advantage of this approach is that the set of prime disjunctions is built for one pass through the data set.

The notion of prime disjunction (conjunction, implicant etc.) and algorithms for finding them have received a lot of attention in various fields, especially in classical cybernetics (generation of prime implicants [22]) and combinatorics. In particular, the notion of prime implicant is analogous to that of maximum frequent itemset, which is used for efficient mining association rules [3,12].

Our approach to the interpretation of multidimensional possibility distributions is somewhat similar to that described in [4,5] (or in fuzzy form in [11]). However in these papers the main operation considered is a decomposition of a multidimensional possibility distribution into a number of distributions of lower dimensionality, which more or less exactly approximate it and can be used to learn the possibilistic networks. In this chapter we suppose that the possibility distribution can be represented by a set of prime disjunctions and the problem of rule induction comes down to finding this set given the dual representation of the dataset by means of possibilistic DNF.

The criterion of high surprisingness of conclusion makes this method very similar to that for interesting subgroup discovery [8,9,10,23] since the condition of set-valued rules (intensionally) defines a subset of all objects revealing highly unusual properties in relation to the target attribute.

Our data and knowledge representation language is based on a so-called method of sectioned vectors and matrices, which originates from the paper [24] and later was generalised onto fuzzy case [14,17,16]. The idea of transformation from fuzzy DNF into fuzzy CNF and finding rules was proposed in [15]. A fuzzy version of this rule induction algorithm, which is based on the covering method is described in [18] while crisp version is described in [21]. The most recent version of this approach is presented in [19,20].

2. REPRESENTATION OF MULTIDIMENSIONAL FINITE DISTRIBUTIONS

Let some problem domain at the syntactic level be described by a finite number of *variables* or *attributes* $x_1, x_2, \ldots, x_n$ each of which takes a finite number of *values* and corresponds to one column of a relational table:

$$x_i \in A_i = \{a_{i1}, a_{i2}, \ldots, a_{in_i}\}, \quad i = 1, 2, \ldots, n$$

where n_i is the number of values of the *i*-th variable and A_i is its set of values. For example, a problem domain might be described by three attributes x_1=RESPONSE, x_2=INCOME, and x_3=HOUSEHOLDSIZE taking the values from the sets A_1={No, Yes}, A_2={Low, Average, High}, and A_3={One, Two, Three, Four}, respectively. The *state space* or the *universe of discourse* is defined as the Cartesian product of all sets of the values:

$$\Omega = A_1 \times A_2 \times \ldots \times A_n$$

The universe of discourse is a finite set with a multidimensional structure. Each syntactic object (state) from the universe of discourse is represented by a combination of values of all variables: $\omega = \langle x_1, x_2, \ldots, x_n \rangle \in \Omega$. The number of such objects is equal to the power of the universe of discourse: $|\Omega| = n_1 \times n_2 \times \ldots \times n_n$. In our example the 3-dimensional universe of discourse consists of $A_1 \times A_2 \times A_3 = 2 \times 3 \times 4$ =24 possible objects.

Formally the problem domain *semantics* is represented by a frequency distribution over the state space. In other words, to represent the semantics each combination of attribute values (syntactic object) should be assigned a natural number or 0, which is thought of as its number of observations. For example, the object $\langle$ Yes, High, Three $\rangle$ might be observed 8 times. In principle, it is possible and for many cases does make sense to map this distribution into the interval [0,1]. Yet, for the problem of rule induction it is simpler to work directly with frequencies so we will not use a mapping into [0,1].

The semantics will be represented by *elementary propositions* about individual variables, which are combined with the help of logical connectives $\vee$ and $\wedge$. All semantic constructions will be written in bold and one lower index will always mean the number of the variable this proposition is about, e.g., $\mathbf{u}_i$ is an elementary proposition about the *i*th variable. Elementary proposition assigns frequencies to all values of one variable, i.e., it is represented by a local distribution over the values of this variable. A frequency assigned to one value, i.e., the local distribution value in one point will be referred to as a *component*. We will write the concrete semantics of elementary propositions as a sequence of the corresponding local distribution values (components): $\mathbf{u}_i = \{\mathbf{u}_{i1}, \mathbf{u}_{i2}, \ldots, \mathbf{u}_{in_i}\}$. Thus the elementary proposition $\mathbf{u}_i$ about the *i*th variable is always made up of n_i components which are denoted by the same bold symbol with two lower indexes corresponding to the number of variable and the number of value. For example, $\mathbf{u}_3 = \{3,0,15,7\}$ is the elementary proposition about the 3rd variable (HOUSEHOLDSIZE) which assigns the frequencies 3, 0, 15 and 7 to its 4 values a_{31}=One, a_{32}=Two, a_{33}=Three, and a_{34}=Four, respectively.

Several elementary propositions combined with the connective $\vee$ are said to be a *disjunction* and denoted by bold symbol (without index), e.g., $\mathbf{d} = \mathbf{d}_2 \vee \mathbf{d}_4 \vee \mathbf{d}_6$ is a disjunction consisting of 3 elementary propositions. If elementary propositions are combined with the connective $\wedge$ then such a construction is said to be a *conjunction*, e.g., $\mathbf{k}=\mathbf{k}_1\wedge\mathbf{k}_3\wedge\mathbf{k}_5$ is a conjunction consisting of 3 elementary propositions. A conjunction of disjunctions is said to be a *conjunctive normal form*, while dually a disjunction of conjunctions is said to be a *disjunctive normal form*.

The semantic constructions themselves do not represent anything — their concrete meaning is defined by means of *interpretation rules*. In other words, interpretation rules allow us to find out what concrete distribution one or another semantic language construction defines. The operations maximum and minimum will be used to define interpretation rules. We will add arguments to propositions when we want to show that it is the meaning of the proposition in the form of the corresponding distribution rather than simply non-interpreted proposition. For example, $\mathbf{d}$, $\mathbf{K}$ and $\mathbf{d}_i$ are propositions while $\mathbf{d}(\omega)$, $\mathbf{K}(\omega)$ and $\mathbf{d}_i(x_i)$ are distributions corresponding to these propositions.

Proposition semantics is defined as a distribution over the universe of discourse. Elementary propositions are interpreted by extending their local distribution onto the whole universe of discourse:

$$\mathbf{u}_i(\omega)=\mathbf{u}_i(\langle x_1,x_2,\ldots,x_n\rangle)=\mathbf{u}_i(x_i)$$

i.e., at any point of the universe of discourse the distribution is equal to some elementary proposition component. Obviously, this is a particular case of the operation of *cylindrical extension* or *deprojection*, which allows us to increase the number of dimensions (in our case from 1 to n).

Two propositions $\mathbf{u}$ and $\mathbf{v}$ combined with the connective $\vee$ define the following distribution:

$$(\mathbf{u}\vee\mathbf{v})(\omega)=\max(\mathbf{u}(\omega),\mathbf{v}(\omega))$$

It can be shown that the disjunction $\mathbf{d}$ defines the distribution, which is equal to the maximum of its n components corresponding to the point coordinates:

$$\mathbf{d}(\omega)=\mathbf{d}(\langle x_1,x_2,\ldots,x_n\rangle)=\mathbf{d}_1(x_1)\vee\mathbf{d}_2(x_2)\vee\ldots\vee\mathbf{d}_n(x_n)=\max_{i=1,\ldots,n}\mathbf{d}_i(x_i)$$

For example, the distribution defined by the disjunction $\mathbf{u}=\mathbf{u}_1\vee\mathbf{u}_2\vee\mathbf{u}_3=\{1,3\}\vee\{9,0,4\}\vee\{3,0,15,7\}$ in the point $\omega=\langle a_{11},a_{21},a_{31}\rangle$ has the value $\mathbf{u}(\langle a_{11},a_{21},a_{31}\rangle)=\max(1,9,3)=9$. The maximum is taken among n components — one from each elementary proposition.

Conjunctions are interpreted dually, i.e., the conjunction **k** defines the distribution, which is equal to the minimum of the components corresponding to the point coordinates:

$$\mathbf{k}(\omega)=\mathbf{k}(\langle x_1,x_2,\ldots,x_n\rangle)=\mathbf{k}_1(x_1)\wedge\mathbf{k}_2(x_2)\wedge\ldots\wedge\mathbf{k}_n(x_n)=\min_{i=1,\ldots,n}\mathbf{k}_i(x_i)$$

Formally semantics is represented by the corresponding distribution over the universe of discourse, however, for complete certainty we have to define its modality, i.e., what we concretely mean by these numbers assigned to combinations of values. In this chapter we consider possibility and necessity distributions, i.e., only two dual modalities. Below in this section we suppose that distributions take values from [0,1].

For possibility distributions the semantic value 0 is interpreted as an absolute impossibility of the corresponding object while all positive numbers are interpreted as various degrees of possibility so that 1 is conventionally interpreted as an uncertainty, complete possibility. Then the absence of information means that the distribution is equal to 1 at any point of the universe of discourse, while the presence of information means that some points are disabled, prohibited with some degree so that the more information, the lower the distribution. With the help of possibility distributions we can represent only negative information about impossibility and are not able to represent information about the necessity of states.

Dually, for necessity distributions the semantic value 0 means absolute uncertainty and positive numbers are interpreted as degrees of necessity. Thus the absence of information is represented by constant 0 distribution (the weakest proposition) while the more information, the higher the distribution. With the help of necessity distributions we can represent positive information, i.e., information describing what certainly has happened. In particular, we are not able to explicitly represent knowledge about impossible events.

Possibility and necessity distributions are supposed to be represented by a number of disjunctions and conjunctions, respectively. In other words, possibility distributions are represented by CNF while necessity distributions are represented by DNF. Thus a disjunction is an elementary piece of possibilistic information, which represents constraints on combinations of values by defining an interval of impossibility. It is important that the degree of possibility can be only decreased, in particular, if some point is impossible then no one additional proposition can make it possible (the property of monotonicity). Several disjunctions combined with $\wedge$ can represent any possibility distribution by upper bounding its surface (Fig. 2). The most specific disjunction is called *singular disjunction* and is made up of all 1's except for one component in each proposition, e.g.,

$\{1,0\} \vee \{1,0,1\} \vee \{1,0,1,1\}$. Such a disjunction pricks a hole in one point of the distribution surface (in our example in the point $\omega = \langle a_{12}, a_{22}, a_{32} \rangle$). The CNF is said to represent the distribution *extensionally* if it contains only singular disjunctions.

Dually, to represent necessity distributions we use conjunctions, which lower bound its surface (Fig. 2). *Singular conjunction* consists of all 0's except for one component in each proposition, e.g., $\{0,1\} \wedge \{0,1,0\} \wedge \{0,0,0,1\}$, and adding it to DNF results in a peak in the distribution surface. DNF is said to represent the distribution *extensionally* if it contains only singular conjunctions. Data can be extensionally represented in the form of DNF so that each singular conjunction represents one record along with the number of its occurrences in the data set.

One distribution is said to be a (possibilistic) *consequence* of another if its values in all points of the universe of discourse are greater or equal to the values of the second distribution. We will also say that the first distribution covers the second one. The consequence relation for conjunctions, disjunctions, DNF and CNF is defined as the consequence relation for the corresponding distributions.

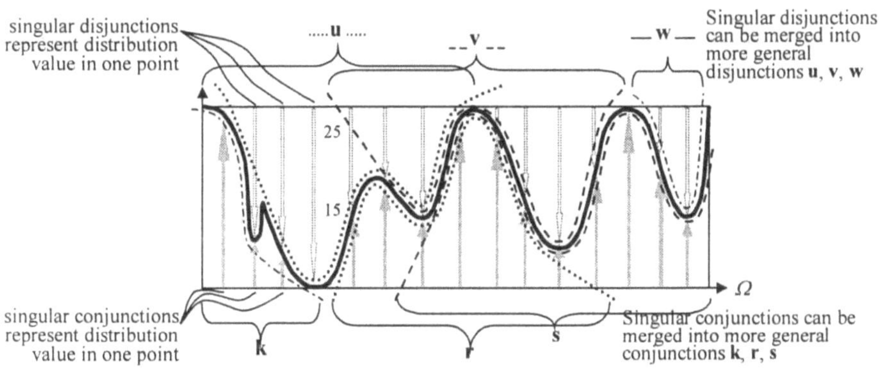

Figure 2. Representation of multidimensional frequency distribution by a number of (upper bounding) disjunctions or a number of (lower bounding) conjunctions.

3. DEPENDENCIES IN MULTIDIMENSIONAL SPACE

By dependency we mean *any information restricted by a simple form (language) of representation*, i.e., the more information can be put into some fixed simple structure the more interesting and significant the dependency is. Thus we suppose that semantics does not exist by itself and it is a *representation language* or other representation mechanism that allows us to express and store it. However, when we deal with dependencies we are interested in simple forms of semantics representation even if simplifying results in some loss of information. For example, a database is a representation of some semantics but the problem then is to transform this representation into a simpler form. A function written as a table of its values in all points is not considered simply represented while its transformation into another basis (e.g., Fourier transformation) may produce more descriptive representation. The notion of simplicity is determined by the representation language chosen for one or another problem domain and, generally, the same information may be simpler in one representation and more complex in another representation language.

Thus to find dependencies first we have to fix our representation language. In this chapter to represent information we use the formalism of CNF and DNF generalised onto the case of (i) finite-valued variables (instead of only 0 and 1 in the conventional boolean approach), and (ii) continuous-valued semantics (instead of only true and false). We also explicitly separate two cases of semantics representation: by means of CNF to represent possibility distributions, and by means of DNF to represent necessity distributions. The structure we use to express dependencies is either disjunction or conjunction, i.e., this is our restriction on the knowledge representation language (e.g., two disjunctions are already not a dependency). Thus in our case by dependency we mean any information represented in the form of either disjunction or conjunction.

Initially, the database can be represented in the form of DNF. However, each singular conjunction of such a representation is the most specific one since it represents the distribution value in one point. One conceptually traditional approach to inducing dependencies consists in merging such point-wise constructions into more general ones so that the resulted conjunction represents the distribution in several points (Fig. 2). Then new more general conjunction absorbs more specific ones, which produced it. For example, in our representation language two singular conjunctions $\{0,0,3\} \wedge \{0,0,0,3\}$ and $\{0,0,3\} \wedge \{0,0,3,0\}$ can be merged into one more general conjunction $\{0,0,3\} \wedge \{0,0,3,3\}$, which however cannot be generalised to cover also the conjunction $\{1,0,0\} \wedge \{5,0,0,0\}$. This process of merging singular conjunctions results in a number of more general conjunctions, each of which represents the initial distribution values within some (positive) interval. The wider this interval, the more general the dependency represented by the conjunction. There is always some limit up to which conjunctions can be generalised due to their fixed structure. If the conjunction cannot be generalised, i.e., no more additional points (singular conjunctions) can be included into it, then it is said to be a prime one. There may be different measures of the interval wideness, i.e., the quantity of information in the conjunction. For many approaches this quantity is calculated as the integral of the distribution represented by the conjunction (the area under the distribution), i.e., essentially, the number of positive examples it covers.

One feature of our approach described in the chapter and distinguishing it from others is that we find dependencies in the dual form as disjunctions Fig. 3. It can be realised as merging singular disjunctions representing negative information about one point (essentially, such a singular disjunction says that some combination of values is impossible). Then more general disjunctions resulted from merging more specific ones describe wider (negative) intervals of impossible combinations of values. Disjunctions, which cannot be generalised, are said to be prime ones. Thus a prime disjunction is a consequence of the initial distribution it represents but is not a consequence of any other disjunction that can be used to represent this distribution. Another characteristic of prime disjunction is that if any component is decreased then it is already not a consequence of the initial distribution. On the other hand, any disjunction that follows from the initial distribution can be obtained by weakening some prime disjunction. Naturally we are interested in finding dependencies in the form of prime disjunctions since they impose the strongest constraints on the possible combinations of attribute values. The problem then is to find an efficient algorithm to transform the initial representation by singular conjunctions into a number of interesting prime disjunctions.

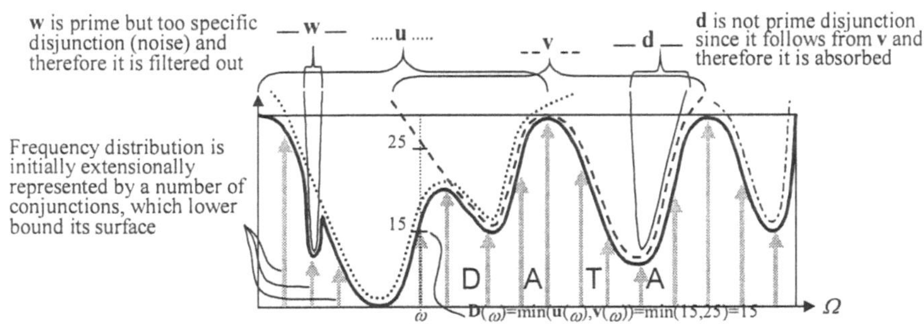

Figure 3. Finding dependencies by transforming singular conjunctions into more general disjunctions.

The above-described conception of finding general dependencies in the form of either disjunctions or conjunction (depending on the chosen modality) generates a lot of interesting information, which however is not very suitable for perception. The result of this data mining process can be thought of as raw unprocessed knowledge resulted from transformation of data from its initial extensional representation into an intensional form. The general goal of this step is to obtain *any* dependencies of certain form. The next important step that should be done is to find something really interesting within this intensional representation. This process is referred to as *knowledge mining* since it processes existing knowledge (raw knowledge) rather than data. In this chapter we focus on mining interesting rules from prime disjunctions. Once interesting rules have been found the final step is to visualise them so that their content becomes clear to the user.

In contrast to dependencies, which can be anything, e.g., a differential equation or neural network, rules should be interesting, easily perceivable and expressive since they are usually used by humans. Traditional for rules is the criterion of classification power according to which the rule significance increases along with its ability to classify some objects (usually the values of the target attribute). Another wide spread criterion is the number of examples the rule covers. Since we use possibilistic modality, i.e., represent the initial semantics by a number of disjunctions, this criterion should be reformulated in its dual form as a size of the impossibility interval instead of the necessity interval. In other words, the wider the impossibility interval, i.e., the more points have the less distribution values, the more general the disjunction and the rule it produces are. Thus the only requirement for rule quality is high quantity of information it holds rather than the form it is expressed or the way it works. In particular, such a criterion allows us to solve the problem of the trade off between the generality of rule condition and specificity of rule conclusion.

The main problem however in rule induction consists in generating the intervals of either necessary or impossible combinations of values, i.e., finding their conjunctive or disjunctive representations. The performance of an algorithm depends on the richness of dependency representation language and the strategy used to search through the space of all dependencies. In the case of disjunctive dependencies discussed in this chapter the space of all disjunctions has to be searched for points satisfying certain conditions. One component of disjunction is responsible for one dimension in this space. The disjunction consisting of all 0's corresponds to the origin of the coordinate system while increasing one component which is referred to as an operation of *elementary induction* moves the disjunction along one axis. Thus the space of disjunctions can be searched by applying elementary induction to different components until the disjunction satisfies the conditions. There are two main strategies how it can be done. The generalised covering method described in Section 4 generates one prime disjunction for several passes through the data set by means of gradual weakening the current disjunction until it covers all points (conjunctions in DNF). The second strategy described in Section 5 builds prime disjunctions in parallel for one pass through the data set. It consists in updating the current set of disjunctions each time a new record (conjunction from DNF) is processed.

4. GENERALISED COVERING METHOD

Let us suppose that a database is written as the matrix of singular conjunctions $\mathbf{K}$, which represents some distribution over the universe of discourse Ω. The problem is to transform this distribution into the form of dependencies between attributes represented by a set of prime disjunctions written as a matrix $\mathbf{D}$ (knowledge). In this section, a generalised covering method is described which can be used to carry out such a transformation into a dual form. This method just like the whole approach generalises the conventional boolean approach in two directions: (i) it allows using finite-valued variables, and (ii) continuous-valued distributions can be used (although we use only integer values for representing frequencies).

Prime disjunctions are built gradually by means of the depth-first search through the space of all possible disjunctions starting from the disjunction consisting of all 0's and then increasing its different components (Fig. 4). As components increase they influence corresponding columns of the matrix and some of its lines are *covered.* Formally it means that the disjunction becomes their consequence (in possibilistic sense). When all lines of the matrix are covered we say that the obtained disjunction is a prime one. An order in which the procedure searches the space of disjunctions is determined by the order in which we increase the components.

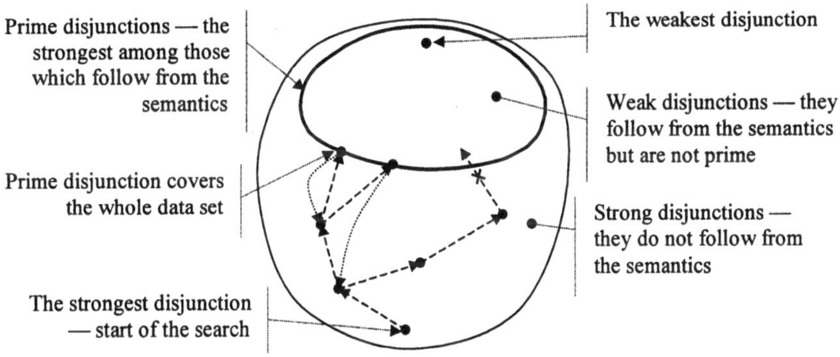

Figure 4. Search through the space of all disjunctions in generalised covering method.

Let us consider an example from the field of screening alcoholism [18] where the goal is to determine a patient alcoholism stage (methods of treatment depend on this diagnosis). The problem domain can be described by the following three attributes:

DEPENDENCE = {Psychical, Psychico-physical, Physical}
LOSS OF SELF-CONTROL = {Quantitative, Situational}
ALCOHOLISM STAGE = {First, Second, Third}

Let us suppose that there is a database of case histories, which can be transformed into the matrix of possibilistic DNF consisting of 14 conjunctions:

$$\mathbf{K} = \left|\begin{array}{lll} \{5,0,0\} & \{5,0\} & \{5,0,0\} \\ \{2,0,0\} & \{2,0\} & \{0,2,0\} \\ \{5,0,0\} & \{0,5\} & \{5,0,0\} \\ \{3,0,0\} & \{0,3\} & \{0,3,0\} \\ \{0,5,0\} & \{5,0\} & \{0,5,0\} \\ \{0,4,0\} & \{4,0\} & \{0,0,4\} \\ \{0,3,0\} & \{0,3\} & \{0,3,0\} \\ \{0,4,0\} & \{0,4\} & \{0,0,4\} \\ \{0,0,5\} & \{5,0\} & \{5,0,0\} \\ \{0,0,5\} & \{5,0\} & \{0,5,0\} \\ \{0,0,5\} & \{5,0\} & \{0,0,5\} \\ \{0,0,5\} & \{0,5\} & \{5,0,0\} \\ \{0,0,3\} & \{0,3\} & \{0,3,0\} \\ \{0,0,5\} & \{0,5\} & \{0,0,5\} \end{array}\right| \begin{array}{l} \mathbf{1} \\ \mathbf{2} \\ \mathbf{3} \\ \mathbf{4} \\ \mathbf{5} \\ \mathbf{6} \\ \mathbf{7} \\ \mathbf{8} \\ \mathbf{9} \\ \mathbf{10} \\ \mathbf{11} \\ \mathbf{12} \\ \mathbf{13} \\ \mathbf{14} \end{array}$$

This matrix represents a 3-dimensional distribution over the space consisting of $3 \times 2 \times 3 = 18$ points. For example, conjunction **1** represents the value 5 in the point $\langle$ Psychical, Quantitative, First$\rangle$. To find a dependency among the attributes we have to build some prime disjunction for this DNF, i.e., the disjunction that is minimal however covers all conjunctions from **K**. It can be noticed that the third column (3-rd value of the 1-st attribute) contains the maximal number of non-0 components therefore it is natural to cover the maximum number of lines by setting the corresponding component in disjunction to 5. Then the disjunction $\{0,0,\underline{5}\} \vee \{0,0\} \vee \{0,0,0\}$ covers 6 last conjunctions **9**, **10**, **11**, **12**, **13**, **14** of the matrix, which after removing them is equal to

$$\mathbf{K}^1 = \left|\begin{array}{lll} \{5,0,0\} & \{5,0\} & \{5,0,0\} \\ \{2,0,0\} & \{2,0\} & \{0,2,0\} \\ \{5,0,0\} & \{0,5\} & \{5,0,0\} \\ \{3,0,0\} & \{0,3\} & \{0,3,0\} \\ \{0,5,0\} & \{5,0\} & \{0,5,0\} \\ \{0,4,0\} & \{4,0\} & \{0,0,4\} \\ \{0,3,0\} & \{0,3\} & \{0,3,0\} \\ \{0,4,0\} & \{0,4\} & \{0,0,4\} \end{array}\right| \begin{array}{l} \mathbf{1} \\ \mathbf{2} \\ \mathbf{3} \\ \mathbf{4} \\ \mathbf{5} \\ \mathbf{6} \\ \mathbf{7} \\ \mathbf{8} \end{array}$$

Now let us choose the second column since it has greater sum of the components. Then the disjunction $\{0,\underline{5},5\} \vee \{0,0\} \vee \{0,0,0\}$ covers 4 last lines **5**, **6**, 7, **8** in the matrix $\mathbf{K}^1$ and we obtain:

$$\mathbf{K}^2 = \left|\begin{array}{lll} \{5,0,0\} & \{5,0\} & \{5,0,0\} \\ \{2,0,0\} & \{2,0\} & \{0,2,0\} \\ \{5,0,0\} & \{0,5\} & \{5,0,0\} \\ \{3,0,0\} & \{0,3\} & \{0,3,0\} \end{array}\right| \begin{array}{l} \mathbf{1} \\ \mathbf{2} \\ \mathbf{3} \\ \mathbf{4} \end{array}$$

Now we see that this matrix can be covered by the third proposition {5,3,0} (the component 5 covers disjunctions **1**, **3**, and the component 3 — **2**, **4**). Thus we obtain that the prime disjunction **d** = {0,5,5} ∨ {0,0} ∨ {5,3,0} covers the whole matrix of conjunctions. This disjunction imposes upper bound on possible frequencies of events. For example, the combination of values $\langle a_{11}, a_{21}, a_{31} \rangle = \langle$ Psychical, Quantitative, First$\rangle$ has maximal frequency $\max(0,0,5) = 5$, while $\mathbf{d}(\langle a_{11}, a_{21}, a_{32} \rangle) = \max(0,0,3) = 3$.

5. METHOD OF DUAL TRANSFORMATION

5.1 Generation of Disjunctions

The method of finding prime disjunctions described in this section is based on the direct transformation from DNF into CNF. On each step of this procedure (Fig. 5) the next conjunction from DNF is added with the help of disjunction (maximum) operation to the current matrix of CNF. This matrix at any moment contains the joint semantics of all processed conjunctions. At the very beginning it is supposed to be 0, i.e., representing the constant 0 distribution. At the next step it is equivalent to one conjunction, i.e., represents the distribution, which is equal to 0 in all except for one point. The procession of each conjunction results in a number of new disjunctions, which are obtained from the old ones with the help of elementary induction. Some of these disjunctions are not prime (i.e., they follow from others) and should be removed from the matrix. In any case the number of disjunctions is extremely high so there should be a mechanism for selecting only the most informative and interesting ones.

To add the conjunction **k** to the matrix of CNF **D** it is necessary to add it to all *m* disjunctions of the matrix:

$$\mathbf{k} \vee \mathbf{D} = \mathbf{k} \vee (\mathbf{d}^1 \wedge \mathbf{d}^2 \wedge \ldots \wedge \mathbf{d}^m) = (\mathbf{k} \vee \mathbf{d}^1) \wedge (\mathbf{k} \vee \mathbf{d}^2) \wedge \ldots \wedge (\mathbf{k} \vee \mathbf{d}^m)$$

Addition of conjunction to disjunction is carried out by the formula:

$$\begin{aligned}
&\mathbf{k} \vee \mathbf{d} = (\mathbf{k}_1 \vee \mathbf{d}) \wedge (\mathbf{k}_2 \vee \mathbf{d}) \wedge \ldots \wedge (\mathbf{k}_n \vee \mathbf{d}) = \\
&(\mathbf{k}_1 \vee (\mathbf{d}_1 \vee \mathbf{d}_2 \vee \ldots \vee \mathbf{d}_n)) \wedge \\
&(\mathbf{k}_2 \vee (\mathbf{d}_1 \vee \mathbf{d}_2 \vee \ldots \vee \mathbf{d}_n)) \wedge \\
&\ldots\ldots\ldots\ldots\ldots\ldots\ldots\ldots \\
&(\mathbf{k}_n \vee (\mathbf{d}_1 \vee \mathbf{d}_2 \vee \ldots \vee \mathbf{d}_n)) = \\
&(\mathbf{k}_1 \vee \mathbf{d}_1 \quad \vee \quad \mathbf{d}_2 \quad \vee \ldots \vee \quad \mathbf{d}_n) \wedge \\
&(\mathbf{d}_1 \quad \vee \quad \mathbf{k}_2 \vee \mathbf{d}_2 \quad \vee \ldots \vee \quad \mathbf{d}_n) \wedge \\
&\ldots\ldots\ldots\ldots\ldots\ldots\ldots\ldots\ldots\ldots\ldots\ldots \\
&(\mathbf{d}_1 \quad \vee \quad \mathbf{d}_2 \quad \vee \ldots \vee \quad \mathbf{k}_n \vee \mathbf{d}_n) \wedge
\end{aligned}$$

and in general case n new disjunctions are generated from one source disjunction by applying the elementary induction, i.e., by increasing one component.

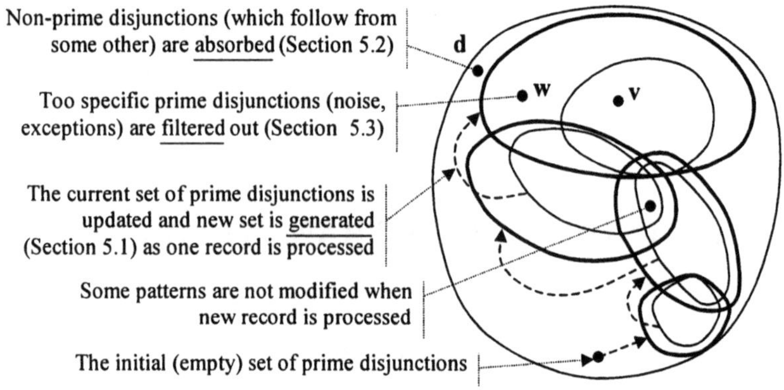

Figure 5. Search in the space of all disjunctions by direct transformation into CNF: the semantics and the corresponding set of prime disjunctions are updated each time new conjunction is processed.

If $\mathbf{k}$ is covered by $\mathbf{d}$ then its addition to $\mathbf{d}$ does not change the semantics: $\mathbf{k} \vee \mathbf{d} = \mathbf{d}$. In this case the disjunction can be simply copied to the new matrix with no modifications. Thus the whole set of new disjunctions can be divided into two subsets: modified and non-modified.

For example, let us suppose that we have two conjunctions $\mathbf{k}^1 =$ 05.005.0005 and $\mathbf{k}^2 =$ 03.003.0030 which have to be transformed into disjunctions. Each new matrix is obtained from the previous one as follows: $\mathbf{D}^i = \mathbf{D}^{i-1} \vee \mathbf{k}^i$, where $i = 1,2$, and $\mathbf{D}^0 = \mathbf{d}^0 =$ 00.000.0000. Thus after processing the first conjunction we obtain:

$$\mathbf{D}^1 = \mathbf{D}^0 \vee \mathbf{k}^1 = 00.000.0000 \vee 05.005.0005 = \left| \begin{array}{c|c} 0\underline{5}.000.0000 & \mathbf{1} \\ 00.00\underline{5}.0000 & \mathbf{2} \\ 00.000.000\underline{5} & \mathbf{3} \end{array} \right.$$

where increased components (to which elementary induction has been applied) are underlined and three resulted disjunctions are denoted with bold numbers. After processing the second conjunction we obtain:

$$\mathbf{D}^2 = \mathbf{D}^1 \vee \mathbf{k}^2 = \left| \begin{array}{c} 05.000.0000 \\ 00.005.0000 \\ 00.000.0005 \end{array} \right| \vee 03.003.0030 = \left| \begin{array}{c|c} 05.000.0000 & \mathbf{1} \\ 00.005.0000 & \mathbf{2} \\ 0\underline{3}.000.0005 & \mathbf{3.1} \\ 00.00\underline{3}.0005 & \mathbf{3.2} \\ 00.000.00\underline{3}5 & \mathbf{3.3} \end{array} \right.$$

where for convenience we separate the lines produced by different parents. Note that the disjunctions **1** and **2** are not modified since they cover $\mathbf{k}^2$ and only three new disjunctions **3.1**, **3.2**, and **3.3** have been generated from their parent **3**.

5.2 Absorption of Disjunctions

As new disjunctions are generated and added to the new matrix the absorption procedure should be carried out to remove weak lines, i.e., the lines which are not prime and follow from others, e.g., **d** in Fig. 5. In general, each new disjunction can either be absorbed itself or absorb other lines. Thus the comparison of lines has to be fulfilled in both directions. To check for the consequence relation between two disjunctions we have to reduce them (see [16,17] for more information about reduced forms) and then compare all their components.

For example, if we add new conjunction $\mathbf{k}^3 = 05.005.0500$ to the matrix $\mathbf{D}^2$ (section 5.1) then we obtain

$$
\mathbf{D}^3 = \mathbf{D}^2 \vee \mathbf{k}^3 = \left|\begin{array}{c} 05.000.0000 \\ 00.005.0000 \\ 03.000.0005 \\ 00.003.0005 \\ 00.000.0035 \end{array}\right| \vee 05.005.0500 = \left|\begin{array}{c|l} 05.000.0000 & \mathbf{1} \\ \hdashline 00.005.0000 & \mathbf{2} \\ \hdashline 0\underline{5}.000.0005 & \mathbf{3.1.1} \supseteq \mathbf{1} \\ 03.00\underline{5}.0005 & \mathbf{3.1.2} \supseteq \mathbf{2} \\ 03.000.0\underline{5}05 & \mathbf{3.1.3} \\ \hdashline 0\underline{5}.003.0005 & \mathbf{3.2.1} \supseteq \mathbf{1} \\ 00.00\underline{5}.0005 & \mathbf{3.2.2} \supseteq \mathbf{2} \\ 00.003.0\underline{5}05 & \mathbf{3.2.3} \\ \hdashline 0\underline{5}.000.0035 & \mathbf{3.3.1} \supseteq \mathbf{1} \\ 00.00\underline{5}.0035 & \mathbf{3.3.2} \supseteq \mathbf{2} \\ 00.000.0\underline{5}35 & \mathbf{3.3.3} \end{array}\right.
$$

where 6 lines are absorbed and therefore the final matrix is:

$$
\mathbf{D}^3 = \left|\begin{array}{c|l} 05.000.0000 & \mathbf{1} \\ \hdashline 00.005.0000 & \mathbf{2} \\ \hdashline 03.000.0505 & \mathbf{3.1.3} \\ \hdashline 00.003.0505 & \mathbf{3.2.3} \\ \hdashline 00.000.0535 & \mathbf{3.3.3} \end{array}\right.
$$

Several properties, which are formulated below, can significantly simplify the absorption process.

Property 1. The disjunctions, which cover the current conjunction and hence are not changed, cannot be absorbed by any other disjunction.

This property follows from the fact that the matrix of disjunctions is always maintained in the state where it contains only prime disjunctions, which do not absorb each other. Hence if some disjunction has not been absorbed earlier then without modifications it will not be absorbed in the new matrix as well since this new matrix contains only the same or weaker disjunctions.

Let us suppose that $\mathbf{u}$ is a non-modified disjunction while $\mathbf{v}$ was modified on the component $\mathbf{v}_{rs}$, and $\mathbf{v}'_{rs}$ is an old value of a modified component ($\mathbf{u}_{ij} = \mathbf{u}'_{ij}$ since $\mathbf{u}$ was not modified). Then the following property takes place.

Property 2. If $\mathbf{u}_{rs} \leq \mathbf{v}'_{rs}$ then $\mathbf{v}$ does not follow from $\mathbf{u}$. (This property is valid only if the constant of $\mathbf{v}$ was not changed. More about constants and reduced forms read in [17].)

To use this property each line has to store information on the old value $\mathbf{v}'_{rs}$ of the modified component and its number (r and s). There are analogous (more complicated) properties for comparing two modified disjunctions, which are not formulated here. These properties are valuable since frequently they allow us to say that one line is not a consequence of another by comparing only one pair of components.

Property 3. If the sum of components in $\mathbf{v}$ or in any of its propositions $\mathbf{v}_i$ is less than the corresponding sum in the disjunction $\mathbf{u}$ then $\mathbf{v}$ does not follow from $\mathbf{u}$.

To use this property we have to maintain the sums of the disjunction and all its elementary proposition components in the corresponding headers. If all these necessary conditions are satisfied then we have to carry out a component-wise comparison of two vectors in the loop consisting of $n_1 + n_2 + \ldots + n_n$ steps.

5.3 Filtration of Disjunctions

In spite of using various methods to increase performance of the generalised transformation from possibilistic DNF into CNF, it is too computationally difficult for real world problems. However for the task of rule induction it is not necessary to carry out this transformation in complete form since usually it is required to find only the strongest dependencies among the attributes. Thus only a limited number of the most interesting disjunctions can be stored and processed whereas those disjunctions, which according to their criterion do not go into it (e.g., $\mathbf{w}$ in Fig. 5) are removed.

The procedure is organised as follows. Before a new disjunction is to be generated we calculate its degree of interestingness, which is compared with that of the last line of the matrix. If the new disjunction does not go into the matrix, it is simply not generated. Otherwise, if it is interesting enough, it is first generated, then checked for absorption, and finally inserted into the corresponding position in the matrix (the last line is removed).

There may be different criteria for ordering disjunctions determining the induction process direction. In our algorithm the criterion of interestingness in the form of the impossibility interval size is used. Informally, the more points of the distribution have smaller values, the more general and strong the corresponding disjunction is. Formally the following formula is used to calculate this parameter:

$$H = \frac{1}{n_1}\sum_{j=1}^{n_1}\mathbf{d}_{1j} + \frac{1}{n_2}\sum_{j=1}^{n_2}\mathbf{d}_{2j} + \ldots + \frac{1}{n_n}\sum_{j=1}^{n_n}\mathbf{d}_{nj}$$

according to which H is equal to the weighted sum of components, and the less this value, the stronger the disjunction. In particular, changing one component from 0 to 1 in two-valued proposition is equivalent to changing three components from 0 to 1 in six-valued proposition. For example, in the matrix $\mathbf{D}^3$ (Section 5.2) two disjunctions **3.1.3** and **3.2.3** can be transformed into rules (three other disjunctions are degenerated and represent the distribution projection on individual variables). However, the second of them is more interesting (informative) since it has larger interval of impossibility:

$$H(\mathbf{3.1.3}) = \tfrac{1}{2}3 + \tfrac{1}{3}0 + \tfrac{1}{4}10 = 4 > 3.5 = \tfrac{1}{2}0 + \tfrac{1}{3}3 + \tfrac{1}{4}10 = H(\mathbf{3.2.3})$$

Generally, each attribute or even each attribute value may have their own user-defined weights, which reflect their informative importance or subjective interestingness for the user. This mechanism provides the capability of more flexible control over the process of rule induction. There may be also other mechanisms of filtration. For example, restricting the number of non-0 components for the target attribute allows us to find only the rules with one value in conclusion and set-valued conditions.

6. GENERATION OF RULES FROM DISJUNCTIONS

The transformation from possibilistic DNF into CNF is the most difficult part of the algorithm and its goal is to generate as many general patterns as possible. The set of patterns is approximately semantically equivalent to the original data and once these prime disjunctions have been generated they can be translated into different forms representing the most interesting information. This process is referred to as knowledge mining. In this section we describe how possibilistic patterns (prime disjunctions) can be translated into interesting set-valued rules. Since the number of rules may be very high and many of them are very similar we also consider the problem of rule clustering and finding the most representative and interesting rules.

Formally, rules are obtained in the conventional way by negating the propositions which should be in the condition and thus obtaining an implication, for example:

$$\mathbf{d}_1 \vee \mathbf{d}_3 \vee \mathbf{d}_5 \quad \Leftrightarrow \quad \overline{\mathbf{d}}_1 \wedge \overline{\mathbf{d}}_3 \rightarrow \mathbf{d}_5$$

As usual, the propositions, which consist of all 0's, should not be considered since they are equivalent to the absence of proposition about the corresponding attribute. The main problem here is that we would like to have crisp conditions instead of possibilistic ones resulted from the negation. Thus we need a mechanism for a meaningful removing uncertainty from the negated elementary propositions.

Since we work with prime disjunctions the only way to obtain a crisp proposition from an uncertain one is to increase some components to the maximal value (corresponding to 1 when mapped into [0,1]). Let us suppose that $\mathbf{d}_{\min} = \max_i \min_j \mathbf{d}_{ij}$, and $\mathbf{d}_{\max} = \max_{i,j} \mathbf{d}_{ij}$ are minimal and maximal components of the disjunction, respectively. (The maximal component is the same for all disjunctions of CNF.) Then the most straightforward way to a obtain crisp proposition from $\mathbf{d}_i$ is as follows:

$$\mathbf{d}_{ij} = \begin{cases} \mathbf{d}_{\max}, & \text{if } \mathbf{d}_{ij} > \mathbf{d}_{\min} \\ \mathbf{d}_{\min}, & \text{otherwise} \end{cases}$$

Here all components, which are greater than $\mathbf{d}_{\min}$ are mapped into $\mathbf{d}_{\max}$ while the components, which are less than or equal to $\mathbf{d}_{\min}$, are set to $\mathbf{d}_{\min}$. The new proposition is crisp since it involves only two values and its negation produces also crisp a condition. For example, the disjunction $\mathbf{d} = \{0,1\} \vee \{0,6,0\} \vee \{0,2,9,5\}$ with $\mathbf{d}_{\min} = 0$ and $\mathbf{d}_{\max} = 9$ can be weakened on the first and second propositions so that we obtain $\{0,9\} \vee \{0,9,0\} \vee \{0,2,9,5\}$, which after transforming into the implication $\{9,0\} \wedge \{9,0,9\} \rightarrow \{0,2,9,5\}$ is interpreted as the following possibilistic rule

IF $x_1 = \{a_{11}\}$ AND $x_2 = \{a_{21}, a_{23}\}$
THEN $x_3 = \{a_{31}: 0, a_{32}: 2, a_{33}: 9, a_{34}: 5\}$

where weights in the conclusion are interpreted in a possibilistic sense, i.e., they define maximal possible values of the distribution within this interval (since we used prime disjunction it is guaranteed that these values cannot be decreased).

Generally, instead of $\mathbf{d}_{\min}$ we can use any user-defined or pattern-dependent (automatically calculated) threshold. It is especially useful when there is a proposition with small maximal component. In the above example, we see that the pattern $\mathbf{d}$ involves a weak proposition $\mathbf{d}_1$. In this case we can obtain a more simple condition by increasing the disjunction minimal value $\mathbf{d}_{\min}$ up to the value of maximal component in the weak proposition (in our example $\mathbf{d}_{\min} = 1$) so that it becomes constant and therefore is not included into condition at all. Then we apply negation to desired propositions of this weaker disjunction and in our case obtain the implication $\{9,1,9\} \rightarrow \{1,2,9,5\}$ which is interpreted as the rule

IF $x_2 = \{a_{21}, a_{23}\}$ THEN $x_3 = \{a_{31}: 1,\ a_{32}: 2,\ a_{33}: 9,\ a_{34}: 5\}$

Note that this rule is more general in the condition but less informative (specific) in the conclusion. It is important that when generating rules in such a way we always loose some information present at the original disjunction and it is why, in particular, different patterns may produce the same rules. However, the optimality of the rules in the sense that the conclusion cannot be strengthened without weakening the condition remains (i.e., maximal frequencies in conclusion are exact).

To make rules more expressive they can be easily filled in with statistical information in the form of the sum of occurrences within the rule condition interval (for one additional pass through the data set). Then we might obtain the rule like the following one:

IF $x_2 = \{a_{21}, a_{23}\}$
THEN $x_3 = \{a_{31}\ (\text{Max} = 1,\ \text{Sum} = 2),\ a_{32}\ (\text{Max} = 2,\ \text{Sum} = 4), \ldots\}$

In addition, other aggregated characteristics like rule confidence can be calculated. General idea is that once rules have been found it would be interesting to find more information on their relation to the data and other rules.

One rather helpful function consists in finding rules, which can be translated into the form of association rules. The condition of such rules must involve only one value for each attribute while the local distribution in its conclusion should be close to the singular form so that when its tails are cut the confidence remains high enough. Obviously, our algorithm has less performance in finding association rules since it is intended for inducing more general class of rules (except for the case of 100% confidence association rules). However, this shows that these approaches are comparable. In particular, when the attribute in conclusion has large number of values and many of them have low frequencies (within rule condition interval) it is much more convenient to cut such infrequent values (items) and then calculate the confidence for frequent values. For example, the rule

IF $x_2 = \{a_{21}, a_{23}\}$
THEN $x_3 = \{a_{31} : 2,\ a_{32} : 216,\ a_{33} : 6,\ a_{34} : 184,\ a_{35} : 1,\ a_{36} : 3,\ a_{37} : 8\}$

(here value weights are sums of frequencies) can be rewritten in the form of an association rule as follows:

IF $x_2 = \{a_{21}, a_{23}\}$
THEN $x_3 = \{a_{32} : 216,\ a_{34} : 184\}$ WHERE $c = 95\%, s = [400/N] \cdot 100\%$

Here N is the total number of instances in the database, s is the rule support, and c is the confidence of conclusion calculated as

$$[(216+184) / (2+216+6+184+1+3+8)]\ 100\% = [400/420]\ 100\% = 95\%$$

Another interesting filtration function, which can be used in knowledge mining, might be called Like and Unlike. This mechanism allows the user to find (select) only the rules or patterns, which are similar to one or several specified rules. It means that the specified rules are declared the most interesting (or the least interesting for Unlike function) and then it is the task of the system to find everything that is close to Like and far from Unlike patterns. In contrast to attribute and value weighting this mechanism allows expressing user preferences directly in the semantic space. Generally, all user preferences induce a function of interestingness over the set of disjunctions. Initially this function is generated by some more or less natural measure of information value in disjunction (e.g., the size of the interval of impossibility). Then it can be updated according to attribute and value weights. Usually, there are some strict constraints on the rule structure, e.g., the number of attributes in condition. Rules that do not satisfy them are considered absolutely uninteresting. And finally, direct semantic preference can be imposed, e.g., by means of Like and Unlike functions. All these preferences make up pattern and rule filters, which calculate the degree of interestingness according to the current set of user preferences.

The described approach can be also used to find interesting subgroups, i.e., a subset of a whole population with some unusual, unexpected or deviating distribution of a target variable [8,9,10,23]. The subgroup description is then just a possibilistic rule condition part. As the rules found by our method are guaranteed to have unusual (informative) conclusion the subgroup is also considered interesting. In such an approach subgroup mining is viewed as a particular case of possibilistic pattern (prime disjunction) interpretation where we are more aimed at the condition description rather than conclusion part.

As we already have noticed the number of rules may be rather high and many of them may be very similar. In this case even if we order all rules according to their interestingness many top level rules will be similar and it will be difficult to find interesting but at the same time significantly different rules. In the previous example, we obtained two interesting rules from one pattern, which however are very semantically close to each other. Then an interesting problem of knowledge mining is finding clusters of similar rules where cluster centre is an existing or new rule representing more or less well all rules in the cluster. Using this mechanism it is possible to view only rule cluster representatives and only if the cluster is really interesting it can be expanded showing all its member rules. Moreover, the clusters may be hierarchical with user-defined levels providing much more flexibility in rule base analysis. One important application of rule clustering is removing overfitting when a fine distribution surface is approximated too exactly by a large number of disjunctions.

The most important thing to carry out cluster analysis is to define an exact distance measure in the space of these objects. In other words, for any two objects, in our case for any two rules or disjunctions, we need to have the measure expressing the difference between them in the space of objects. Since the space of disjunctions is a multidimensional space (one disjunction component corresponds to one dimension) we can use any metric used in traditional cluster analysis. In particular, we can take the sum of differences between components:

$$\delta(\mathbf{u},\mathbf{v}) = \sum_{i=1,\ldots,n} \sum_{j=1,\ldots,n_i} |\, \mathbf{u}_{ij} - \mathbf{v}_{ij} \,|$$

According to this formula the distance between two disjunctions $\{0,9\} \vee \{0,9,0\} \vee \{0,2,9,5\}$ and $\{1,1\} \vee \{1,9,1\} \vee \{1,2,9,5\}$ is 12 while their distance from, say, $\{9,0\} \vee \{9,0,0\} \vee \{9,7,0,3\}$ is much larger and is equal to 61 and 51, respectively. Once the space metric has been defined we can use any existing clustering algorithm.

One specific feature of metric for rules and patterns is that we would not like to significantly distinct disjunctions one of which is a direct generalisation of another, i.e., if we take a disjunction and increase its components then the new disjunction should not be too far from the original. In particular, absolutely valid disjunction (with all components equal to

$\mathbf{d}_{max}$) should be close to any other disjunction. Thus for strong disjunctions such metrics should be much more distinctive and sensitive than for weak disjunctions. It is very natural that if we take some disjunction as a cluster centre then all its generalisations are also included in this very cluster since their distance from the centre is not too large. Although as a whole this problem is opened, one simple mechanism for such type of clustering could be implemented as follows. There is an ordered list of disjunctions. As a new cluster centre we take the next free disjunction in the list, i.e., the disjunction that has not been included into any cluster. Then the cluster members are chosen as those, which are close enough to the centre among free disjunctions. This procedure guarantees that cluster centres will always be strong and real disjunctions although it is possible to recalculate them as some new disjunction (this is a separate problem).

Another interesting procedure to cluster rules is to weaken them gradually until some of them become identical. The degree of weakening defines the level in cluster hierarchy while identical disjunctions form one cluster. In particular, complete weakening results in the only cluster involving all disjunctions. The problem in this approach is to meaningfully define the operation of weakening disjunctions. Formally, it means that some constant value is added to disjunction but the operation used to add this constant may be different, e.g., maximum or arithmetic sum.

7. CONCLUSION

The described algorithm for the case of multi-valued attributes and two-valued semantics has been implemented in the Chelovek for Windows 95 rule induction system* (Fig. 6). Since the number of patterns is limited by a special parameter the processing time is linear to the table length. For several hundreds of patterns the processing time per record is rather low (less than the overheads such as loading the record and parsing the attribute values). When the number of patterns exceeds 1000 the combinatorial part becomes significant. We also noticed that the algorithm works much better with ordered data.

Below we summarise our rule induction algorithm characteristic features, advantages and disadvantages.

- An important conceptual characteristic of the approach is that we find disjunctive patterns rather than conjunctive in traditional algorithms. In particular, since disjunction by definition covers all examples the

* Developed at the Institute of Mathematics & Informatics, Moldavian Academy of Sciences. Available in http://www.geocities.com/ResearchTriangle/7220/edip.zip

quantity of information in it is measured dually and is proportional to the width of its negative interval.

- We use an original formal framework generalising the conventional boolean approach on the case of (i) finite-valued variables, and (ii) continuous-valued semantics. This allows us to efficiently generate set-valued rules with possibilistic conclusions.
- The notion of prime disjunction as the most general and informative one allows us to find the most interesting (surprising) rules. Moreover, the algorithm guarantees finding all such rules while the rules are optimal in the sense that they have the widest condition and the narrowest conclusion (an attempt to generalise condition or to narrow conclusion results in a wrong rule).
- Most rule induction algorithms explicitly separate rule condition and conclusion parts and search through the space of all conditions for those satisfying certain criteria in conclusion. In our approach we use the unified language for representing pieces of information in the form of disjunctions the most informative of which once found can be represented as interesting rules.

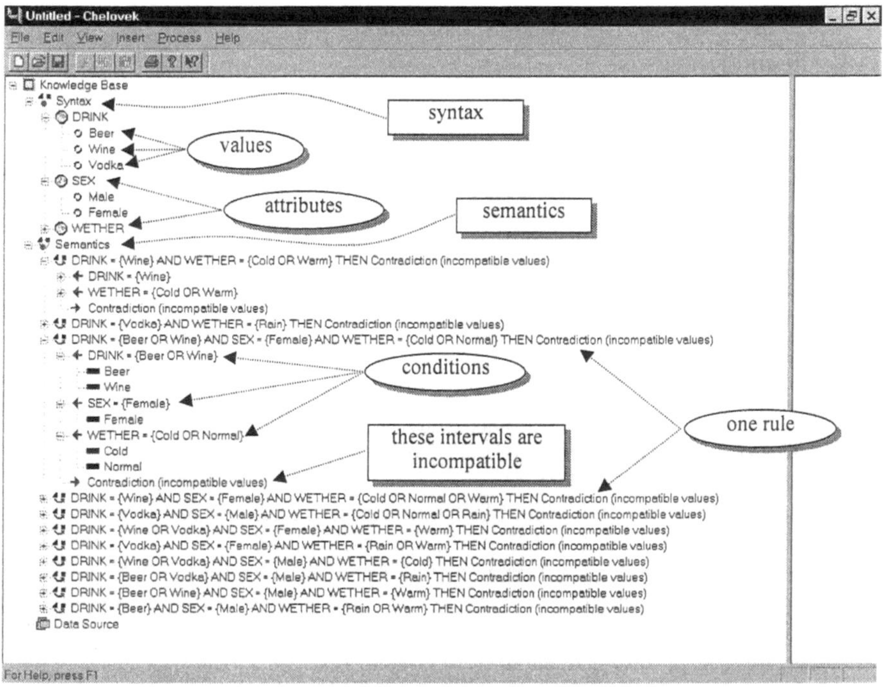

Figure 6. Screenshot of the Chelovek rule induction system.

- The algorithm processes all records for one pass through the database what allows applying it to large databases.
- The rules can be easily filled in with statistical information in the form of the sum of records within the rule condition interval. Then some rules can be represented as (set-valued) association rules.
- The method can be used to find interesting subgroups, i.e., a subset of objects revealing highly unusual properties in relation to the whole data set.
- All rules have equal rights, i.e., the whole semantics and the rule interpretation do not depend on their order (e.g., for CN2 it is not so [7,6]).
- The knowledge base in the form of a number of the most strong prime disjunctions is approximately equivalent to the database and therefore it can be easily used for prediction purposes when it is necessary to determine the value of one attribute given (constraints on) the values of other attributes.
- One minus of the algorithm is a large number of generated rules especially in the case of dense distributions with fine surface (overfitting). This problem can be solved with the help of more powerful search, filtration and rule clustering mechanisms used in the rule induction system.

REFERENCES

1. R. Agrawal, T. Imielinski, and A. Swami, Mining association rules between sets of items in large databases, Proc. of the ACM SIGMOD Conference on Management of Data, Washington, D.C., May 1993, 207–216.
2. R. Agrawal and R. Srikant, Fast algorithms for mining association rules in large databases, Proceedings of the 20th International Conference on Very Large Data Bases, pp. 487–499, 1994.
3. R.J. Bayardo Jr., Efficiently mining long patterns from databases, In Proc. of the 1998 SIGMOD Conf. on the Management of Data, 1998.
4. C. Borgelt and R. Kruse, Efficient maximum projection of database-induced multivariate possibility distributions, Proc. 7th IEEE International Conference on Fuzzy Systems (FUZZ-IEEE'98), Vol. 1, pp. 663–668, Anchorage, Alaska, 1998.
5. C. Borgelt and R. Kruse, Probabilistic and possibilistic networks and how to learn them from data, in: O. Kaynak, L. Zadeh, B. Turksen, and I. Rudas (eds.), Computational Intelligence: Soft Computing and Fuzzy-Neuro Integration with Applications, NATO ASI Series F, Vol. 162, pp. 403–426, Springer, New York, 1998.
6. P. Clark and R. Boswell. Rule induction with CN2: some recent improvements. In Y. Kodratoff, editor, Machine Learning — EWSL-91, pages 151–163, Berlin, 1991. Springer-Verlag.
7. P. Clark and T. Niblett. The CN2 induction algorithm. Machine Learning, **3**(4):261–283, 1989.

8. W. Klösgen, Explora: A multipattern and multistrategy discovery assistant, In: Advances in Knowledge Discovery and Data Mining, U. Fayyad, G. Piatetsky-Shapiro, P. Smyth, and R. Uthurusamy (eds.), MIT Press, Cambridge MA, 249–271, 1996.
9. W. Klösgen, Applications and research problems of subgroup mining, In: Foundations of Intelligent Systems, Z. Ras and A. Skowron (eds.), Springer, Heidelberg, 1–15, 1999.
10. W. Klösgen, Subgroup patterns. In: Handbook of Data Mining and Knowledge Discovery, W. Klösgen and J. Zytkow (eds.), Oxford University Press, New York, 1999 (to appear).
11. R. Kruse and E. Schwecke, Fuzzy reasoning in a multidimensional space of hypotheses, Int. J. of Approximate Reasoning 4, 47–68, 1990.
12. D. Lin and Z.M. Kedem, Pincer-Search: A new algorithm for discovering the maximum frequent set. In Proc. of the Sixth European Conf. on Extending Database Technology, 1998.
13. H. Mannila, H. Toivonen, and A.I. Verkamo, Efficient algorithms for discovering association rules, In KDD-94: AAAI Workshop on Knowledge Discovery in Databases, Seattle, Washington, July 1994, 181–192.
14. A.A. Savinov, Matrix representation of fuzzy knowledge in attribute models, Preprint, Institute of Mathematics and CC, AS Moldavia, Kishinev, Shtiintsa, 1991 (in Russian).
15. A.A. Savinov, Forming knowledge by examples in fuzzy finite predicates, Proc. conf. "Hybrid Intellectual Systems", Part 1, Rostov-na-Donu—Terskol, 177–179, 1991 (in Russian).
16. A.A. Savinov, Fuzzy multi-dimensional analysis, online paper, 1997, http://www.geocities.com/ResearchTriangle/7220/fmda_0.html.
17. A.A. Savinov, Fuzzy multi-dimensional analysis and resolution operation, Computer Sci. J. of Moldova **6**(3), 252-285, 1998.
18. A.A. Savinov. Application of multi-dimensional fuzzy analysis to decision making. In: Advances in Soft Computing — Engineering Design and Manufacturing, R. Roy, T. Furuhashi and P.K. Chawdhry (eds.), Springer-Verlag, London, 1999.
19. A.A. Savinov, Mining possibilistic set-valued rules by generating prime disjunctions, Proc. 3rd European Conference on Principles and Practice of Knowledge Discovery in Databases — PKDD'99, Prague, Czech Republic, September 15-18, 1999, 536-541.
20. A.A. Savinov, An algorithm for induction of possibilistic set-valued rules by finding prime disjunctions, 4th On-line World Conference on Soft Computing in Industrial Applications (WSC4), 21-30 Sept. 1999.
21. A.A. Savinov, An algorithm for finding logical dependencies among multivalued attrbiutes, Perner, P. (Ed.), Proc. Fachgruppentreffen Maschinelles Lernen (FGML'99), Magdeburg, September 27-29, Institut fuer Bildverarbeitung und angewandte Informatik e.V., Leipzig, 8–14, 1999.
22. J.R. Slagel, C.-L. Chang, and R.C.T. Lee, A new algorithm for generating prime implicants, *IEEE Trans. on Computers*, C-19(4):304–310, 1970.
23. S. Wrobel, An algorithm for multi-relational discovery of subgroups, Proc. 1st European Symposium on Principles of KDD, eds. J. Komorowski, and J. Jytkow, Springer, Heidelberg, 78–87, 1997.
24. A.D. Zakrevsky, Logical inference in finite predicates, Preprint No.6, Institute of Technical Cybernetics, AS Belorussia, Minsk, 1989 (in Russian).

Chapter 7

COMPLEXITY REDUCTION OF A GENERALISED RATIONAL FORM

Peter Baranyi[1] and Yeung Yam[2]
[1]*Research Group for Mechanics of the Hungarian Academy of Science and Dept. Telecommunication and Telematics, Technical University of Budapest, Sztoczek u.2., Budapest,H-1111,Hungary,baranyi@ttt-202.ttt.bme.hu*
and
[2]*Department of Mechanical and Automation Engineering, Chinese University of Hong Kong, Shatin, N.T., Hong Kong, yyam@mae.cuhk.edu.hk*

Abstract: This chapter is motivated by the fact that though fuzzy techniques are popular engineering tools, their utilisation is being restricted by their exponential complexity property. The objectives of this chapter are twofold: one is to find a general form for fuzzy system output covering the widest possible areas of applications, and the other is to present a complexity reduction algorithm for the general form.

1. INTRODUCTION

The advantage of fuzzy control lies in its ability to mimic and implement the actions of expert operator(s) without the need of accurate mathematical models. The drawback, however, is that there is no standardised framework regarding the design, optimality, reducibility, and partitioning of a fuzzy rule set. A fuzzy rule base, be it generated from expert operators or by some learning or identification schemes, may contain redundant, weakly contributing, or outright inconsistent components. Moreover, in pursuit of good approximation, one may be tempted to overly assign the number of antecedent sets, thereby resulting in large fuzzy rule bases and much problems in computation time and storage space. A formal approach to extracting the more pertinent elements of a given rule set is, hence, highly desirable. The present chapter is an attempt in this direction.

The quest to minimise the complexity of fuzzy rule-bases for reduced computation time and the number of rules is not new. In 1974, *Mamdani* the first to modify the CRI algorithm to obtain considerable calculation time reduction. His inference algorithm is based on the cylindrical approximation of fuzzy relations. Other algorithms, such as *Sugeno-*, *Takagi-Sugeno-*, *Larsen-*, Product-Sum-Gravity, etc., are subsequently developed following the same key idea, by simply operating the fuzzy sets along each input dimension. However, they still suffer from the exponential calculation-time and storage-space complexity that has been demonstrated by *Kóczy* and *Hirota* through the uniform complexity expression of [1,2], namely, the number of rules grows exponentially with the number of antecedent fuzzy terms and the number of input variables (*Kóczy* and *Hirota* [1]). This is the reason why the number of inputs is usually not more than five in fuzzy applications. In this regards, *Moser*, *Klement* and *Tikk* have shown that most of the fuzzy algorithms do not have universal approximation property if the number of antecedent sets is limited (*Klement*, *Moser* [3], *Tikk* [4]).

A fuzzy rule-base design, hence, has two important objectives. One is to achieve a good approximation. The other is to reduce the number of rules. The main difficulty is that these two objectives are contradictory. Complexity reduction is therefore becoming a pertinent research topic of fuzzy theory.

Reduction methods can be classified according to their underlying concept. One group uses a new or modified inference algorithm to induce reduction. For example: inference paradigms based on uniform complexity-expression of *Kóczy* and *Hirota* [2], α-cut mapping method of *Stoica* [5], and the minimum-distance-inference method of *Yu* and *Bien* [6] can be viewed as belonging to this group. The other group contains rule-base reduction methods, which works with the same inference algorithm, but reduces the size of the given rule base. This includes the reduction method for input variables by *Titli et al.* [7], *Sugeno*'s hierarchical rule-structure [8], and singular value based (SVD) reduction methods [9-16]. SVD has also been applied by *Wang et al.* [20] to reduce the dimensionality of the input space, after which the system is reduced via optimisation of an objective function. It's worth pointing out that fuzzy interpolation methods can be included in both groups. One can find different types of fuzzy interpolation methods in *Kóczy, Hirota, Gedeon, Kovács and Kawaguchi* [1,21-24].

The present chapter focuses on the SVD based reduction methods of [9-16], which are based on conducting singular value decomposition of the rule consequents and then generating proper linear combinations of the original membership functions to form new ones for the reduced set. The first work is published in 1996 for rule bases using product-sum-gravity inference algorithm, piece-wise linear antecedents, and essentially, singleton consequents [9]. Shortly after the method is extended to non-singleton consequents [10,17], *Takagi-Sugeno* type rule bases [10,18], and to a special fuzzy rule interpolation method [19]. The method is further generalised in [12] and [13] to using SVD for reducing linguistic symbol arrays. A practical implementation technique is also proposed in [14] for extremely large rule bases.

An important advantage of the SVD reduction techniques is that there is a formal measure to filtering out the redundant and weakly contributing components. This implies that the degree of reduction can be applied according to the maximum acceptable error. For various cases, output error bound between the original set and the reduced set is readily expressible based on the sum of discarded singular values [10,17,18].

On the other hand, techniques using B-spline face the same exponentially growing computational and parameter requirement [25,26]. Non-Uniform Rational B-Spline (NURBS) technique has become industry standard throughout CAD/CAM/CAE systems for the representation, design, and data exchange of geometric information processing by computers. It has also been incorporated into several international and American national standards such as IGES (Initial Graphics Exchange Specification), STEP (Standard for the Exchange of Product Model Data) and PHIGS (Programmer's Hierarchical Interactive Graphics System). These systems use piecewise polynomial or rational curves and surfaces to represent complex geometry. Regardless of their type, they can be simply referred to as splines, for spline surfaces are just splines patching together. B-spline surfaces grew out of the work of Bézier in 1970 and formed the basic concept of non-uniform rational B-splines. The use of Spline, Bézier-, B-spline-techniques, Monomial- or Hermite-form etc., are included in NURBS. It turns out that the polynomial and the rational B-spline functions have the same form as the product-sum-gravity inference based algorithms [25,26], which motivates our work to extend the SVD based reduction techniques to a general rational form. The general form to be introduced in this chapter is family of functions containing the product-sum-gravity inferred output and the explicit functional form of B-spline as special cases, with corresponding restrictions of their own. The proposed reduction method can hence be used in any kind of numerical algorithms, as long as the prescribed general form is valid.

The present chapter serves to summarise the results of the works of [9-19]. Algorithms to be given related to the technique are mostly developed in [9,10,11] for fuzzy approximation and reduction of general functions.

2. GENERALISED RATIONAL FORM

This section gives a general form that includes as special cases output formulas used in B-spline techniques and product-sum-gravity based fuzzy systems.

Algorithm 1: Product-sum-gravity inference [26]

i) Characterisation of input - output

Let the input and output universes X_i, $i=1\ldots n$ and Y, respectively.

ii) Characterisation of fuzzy sets

Antecedents: The antecedent fuzzy sets $A_{i,j} : \mu_{A_{i,j}}(x_i)$, $x_i \in X_i$, $j=1\ldots n_i$, are defined on universe X_i, where n_i is the number of antecedent sets.

Consequents: The consequent fuzzy sets are defined on universe Y as $B_{t_1,\ldots,t_n} : \mu_{B_{t_1,\ldots,t_n}}(y)$, $y \in Y$, where $t_i = 1\ldots n_i$.

Observation: Let the input value x^* be *fuzzificated* into a singleton observation fuzzy set A^* such as: $A^* : \mu_{A^*}(x) = \delta(x^*)$.

Rules: If A_{1,t_1} and ... and A_{n,t_n} then $B_{t_1,\ldots,t_n}$

iii) Characterisation of the inference

The inference technique is based on product-sum-gravity [26].

Product: This step gives the contribution of the consequent sets to the conclusion according to the degree of matching among the observation and the antecedent sets. Here, the contribution of consequent $B_{t_1,\ldots,t_n}$ is:

$$\prod_{i=1}^{n} \mu_{A_{i,t_i}}(x_i),$$

where x_i is the input value.

Sum-gravity: This step follows the centre of gravity defuzzification technique. Applying *Larsen* method the contributing sets are:

$$B^c_{t_1,\ldots,t_n} : \mu_{B_{t_1,\ldots,t_n}}(y) \prod_{i=1}^{n} \mu_{A_{i,t_i}}(x_i).$$

The defuzzificated output value is then the weighted average of the contributing sets:

$$y = F(x_1,\cdots,x_n) = \frac{\sum_{t_1=1}^{n_1}\cdots\sum_{t_n=1}^{n_n}\left(w_{t_1,\cdots,t_n} b_{t_1,\cdots,t_n} \prod_{i=1}^{n} \mu_{A_{i,t_i}}(x_i)\right)}{\sum_{t_1=1}^{n_1}\cdots\sum_{t_n=1}^{n_n}\left(w_{t_1,\cdots,t_n} \prod_{i=1}^{n} \mu_{A_{i,t_i}}(x_i)\right)}, \tag{1}$$

where values $w_{t_1,\cdots,t_n}$ are the weightings of the contributing sets:

$$w_{t_1,\cdots,t_n} = \int_Y \mu_{B_{t_1,\cdots,t_n}}(y)dy,$$

and value $b_{t_1,\cdots,t_n}$ is the centre-of-gravity point of the fuzzy set $B_{t_1,\cdots,t_n}$.

It's worth mentioning that the rational form (1) is also the basic form in B-spline techniques. See [25] for more details.

Definition 1: Generalised rational form.

The algorithm above can be extended to a general form that also includes the generalised form of *Takagi-Sugeno-Kang* fuzzy inference algorithm, where consequent functions $f_{t_1,\cdots,t_n}(x_1,\cdots,x_n)$ are used instead of consequent sets $B_{t_1,\cdots,t_n}$:

$$y = F(x_1,\cdots,x_n) = \frac{\sum_{t_1=1}^{n_1}\cdots\sum_{t_n=1}^{n_n}\left(w_{t_1,\cdots,t_n} f_{t_1,\cdots,t_n}(x_1,\cdots,x_n) \prod_{i=1}^{n} \mu_{A_{i,t_i}}(x_i)\right)}{\sum_{t_1=1}^{n_1}\cdots\sum_{t_n=1}^{n_n}\left(w_{t_1,\cdots,t_n} \prod_{i=1}^{n} \mu_{A_{i,t_i}}(x_i)\right)}. \tag{2}$$

In practice, the following function type is usually used:

$$f_{t_1,\cdots,t_n}(x_1,\cdots,x_n) = \sum_{k=1}^{m} p_{k,t_1,\cdots,t_n} \Phi_k(x_1,\cdots,x_n).$$

For implementation (3) is used instead of (2) as:

$$y = F(x_1,\cdots,x_n) = \frac{\sum_{t_1=1}^{n_1}\cdots\sum_{t_n=1}^{n_n}\left(q_{t_1,\cdots,t_n}(x_1,\cdots,x_n)\prod_{i=1}^{n}\mu_{A_{i,t_i}}(x_i)\right)}{\sum_{t_1=1}^{n_1}\cdots\sum_{t_n=1}^{n_n}\left(w_{t_1,\cdots,t_n}\prod_{i=1}^{n}\mu_{A_{i,t_i}}(x_i)\right)}, \tag{3}$$

where

$$q_{t_1,\cdots,t_n}(x_1,\cdots,x_n) = w_{t_1,\cdots,t_n} f_{t_1,\cdots,t_n}(x_1,\cdots,x_n) = \sum_{k=1}^{m} b_{k,t_1,\cdots,t_n}\Phi_k(x_1,\cdots,x_n). \tag{4}$$

3. COMPLEXITY INVESTIGATION OF THE GENERALISED RATIONAL FORM

This section investigates the calculation and parameter complexity of (3).

Lemma 1: (3) has exponential calculation and parameter complexity:

$$T_G = \prod_{i=1}^{n} n_i\left(t_p(n-1+r+m)+t_s(m-1)\right)+mT_B+T_A\sum_{i=1}^{n} n_i+t_d, \tag{5}$$

where T_A and T_B are the calculation time of $\mu_{A_{i,t_i}}(x_i)$ and $f_{t_1,\cdots,t_n}(x_1,\cdots,x_n)$, respectively, and t_p, t_s, and t_d are those for the product, addition, and division operation, respectively.

To arrive at (5), one notes that calculating y requires multiplying functions $\mu_{A_{i,t_i}}(x_i)$ and $f_{t_1,\cdots,t_n}(x_1,\cdots,x_n)$ together. Hence, the calculation time is $\prod_{i=1}^{n} n_i$ times $(n-1)t_p$ for $\prod_{i=1}^{n}\mu_{A_{i,t_i}}(x_i)$ and $mt_p+(m-1)t_s$ for $f_{t_1,\cdots,t_n}(x_1,\cdots,x_n)$, and consequently, the time for calculating the whole is (5).

The parameter complexity can be similarly deduced. The number of values $b_{k,t_1,\cdots,t_n}$ grows exponentially with n and n_i.

Lemma 1 is in full accordance with [1,2], namely, the computation complexity grows exponentially with the number of antecedent sets and inputs.

4. COMPLEXITY REDUCTION OF THE GENERALISED RATIONAL FORM

This section is to introduce a method capable of generating a minimal form of (3), in the sense that the resultant function utilises the minimal number of antecedent functions for representation. The section considers only exact reductions, i.e., that there is no error between the original function and its reduced form. Inexact reduction will be treated in Section 8.

Lemma 2: The function (3) can always be transformed into the following form:

$$y = F(x_1,\cdots,x_n) = F^r(x_1,\cdots,x_n)$$
$$= \frac{\sum_{t_1=1}^{n_1^r}\cdots\sum_{t_n=1}^{n_n^r}\left(q^r_{t_1,\cdots,t_n}(x_1,\cdots,x_n)\prod_i^n \mu_{A^r{}_{i,t_i}}(x_i)\right)}{\sum_{t_1=1}^{n_1^r}\cdots\sum_{t_n=1}^{n_n^r}\left(w^r_{t_1,\cdots,t_n}\prod_i^n \mu_{A^r{}_{i,t_i}}(x_i)\right)}, \tag{6}$$

where $\forall i : n_i^r \le n_i$ and

$$q^r_{t_1,\cdots,t_n}(x_1,\cdots,x_n) = \sum_{k=1}^{m} b^r_{k,t_1,\cdots,t_n}\Phi_k(x_1,\cdots,x_n). \tag{7}$$

The superscript "*r*" here denotes "reduced".

Proof: To effectuate an easier understanding, the reduction algorithm will be introduced in three steps. The first outlines the main steps of reducing the number of parameters for the numerator of (3) via SVD. The second shows the construction of the functions $\mu_{A^r{}_{i,t_i}}(x_i)$ from the reduced parameters, which would result in the complexity reduction form for the numerator. Step 3 extends steps 1 and 2 to conduct reduction accounting also the effect of the denominator of (3) to generate the values $w^r_{t_1,\cdots,t_n}$ for (6). Expressions for the reduced functions of $f^r_{t_1,\cdots,t_n}(x_1,\cdots,x_n)$, $\mu_{A^r{}_{i,t_i}}(x_i)$, as well as for the values $w^r_{t_1,\cdots,t_n}$ and n_i^r, are given in Method 1 to be introduced below.

For convenience, the reduction method will be demonstrated here with the example of a three input variable case ($n=3$). Let's first introduce the SVD reduction as:

Definition 3: SVDR: Singular Value Based Reduction. Suppose that matrix $\underline{\underline{\mathbf{B}}}_{(n_1 \times n_2)} = [y_{i,j}]$ is given. Applying singular value decomposition yields:

$$\underline{\underline{\mathbf{B}}} = \underline{\underline{\mathbf{A}}}_1 \underline{\underline{\mathbf{D}}} \underline{\underline{\mathbf{A}}}_2^{\mathrm{T}} = [\underline{\underline{\mathbf{A}}}_1^r \mid \underline{\underline{\mathbf{A}}}_1^d] \begin{bmatrix} \underline{\underline{\mathbf{B}}}^r & \underline{\underline{\mathbf{O}}} \\ \underline{\underline{\mathbf{O}}} & \underline{\underline{\mathbf{B}}}^d \end{bmatrix} [\underline{\underline{\mathbf{A}}}_2^r \mid \underline{\underline{\mathbf{A}}}_2^d]^{\mathrm{T}}. \tag{8}$$

Here, matrix $\underline{\underline{\mathbf{O}}}$ contains zero elements, matrices $\underline{\underline{\mathbf{A}}}_1$ and $\underline{\underline{\mathbf{A}}}_2$ are orthogonal, and diagonal matrix $\underline{\underline{\mathbf{D}}}$ contains the singular values in decreasing magnitude. The zero or the smallest of singular values (smaller that singular value threshold T_0, say) can be discarded to yield a simpler system. Matrix $\underline{\underline{\mathbf{B}}}^r$ thus contains the retained singular values and $\underline{\underline{\mathbf{B}}}^d$ contains the discarded ones. The result of SVDR is therefore:

$$\hat{\underline{\underline{\mathbf{B}}}} = \underline{\underline{\mathbf{A}}}_1^r \underline{\underline{\mathbf{B}}}^r \underline{\underline{\mathbf{A}}}_2^{r\,\mathrm{T}}.$$

Method 1: Reduction of the general form (3).

Step 1:

Let the 3-dimensional matrices $\underline{\underline{\mathbf{B}}}_k$, $k=1\ldots m$ be constructed with the elements b_{k,t_1,t_2,t_3} as given in (3) and (4).

(i) **Calculating matrix $\underline{\underline{\mathbf{A}}}_1^r$:** Construct the 2-dimensional matrix

$\underline{\underline{\mathbf{S}}}_{k,(n_1 \times n_2 n_3)} = \begin{bmatrix} \underline{\underline{\mathbf{H}}}_{k,1} \cdots \underline{\underline{\mathbf{H}}}_{k,n_3} \end{bmatrix} = [s_{k,l,o}]$, where

$$\underline{\underline{\mathbf{H}}}_{k,t_3} = \begin{bmatrix} b_{k,1,1,t_3} & \cdots & b_{k,1,n_2,t_3} \\ \vdots & \ddots & \vdots \\ b_{k,n_1,1,t_3} & \cdots & b_{k,n_1,n_2,t_3} \end{bmatrix}, \text{ and } t_3 = 1\ldots n_3.$$

The elements of matrix $\underline{\underline{\mathbf{S}}}_k$ are hence: $s_{k,t_1,t_2+(t_3-1)n_2} = b_{k,t_1,t_2,t_3}$. Furthermore, construct $\underline{\underline{\mathbf{C}}}_{1,(n_1 \times n_2 n_3 m)} = \lfloor \underline{\underline{\mathbf{S}}}_1 \cdots \underline{\underline{\mathbf{S}}}_m \rfloor$ and apply SVDR to $\underline{\underline{\mathbf{C}}}_1$:

$$\underline{\underline{\mathbf{C}}}_1 = \underline{\underline{\mathbf{A}}}_1^r \underline{\underline{\mathbf{D}}}_1 \underline{\underline{\mathbf{V}}}_1 = \underline{\underline{\mathbf{A}}}^r_{1,(n_1 \times n_1^r)} \underline{\underline{\mathbf{C}}}'_{1,(n_1^r \times n_2 n_3 m)}.$$

With $\underline{\underline{\mathbf{C}}}'_1$ partitioned as $\underline{\underline{\mathbf{C}}}'_1 = \lfloor \underline{\underline{\mathbf{G}}}'_1 \cdots \underline{\underline{\mathbf{G}}}'_m \rfloor$, one then obtains $\underline{\underline{\mathbf{B}}}'_k = [b'_{k,t_1,t_2,t_3}]$ from $\underline{\underline{\mathbf{G}}}'_{k,(n_1^r \times n_2 n_3)} = [g'_{k,l,o}]$ with the assignment: $b'_{k,t_1,t_2,t_3} = g'_{k,t_1,t_2+(t_3-1)n_2}$, $t_1 = 1 \ldots n_1^r$.

(ii) **Calculating matrix $\underline{\underline{\mathbf{A}}}_2^r$:** similar to (i), let

$$\underline{\underline{\mathbf{S}}}'_{k,(n_2 \times n_3 n_1^r)} = \begin{bmatrix} \underline{\underline{\mathbf{H}}}'_{k,1} & \cdots & \underline{\underline{\mathbf{H}}}'_{k,n_1^r} \end{bmatrix} = [s'_{k,l,o}], \text{ where}$$

$$\underline{\underline{\mathbf{H}}}'_{k,t_1} = \begin{bmatrix} b'_{k,t_1,1,1} & \cdots & b'_{k,t_1,1,n_3} \\ \vdots & \ddots & \vdots \\ b'_{k,t_1,n_2,1} & \cdots & b'_{k,t_1,n_2,n_3} \end{bmatrix}, \text{ and } t_1 = 1 \ldots n_1^r.$$

The elements of matrix $\underline{\underline{\mathbf{S}}}'_k$ is then $s'_{k,t_2,t_3+(t_1-1)n_3} = b'_{k,t_1,t_2,t_3}$, for $t_1 = 1 \ldots n_1^r$. Letting $\underline{\underline{\mathbf{C}}}_{2,(n_2 \times n_3 n_1^r m)} = \lfloor \underline{\underline{\mathbf{S}}}'_1 \cdots \underline{\underline{\mathbf{S}}}'_m \rfloor$ and applying SVDR to $\underline{\underline{\mathbf{C}}}_2$, one has:

$$\hat{\underline{\underline{\mathbf{C}}}}_2 = \underline{\underline{\mathbf{A}}}_2^r \underline{\underline{\mathbf{D}}}_2 \underline{\underline{\mathbf{V}}}_2 = \underline{\underline{\mathbf{A}}}^r_{2,(n_2 \times n_2^r)} \underline{\underline{\mathbf{C}}}'_{2,(n_2^r \times n_3 n_1^r m)}.$$

Again, with $\underline{\underline{\mathbf{C}}}'_2$ partitioned as: $\underline{\underline{\mathbf{C}}}'_2 = \lfloor \underline{\underline{\mathbf{G}}}''_1 \cdots \underline{\underline{\mathbf{G}}}''_m \rfloor$, the elements of $\underline{\underline{\mathbf{B}}}''_k = [b''_{k,t_1,t_2,t_3}]$ can be obtained from $\underline{\underline{\mathbf{G}}}''_{k,(n_2^r \times n_3 n_1^r)} = [g''_{k,l,o}]$ with the assignment $b''_{k,t_1,t_2,t_3} = g''_{k,t_2,t_3+(t_1-1)n_3}$, $t_1 = 1 \ldots n_1^r$, $t_2 = 1 \ldots n_2^r$.

(iii) **Calculating matrix $\underline{\underline{\mathbf{A}}}_3^r$:** similar to above, let

$$\underline{\underline{\mathbf{S}}}''_{k,(n_3 \times n_1^r n_2^r)} = \begin{bmatrix} \underline{\underline{\mathbf{H}}}''_{k,1} \cdots \underline{\underline{\mathbf{H}}}''_{k,n_2^r} \end{bmatrix} = [s''_{k,l,o}], \text{ where}$$

$$\underline{\underline{\mathbf{H}}}''_{k,t_2} = \begin{bmatrix} b''_{k,1,t_2,1} & \cdots & b''_{k,n^r_1,t_2,1} \\ \vdots & \ddots & \vdots \\ b''_{k,1,t_2,n_3} & \cdots & b''_{k,n^r_1,t_2,n_3} \end{bmatrix}, \text{ and } t_2 = 1 \ldots n_2^r,$$

so that the elements of $\underline{\underline{\mathbf{S}}}''_k$ are: $s''_{k,t_3,t_1+(t_2-1)n_1^r} = b''_{k,t_1,t_2,t_3}$, for $t_1 = 1 \ldots n_1^r$ and $t_2 = 1 \ldots n_2^r$. Applying SVDR to $\underline{\underline{\mathbf{C}}}_{3,(n_3 \times n_1^r n_2^r m)} = \left\lfloor \underline{\underline{\mathbf{S}}}''_1 \cdots \underline{\underline{\mathbf{S}}}''_m \right\rfloor$ then yields

$$\hat{\underline{\underline{\mathbf{C}}}}_3 = \underline{\underline{\mathbf{A}}}_3^r \underline{\underline{\mathbf{D}}}_3 \underline{\underline{\mathbf{V}}}_3 = \underline{\underline{\mathbf{A}}}^r_{3,(n_3 \times n_3^r)} \underline{\underline{\mathbf{C}}}'_{3,(n_3^r \times n_1^r n_2^r m)}.$$

Again, with $\underline{\underline{\mathbf{C}}}'_3 = \left\lfloor \underline{\underline{\mathbf{G}}}'''_1 \cdots \underline{\underline{\mathbf{G}}}'''_m \right\rfloor$, one gets $b'''_{k,t_1,t_2,t_3} = g'''_{k,t_3,t_1+(t_2-1)n_1^r}$, $t_1 = 1 \ldots n_1^r$, $t_2 = 1 \ldots n_2^r$, $t_3 = 1 \ldots n_3^r$, with $\underline{\underline{\mathbf{G}}}'''_{k,(n_3^r \times n_1^r n_2^r)} = [g'''_{k,l,o}]$ and $\underline{\underline{\mathbf{B}}}'''_k = [b'''_{k,t_1,t_2,t_3}]$.

As a result, one obtains $\underline{\underline{\mathbf{B}}}^r_{k,(n_1^r \times \cdots \times n_n^r)} = [b^r_{k,t_1,t_2,t_3}] = \underline{\underline{\mathbf{B}}}'''_k = [b'''_{k,t_1,t_2,t_3}]$ and $\underline{\underline{\mathbf{A}}}^r_{i,(n_i \times n_i^r)} = \left\lfloor a_{i,l,o} \right\rfloor$, $i = 1 \ldots 3$, where $\forall i : n_i^r \leq n_i$.

In short, Step 1 actually transforms the 3-dimensional matrices $\underline{\underline{\mathbf{B}}}_{k(n_1 \times n_2 \times n_3)}$ into a sequence of 2-dimensional matrices along each dimension, where upon the same process of SVDR can be applied. Following this approach one obtains readily that $\underline{\underline{\mathbf{A}}}^r_{i,(n_i \times n_i^r)}$, $i = 1 \ldots n$, and $\underline{\underline{\mathbf{B}}}^r_{k,(n_1^r \times \cdots \times n_n^r)}$, $\forall i : n_i^r \leq n_i$, when $n > 3$.

Step 2:

The parameters of (7) are given by the matrices $\underline{\underline{\mathbf{B}}}^r_{k,(n_1^r \times \cdots \times n_n^r)}$. The reduced antecedents in (6) can be obtained from parameters of the reduced matrices as:

$$\mu_{A^r_{i,j}}(x_i) = \sum_{l=1}^{n_i} \mu_{A_{i,l}}(x_i) a_{i,l,j} \ , \ j = 1..n_i^r \ , \tag{9}$$

where elements $a_{i,l,j}$ consist of matrix $\underline{\underline{\mathbf{A}}}_i^r$.

Remark: The resultant function of (9) may not be interpretable as a fuzzy set, due to the fact that it does not guarantee that $\forall x_i : \mu_{A^r_{i,j}}(x_i) \in [0,1]$. If the reduced form is intended to serve additional purpose(s) other than complexity reduction, then it may be required to accommodate additional characterisation pertaining to the specific purpose(s) in mind. As an example, the transformations are given in Section 7 to enable that $\mu_{A^r_{i,j}}(x_i)$ be interpretable as fuzzy membership functions.

Step 3. Fitting numerator and denominator

The numerator and denominator of (3) are the same type. Therefore, steps 1 and 2 can be applied to both individually. However, this may not result in the same functions $\mu_{A^r_{i,j}}(x_i)$ as required in the numerator and denominator of (6). Let us follow the description with arbitrary number of dimension.

This problem can be alleviated by constructing the n-dimensional matrices $\underline{\underline{\mathbf{B}}}_k$, $k = 1..m$, using elements $b_{k,t_1,\cdots,t_n}$, and with an additional matrix $\underline{\underline{\mathbf{B}}}_{m+1} = [w_{t_1,\ldots,t_n}]$. Then let $\underline{\underline{\mathbf{A}}}^r_{i,(n_i \times n_i^r)}$, $i = 1..n$, and $\underline{\underline{\mathbf{B}}}^r_{k,(n_1^r \times \cdots \times n_n^r)}$, $k = 1\ldots m+1$ (where $\forall i : n_i^r \le n_i$) be generated by applying Method 1 to all matrices $\underline{\underline{\mathbf{B}}}$. As a result, the reduced function $\mu_{A^r_{i,j}}(x_i)$ as calculated from (9) will be common to both numerator and denominator of (6). For the remaining parameters, the functions $q^r_{t_1,\cdots,t_n}(x_1,\cdots,x_n)$ can be formed with elements $b^r_{k,t_1,\cdots,t_n}$ of the matrices $\underline{\underline{\mathbf{B}}}^r_{k,(n_1^r \times \cdots \times n_n^r)}$, $k = 1\ldots m$, and the elements $w^r_{t_1,\cdots,t_n}$ can be given by elements of the matrix $\underline{\underline{\mathbf{B}}}^r_{m+1,(n_1^r \times \cdots \times n_n^r)}$. This yields all necessary parameters for the reduced form (6).

5. COMPLEXITY INVESTIGATION OF THE REDUCED RATIONAL FORM

This section investigates the effectiveness of the reduction method presented in the previous section. It compares (3) and (6) for two cases of antecedent membership functions. The first involves antecedents functions for which the transformed form of (9) cannot be analytically simplified, resulting in an additional computational term over the original sets. The second studies the case where the transformed antecedent can be analytically expressed in the same form as the original ones. For instance, if the antecedent sets are given in form of $\mu_{A_{i,j}}(x_i) = c_{i,j}\sin(ax_i + \gamma_{i,j})$ then $\mu_{A^r{}_{i,j}}(x_i)$ of (9) can still be expressed in the same form of $\mu_{A^r{}_{i,j}}(x_i) = c^r{}_{i,j}\sin(ax_i + \gamma^r{}_{i,j})$. Another example is that the antecedents are piecewise linear, see [9,10,17,18], or that they are in a general form expressible in terms of basis functions $\varphi_{i,u}(x)$:

$$\mu_{A_{i,j}}(x_i) = \sum_{u=1}^{o_i} c_{i,u,j}\varphi_{i,u}(x_i)\ ,\ j = 1 \ldots n_i . \tag{10}$$

The number of basis functions are o_i on universe X_i.

We will now present conditions for complexity reduction for these two cases.

First case: transformed antecedents cannot be analytically simplified

Lemma 3: The reduced form (6) generated by Method 1 has reduced calculation-time and storage-space over (3) if $\exists i : n_i^r < n^t$ and $n > n^t$. The relative reduction is approximated by:

$$R \approx \left(\prod_{i=1}^{n} n_i^r\right) \Big/ \left(\prod_{i=1}^{n} n_i\right). \tag{11}$$

Proof: By Lemma 1, the transformed antecedents $\mu_{A^r{}_{i,j}}(x_i)$ are given by (9) of step 2. Suppose the worse case that $\mu_{A^r{}_{i,j}}(x_i)$ cannot be analytically simplified and they remain expressed in the explicit form of (9). The computational complexity of (6) then becomes:

$$T_R = \prod_{i=1}^{n} n_i^r \left((n-1+r+m)t_p + (m-1)t_s \right) + mT_B + T_A \sum_{i=1}^{n} n_i + \sum_{i=1}^{n} T_{C,i} + t_d,$$

where $T_{C,i} = t_p n_i n_i^r + t_s (n_i - 1) n_i^r$ is the calculation time for all $\mu_{A^r_{i,j}}(x_i)$ on X_i. Hence, T_R has an additional term compared to (5). However, this term is not exponentially growing. Consequently, threshold n^t can always be found when if $\exists i : n_i^r < n^t$ then $T_R < T_G$. The ratio (11) is obtained considering only the dominant terms, i.e., the exponentially growing parts of (5) and (11).

Proof for the storage-space complexity reduction is similar.

Second case: transformed sets are expressible in the same form as the original ones.

Lemma 4: The reduced form (6) generated by Method 1 has reduced calculation-time and storage-space over (3) for the case when $\mu_{A_{i,j}}(x_i)$ are expressible in terms of basis functions, as in (10), if $\exists i : n_i^r < n_i$. The relative reduction is approximated by:

$$R \approx \left(\prod_{i=1}^{n} n_i^r \right) \Big/ \left(\prod_{i=1}^{n} n_i \right).$$

Proof: In this case the reduced antecedents $\mu_{A^r_{i,j}}(x_i)$ are in the same form as $\mu_{A_{i,j}}(x_i)$. Consequently the calculation time for (6) is the same expression of (5), only with n_i replaced by n_i^r. Therefore $T_R < T_G$ if $\exists i : n_i^r < n_i$.

6. IMPROVING THE EFFECTIVENESS OF REDUCTION

The reduction in the case of Lemma 4 may be further improved by the following step.

Method 2: Further improving the reduction

From (10), functions $\mu_{A_{i,j}}(x_i)$ can be put into the following form:

$$\begin{bmatrix}\mu_{A_{i,1}}(x_i) & \cdots & \mu_{A_{i,n_i}}(x_i)\end{bmatrix} = \begin{bmatrix}\varphi_{i,1}(x_i) & \cdots & \varphi_{i,o_i}(x_i)\end{bmatrix}\underline{\underline{\mathbf{C}}}_i,$$

where $\underline{\underline{\mathbf{C}}}_i = [c_{i,u,j}]$. As such, $\mu_{A^r_{i,j}}(x_i)$ in (9) can therefore be expressed as

$$\begin{bmatrix}\mu_{A^r_{i,1}}(x_i) & \cdots & \mu_{A^r_{i,n^r_i}}(x_i)\end{bmatrix} = \begin{bmatrix}\varphi_{i,1}(x_i) & \cdots & \varphi_{i,o_i}(x_i)\end{bmatrix}\underline{\underline{\mathbf{C}}}_i\underline{\underline{\mathbf{A}}}^r_i.$$

Hence, one can actually bypass $\underline{\underline{\mathbf{C}}}_i$ in the expressions by apply Method 1 on matrices $\underline{\underline{\mathbf{B}}}^{\mathrm{o}}_k = [b^{\mathrm{o}}_{k,z_1,\ldots,z_n}]$, with:

$$b^{\mathrm{o}}_{k,z_1,\ldots,z_n} = \sum_{t_1=1}^{n_1} \cdots \sum_{t_n=1}^{n_n} b_{k,t_1,\ldots,t_n} \prod_{i=1}^{n} c_{i,z_i,t_i}.$$

This shows that in this case one can use in (3) values $b^{\mathrm{o}}_{k,z_1,\ldots,z_n}$ and $\varphi_{i,u}(x_i)$ instead of $b_{k,z_1,\ldots,z_n}$ and $\mu_{A_{i,j}}(x_i)$, respectively. Consequently, if the rank of $\underline{\underline{\mathbf{B}}}^{\mathrm{o}}_k$ in the 2 dimensional sequences (see Method 1) is smaller than the rank of the 2 dimensional sequences of $\underline{\underline{\mathbf{B}}}_k$ then improved reduction is attained by Method 2.

7. SPECIALISATION FOR FUZZY INFERENCE CASES

A most popular scheme in many fuzzy applications is the utilisation of product-sum-gravity inference algorithm, together with piece-wise linear antecedent sets in the *Ruspini* partition:

$$\forall i: \sum_{j=1}^{n_i} \mu_{A_{i,j}}(x_i) = 1, \tag{12}$$

and also singleton consequent sets

$$B_{t_1,\cdots,t_n} : \mu_{B_{t_1,\cdots,t_n}}(y) = \delta(b_{t_1,\cdots,t_2}). \tag{13}$$

Substituting (12) and (13) into (3) yields the following simplified form:

$$y = F(x_1,\cdots,x_n) = \sum_{t_1=1}^{n_1} \cdots \sum_{t_n=1}^{n_n} \left(b_{t_1,\cdots,t_n} \prod_{i=1}^{n} \mu_{A_{i,t_i}}(x_i) \right). \tag{14}$$

where we have $\sum_{t_1=1}^{n_1} \cdots \sum_{t_n=1}^{n_n} \left(b_{t_1,\cdots,t_n} \prod_{i=1}^{n} \mu_{A_{i,t_i}}(x_i) \right) = 1$ because of (12) and that $\forall t_1,\cdots,t_2 : w_{t_1,\cdots,t_2} = 1$.

The piece-wise linear antecedent sets are defined by their characteristic points as in Figure 1:

$$c_{i,j,h} = \mu_{A_{i,j}}(a_{i,h}),\ a_{i,h} \in X_i,\ c_{i,j,h} \in [0,1],\ j = 1 \ldots n_i.$$

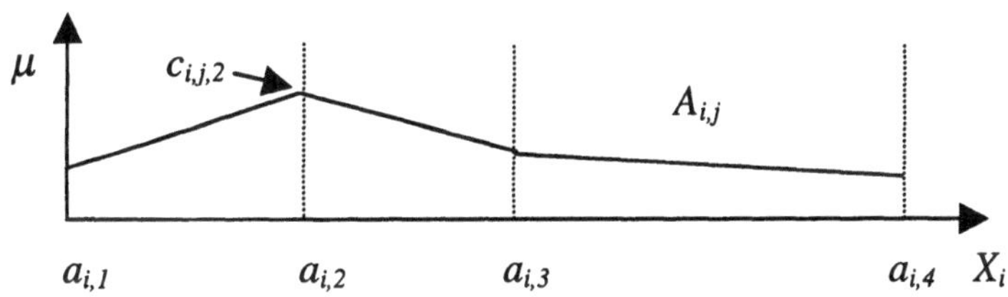

Figure 1. Piece-wise antecedent sets defined by characteristic points

A special case of this is when $\underline{\underline{\mathbf{C}}}_i = \lfloor c_{i,j,h} \rfloor$ are identity matrices. This implies that the antecedent sets are triangular shaped. In this case $F(x_1,\cdots,x_n)$ is a piece-wise linear interpolation that yields $b_{t_1,\cdots,t_n} = F(a_{1,t_1},\cdots,a_{n,t_n})$. Applying Method 1, the reduced antecedent and consequent sets obtained is then $\mu_{A^r_{i,j}}(x_i)$ and $\mu_{B^r_{t_1,\cdots,t_n}}(y) = \delta(b^r_{t_1,\cdots,t_2})$. The elements of $\underline{\underline{\mathbf{A}}}_i^r = \underline{\underline{C}}_i \underline{\underline{\mathbf{A}}}_i^r$ ($\underline{\underline{C}}_i$ is an identity matrix) now define the characteristic points of the reduced sets $\mu_{A^r_{i,j}}(x_i)$ as $c_{i,j,h}$ for $\mu_{A_{i,j}}(x_i)$. This process is illustrated in Figure 2 for the two-variable case. The first co-ordinate system shows the sampled values $b_{i,j}$ of a given function. The second co-ordinate system is the approximation of the given function by

Product-Sum-Gravity inference algorithm (14). The last one depicts the reduced rule-base defining the same approximation.

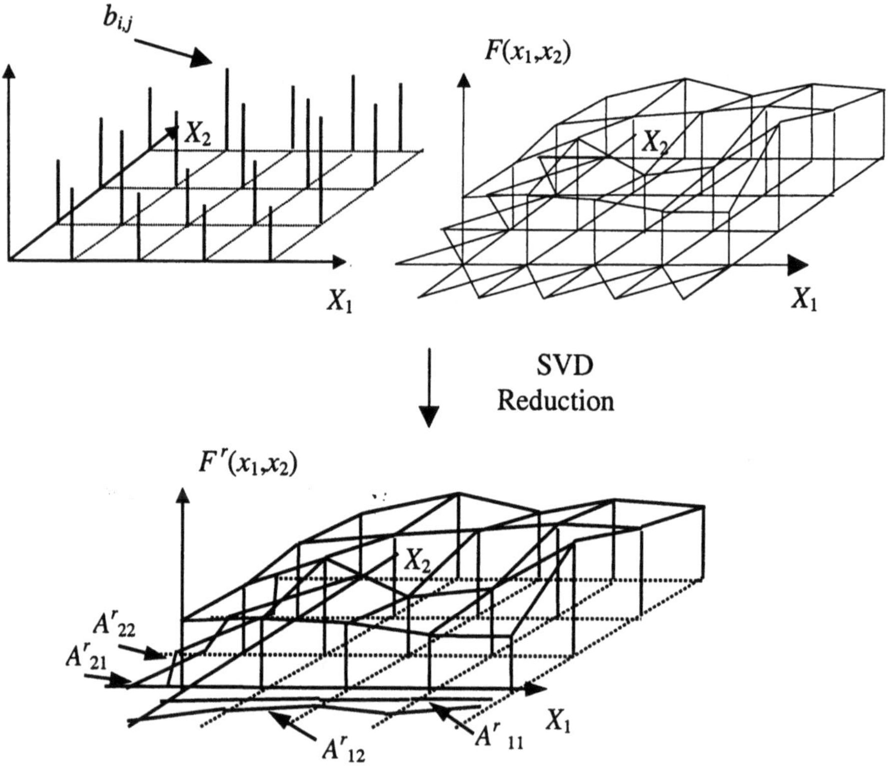

Figure 2: Illustration of the reduction for the two-variable case

If $\underline{\underline{\mathbf{C}}}_i = \lfloor c_{i,j,h} \rfloor$ are not identity matrices Method 2 may be applied to improve the reduction.

Note that the transformed antecedent sets in (9) may not maintain the *Ruspini* partition here as values $a_{i,l,j}$ may be negative and there is no guarantee that $\forall i: \sum_{j=1}^{n_i^r} \mu_{A^r_{i,j}}(x_i) = 1$, see (12). Moreover, the antecedent membership functions may not be bounded by $[0,1]$ at all. Function $F(x_1, \cdots, x_n)$, however, can still be implemented by the reduced form of

$$y = F(x_1, \cdots, x_n) = \sum_{t_1=1}^{n_1^r} \cdots \sum_{t_n=1}^{n_n^r} \left(b^r_{t_1,\cdots,t_n}(x_1, \cdots, x_n) \prod_i^n \mu_{A^r_{i,t_i}}(x_i) \right) \tag{15}$$

even when the *Ruspini* partition is lost.

If the reduced form is intended not only for reducing the computational complexity, but also meant to be interpretable as fuzzy or B-spline quantities as well, it should maintain additional characterisation pertaining to the specific purpose(s). This leads to further transformation, for instance, to obtain matrices $\underline{\underline{\mathbf{A}}}_r^i$ such that the functions $\mu_{A^r_{i,t_i}}(x_i)$ are within the interval [0,1] and also satisfying the *Ruspini* partition. In this regards, algorithms for *nonnegativeness*, *sum-normalisation* and *set normalisation* transformations have been developed in [9,10]. Here, we will briefly discuss two such transformations and extend RSVD to become SVDE.

Transformation 1: SN (*Sum-Normalisation*). Let matrix $\underline{\underline{\mathbf{B}}} = \underline{\underline{\mathbf{A}}}'_1 \underline{\underline{\mathbf{B}}}' \underline{\underline{\mathbf{A}}}'^{\mathrm{T}}_2$ be given. SN transforms $\underline{\underline{\mathbf{A}}}'_1$, $\underline{\underline{\mathbf{B}}}'$, and $\underline{\underline{\mathbf{A}}}'_2$ into $\underline{\underline{\mathbf{A}}}''_1$, $\underline{\underline{\mathbf{B}}}''$, and $\underline{\underline{\mathbf{A}}}''_2$, respectively, while keeping $\underline{\underline{\mathbf{B}}} = \underline{\underline{\mathbf{A}}}''_1 \underline{\underline{\mathbf{B}}}'' \underline{\underline{\mathbf{A}}}''^{\mathrm{T}}_2$ and ensuring that the sum of the elements in each row of $\underline{\underline{\mathbf{A}}}''_1$ and $\underline{\underline{\mathbf{A}}}''_2$ equals 1. Let $\Phi_i = diag(\tilde{sum}(\underline{\underline{\mathbf{A}}}*^{r\,\mathrm{T}}_i))$, where $\tilde{sum}(\underline{\underline{\mathbf{A}}})$ is the column vector summing over the rows of $\underline{\underline{\mathbf{A}}}$. Then according to [9,10],

If $\tilde{sum}\left(\left(\underline{\underline{\mathbf{A}}}*^d_i\right)^{\mathrm{T}}\right) = \underline{0}$ (where vector $\underline{0}$ consists of zero elements) then $\underline{\underline{\mathbf{A}}}''_i$ can be determined as:

$$\underline{\underline{\mathbf{A}}}''_i = \underline{\underline{\mathbf{A}}}'_i \, \Phi_i \, . \tag{16}$$

Or, if $\tilde{sum}\left(\left(\underline{\underline{\mathbf{A}}}*^d_i\right)^{\mathrm{T}}\right) \neq \underline{0}$ then $\underline{\underline{\mathbf{A}}}''_i$ can be determined as:

$$\underline{\underline{\mathbf{A}}}''_i = \left[\underline{\underline{\mathbf{A}}}_1 \quad \underline{\underline{\mathbf{A}}}*^d_i \; \tilde{sum}\left(\left(\underline{\underline{\mathbf{A}}}*^d_i\right)^{\mathrm{T}}\right)\right] \begin{bmatrix} \Phi_i & 0 \\ 0 & 0 \end{bmatrix}. \tag{17}$$

After $\underline{\underline{\mathbf{A}}}''_i$ is determined by (16) or (17), one can always find $\underline{\underline{\mathbf{B}}}''$ so that $\underline{\underline{\mathbf{B}}} = \underline{\underline{\mathbf{A}}}'_1 \underline{\underline{\mathbf{B}}}' \underline{\underline{\mathbf{A}}}'^{\mathrm{T}}_2 = \underline{\underline{\mathbf{A}}}''_1 \underline{\underline{\mathbf{B}}}'' \underline{\underline{\mathbf{A}}}''^{\mathrm{T}}_2$.

Transformation 2: NN (*NonNegativeness*). Let matrix $\underline{\underline{\mathbf{B}}} = \underline{\underline{\mathbf{A}}}''_1 \underline{\underline{\mathbf{B}}}'' \underline{\underline{\mathbf{A}}}''^{\mathrm{T}}_2$ be given. NN transforms $\underline{\underline{\mathbf{A}}}''_1$, $\underline{\underline{\mathbf{B}}}''$, and $\underline{\underline{\mathbf{A}}}''_2$ into $\underline{\underline{\mathbf{A}}}'''_1$, $\underline{\underline{\mathbf{B}}}'''$, and $\underline{\underline{\mathbf{A}}}'''_2$, respectively, while keeping $\underline{\underline{\mathbf{B}}} = \underline{\underline{\mathbf{A}}}'''_1 \underline{\underline{\mathbf{B}}}''' \underline{\underline{\mathbf{A}}}'''^{\mathrm{T}}_2$ and ensuring that the elements of $\underline{\underline{\mathbf{A}}}'''_1$ and $\underline{\underline{\mathbf{A}}}'''_2$ being within the interval [0,1]. According to [9,10] let:

$$\varsigma_i = \begin{cases} 1 & if \quad a_{i\min} \geq -1 \\ \dfrac{1}{|a_{i\min}|} & otherwise \end{cases}$$

where $a_{i\,\min}$ is the minimum element of $\underline{\underline{\mathbf{A}}}''_i$. Then

$$\underline{\underline{\mathbf{A}}}'''_i = \underline{\underline{\mathbf{A}}}''_i \frac{1}{c_i + \varsigma_i}(1_{c_i \times c_i} + \varsigma_i I_{c_i \times c_i}) \tag{18}$$

where c_i is the number of columns of $\underline{\underline{\mathbf{A}}}''_i$. With $\underline{\underline{\mathbf{A}}}'''_i$ as determined by (18), one can always find $\underline{\underline{\mathbf{B}}}'''$ so that $\underline{\underline{\mathbf{B}}} = \underline{\underline{\mathbf{A}}}''_1 \underline{\underline{\mathbf{B}}}'' \underline{\underline{\mathbf{A}}}''^{\mathrm{T}}_2 = \underline{\underline{\mathbf{A}}}'''_1 \underline{\underline{\mathbf{B}}}''' \underline{\underline{\mathbf{A}}}'''^{\mathrm{T}}_2$.

With the above transformations, SVDR is extended to become SVDE, with *non-negativeness* and *sum-normalisation* properties in order to maintain the *Ruspini* partition in the case of piece-wise linear sets. SVDE results in matrices $\underline{\underline{\mathbf{A}}}'''_i$, the elements of which will be within interval [0,1] and the sum of elements of each row is equal to one.

Definition: SVDE (Singular Value Decomposition Extended). SVDE includes the operations of SVDR, SN and NN. For a given matrix $\underline{\underline{\mathbf{B}}}_{(n_1 \times n_2)}$, SVDE results in $\underline{\underline{\mathbf{B}}} = \underline{\underline{\mathbf{A}}}^r_1 \underline{\underline{\mathbf{B}}}^r \underline{\underline{\mathbf{A}}}^{r\,\mathrm{T}}_2$, where $\underline{\underline{\mathbf{A}}}^r_1 = \underline{\underline{\mathbf{A}}}'''_1$, $\underline{\underline{\mathbf{B}}}^r_{(n^r_1 \times n^r_2)} = \underline{\underline{\mathbf{B}}}'''$ $n^r_i \leq n_i : i = 1,2$ and $\underline{\underline{\mathbf{A}}}^r_2 = \underline{\underline{\mathbf{A}}}'''_2$, and that $\underline{\underline{\mathbf{A}}}'''_i$ satisfy SN and NN.

Consequently, if SVDE is used instead of SVDR, the reduced form of (15) also describes a fuzzy rule base, where the antecedent fuzzy sets are defined in the *Ruspini* partition. Moreover, the use of (15) corresponds to adopting the paradigm of product-sum-gravity inference.

8. INEXACT REDUCTION AND ERROR BOUND

Discarding non-zero singular values increases the effectiveness of reduction, but brings in reduction error. In this case, the reduced form (6) is only an approximation of the general form (3), $\hat{F}(x_1, \cdots, x_n) = F^r(x_1, \cdots, x_n)$.

The resulting error is related to the sum of discarded singular values in the reduction process and also the type of antecedent membership functions used in the general forms. This implies that the ratio of compression is dependent on given error threshold. Expressions of the resultant error for different cases can be found in [9,10,17,18].

For the case where $\underline{\underline{\mathbf{C}}}_i$ are identity matrices, i.e., the antecedent sets are triangular shaped, we have the following additional results.

Lemma 5: The maximum reduction error between (14) and (15) in the case of (14) with identity $\underline{\underline{\mathbf{C}}}_i$ is

$$\left|F(x_1,\cdots,x_n)-F^r(x_1,\cdots,x_n)\right| \le E_{\max} = \max_{t_1,\cdots,t_n}\left|b_{t_1,\cdots,t_n} - b^r_{t_1,\cdots,t_n}\right| = \sum_i^{\alpha}\delta_i$$

where δ_i are the discarded singular values during SVDR and α is the number of all discarded singular values.

Proof: The bound of the reduction error measures the maximum difference between the linear interpolated surface by (14) and (15). Again, taking the example of the two variable case. Since columns of $\underline{\underline{\mathbf{A}}}_1^d$ and $\underline{\underline{\mathbf{A}}}_2^d$ of (8) have Euclidean norm of unity, absolute values of their elements must be bounded by 1. Then from (8), the error introduced by SVDR is hence:

$$\left|\underline{\underline{\mathbf{B}}}-\hat{\underline{\underline{\mathbf{B}}}}\right| \le \left|\underline{\underline{\mathbf{A}}}_1^d\,\underline{\underline{\mathbf{B}}}^d\,\underline{\underline{\mathbf{A}}}_2^{d\,\mathrm{T}}\right| \le \underline{\underline{\mathbf{1}}}_{n_1\times(n_1-n^d)}\underline{\underline{\mathbf{B}}}^d\,\underline{\underline{\mathbf{1}}}_{n_2\times(n_2-n^d)} = \left(\sum_{i=1}^{n^d}\delta_{i+n^r}\right)\underline{\underline{\mathbf{1}}}_{n_1\times n_2}$$

where n^d is the number of discarded singular values δ. Following similar approach, it can also be shown that the reduction error remains the same during the whole process of Method 1. The derivation is straightforward but lengthy and will not be shown here. One can refer to [9, 10] for more details.

Finally, one notes that functions (14) and (15) are, respectively, linear interpolations based on values $b_{t_1,\cdots,t_n}$ and $b^r{}_{t_1,\cdots,t_n}$. The maximum error between (14) and (15) is hence expressible as:

$$E_{\max} = \max_{t_1,\cdots,t_n}\left|b_{t_1,\cdots,t_n} - b^r_{t_1,\cdots,t_n}\right| = \sum_i^{\alpha}\delta_i\ ,$$

which completes the proof.

9. EXAMPLES

Two numerical examples are now given to illustrate the effectiveness and application of the present technique. Due to limited space, the first example is discussed more fully than the second. Details of the examples, and other as well, can be found in [9,10,11].

The first example, taken from [9,27], concerns the function $y = f(x_1, x_2) = \left(1 + x_1^{-2} + x_2^{-1.5}\right)^2$ as depicted on Figure 3. To start, let the function be approximated by (14), namely, a product-sum-gravity inference-based fuzzy system with 5 triangular shaped antecedent sets defined in the *Ruspini* partition in each input domain, and that the consequents are singleton functions. By (14), the resultant approximation to the function would be piece-wise planar, as shown in Figure 4. The characteristic points of the antecedents for each input domain are given as $a_i = \{1,2,3,4,5\}$, $i = 1,2$, and hence $b_{t_1,t_2} = f(a_{1,t_1}, a_{2,t_2})$, $t_1, t_2 = 1 \ldots 5$. The fuzzy approximation in this case hence has 25 rules, and we are to apply Method 1 to reduce it. The resulting non-zero singular values in this case are 17.3937, 1.4282 and 0.0045. Because the function $y = f(x_1, x_2)$ can be put into the following form:

$$y = f(x_1, x_2) = \begin{bmatrix} 1 & \dfrac{1}{x_1^2} & \dfrac{1}{x_1^4} \end{bmatrix} \begin{bmatrix} 1 & 2 & 1 \\ 2 & 2 & 0 \\ 1 & 0 & 0 \end{bmatrix} \begin{bmatrix} 1 \\ 1/x_2^{0.5} \\ 1/x_2^3 \end{bmatrix},$$

the rank of the matrix $\underline{\underline{\mathbf{B}}}$ and hence the number of nonzero singular values, is at most 3. Keeping all three non-zero singular values yields three antecedent sets for each input dimension in (15). The reduced rule base thus has 9 rules. Note that in this case there is no SVD reduction error. There is no difference in approximation performance of the original function by the starting fuzzy rule base and the reduced one. And both will have a maximum error of 1.73 relative to the original function due to the linear interpolation.

The next study in this example consider fuzzy approximation of the function with non-singleton consequents, namely, with rational from (3). The results are shown in Figure 5. The resultant interpolation is non-linear and the corresponding error due to such interpolation is 0.116, much reduced from the linear interpolation case. Carrying out reduction in this case yields the rule based in Figure 6, and an error of 0.386 to the full version of fuzzy approximation with non-singleton consequents.

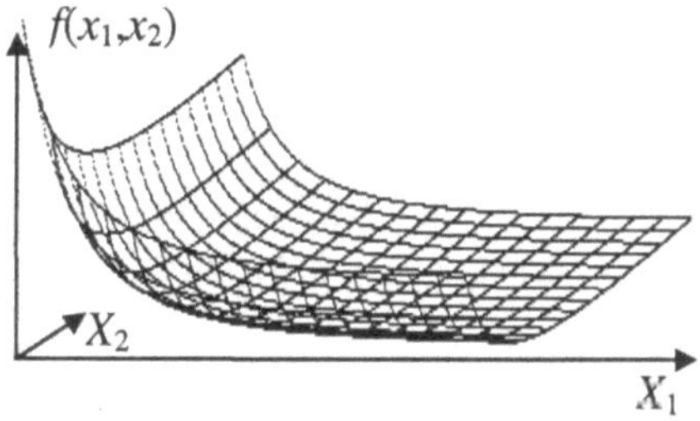

Figure 3: Original function

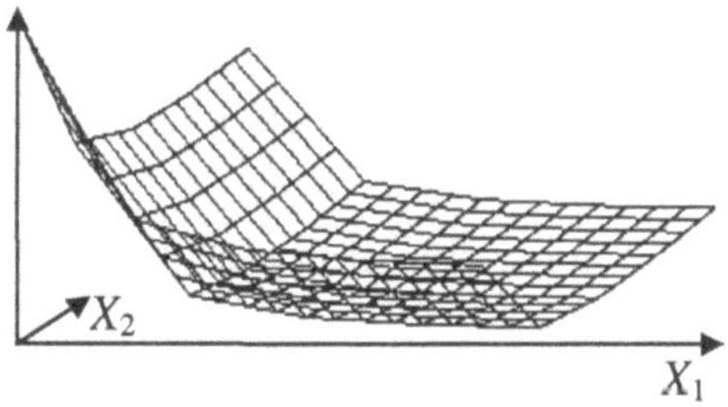

Figure 4: Approximation by (14)

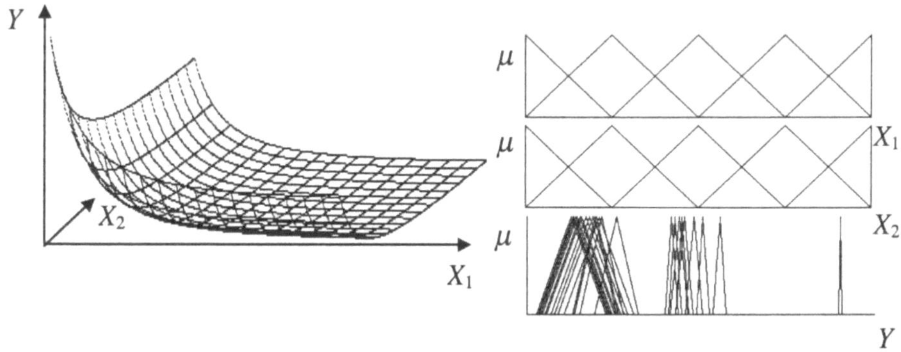

Figure 5: Approximation by the rational form

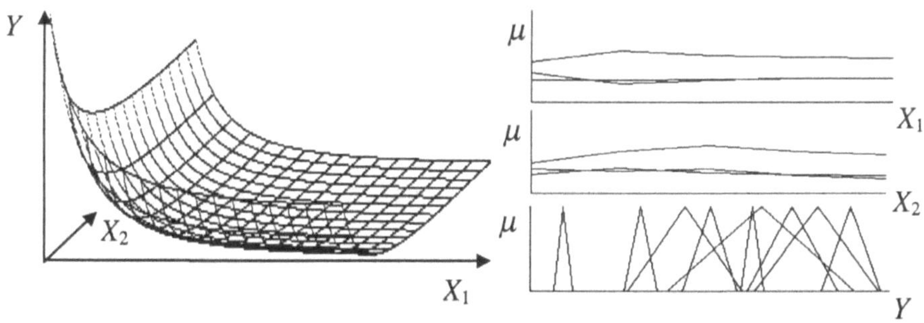

Figure 6: Reduced rule base

Again, the number of the rules is reduced from 25 to 9, which amounts to a reduction of 64%. Comparing Figures 5 and 6 to the original function in Figure 4, one notes that non-singleton fuzzy approximation is very efficient over the singleton case, is mainly due to the reduction in interpolation error.

The second example demonstrates SVD reduction in a B-spline application. Figure 7 shows the figures of three cats. The first cat (a) is generated by B-spline technique [25] using polynomial form, which amounts to having $\forall t_1,\cdots,t_n : w_{t_1,\cdots,t_n} = 1$ and $f_{t_1,\cdots,t_n}(x_1,\cdots,x_n) = b_{t_1,\cdots,t_n}$, i.e., using only the numerator of (1). In this case, three 2-dimensional matrices $\underline{\underline{\mathbf{B}}}$ are defined for generating the 3-dimensional cat (one for each dimension). The number of second order basis functions for the inputs are 9 and 11 and are shown in Figure 8. The number of parameters to specify the control points for each dimension is hence 9*11, so the cat in Figure 7(a) is defined by 3*9*11=297 parameters. With the present method and discarding 4 zero singular values, we determine that the same cat can actually be reproduced by only 3*5*7=105 parameters. The transformed basis functions are the same as the original ones. This, according to Lemma 2, amounts to a complexity reduction of 65%. Any operation with reduced form, for instance, rotating the cat, needs only 35% of the computational effort required by the original form. This is a considerable saving of calculation. Such reduction also facilitates efficient transferral of between B-spline graphic applications.

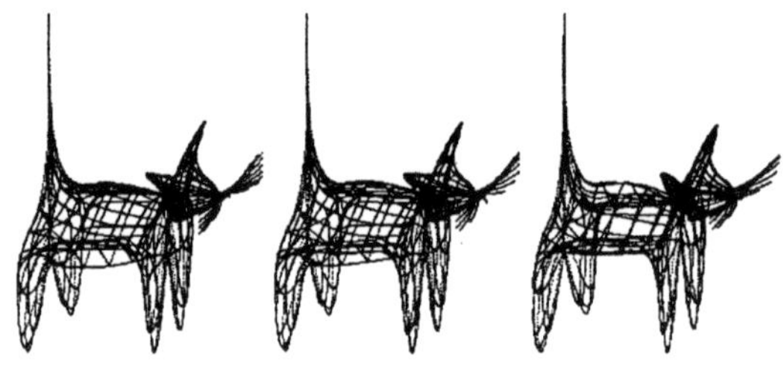

a b c

Figure 7: Three dimensional figures by B-spline technique

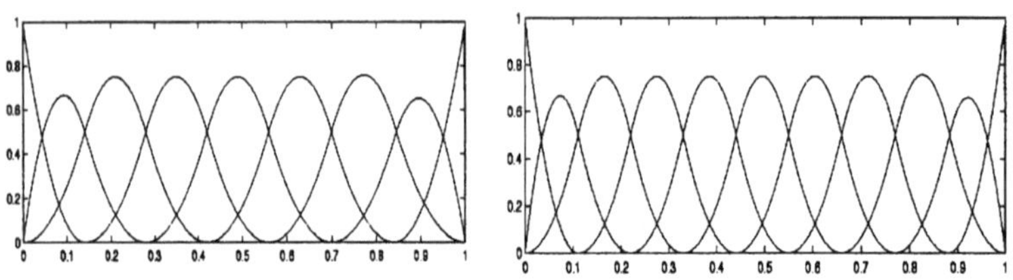

Figure 8: Rational form with 9 and 11 second order basis functions

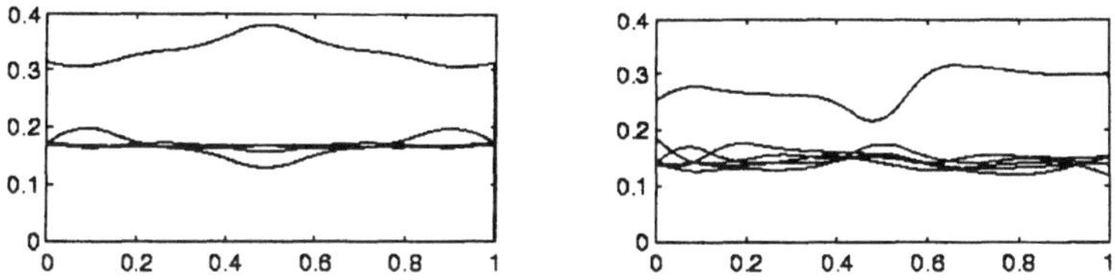

Figure 9: Transformed basis function in the reduced form

Discarding an additional non-zero singular value enhances the reduction but results in a slightly different cat as depicted in Figure 7(b). The transformed basis functions in this case are depicted in Figure 9. The number of parameters is 3*4*6=72, and the reduction is 76%. Discarding one more non-zero singular value results in requiring only 45 parameters and 84% reduction. The result is depicted in Figure 7(c), which still passes looking like a cat.

References

[1] L.T. Kóczy and K. Hirota "Size reduction by interpolation in fuzzy rule bases", *IEEE Trans. on System Man and Cybernetics,* Vol. 27, 1997, pp14-25.

[2] L.T. Kóczy and K. Hirota "Fuzzy inference by compact rules", *Proc. of Int. Conf. on FL & NN* (IIZUKA '90), Iizuka, Fukuoka, 1990, pp. 307-310.

[3] E.P. Klement, L.T. Kóczy and B. Moser "Are fuzzy systems universal approximators?", *Int. Jour. General Systems*, to appear.

[4] D. Tikk "The nowhere denseness of *Takagi-Sugeno-Kang* type fuzzy controllers containing prerestricted number of rules", *Tatra Mountains Mathematical Publications.*

[5] A. Stoica "Fuzzy processing based on α-cut mapping", *5th IFSA World Congress*, Seoul, pp. 1266-1269.

[6] W. Yu and Z. Bien "Design of fuzzy logic controller with inconsistent rule base", *Jour. of intelligent and Fuzzy Systems,* Vol. 2, 1994, pp 147-159

[7] J. Bruinzeel, V. Lacrose, A. Titli and H.B. Verbruggen "Real time fuzzy control of complex systems using rule-base reduction methods", *2nd World Aut. Con.* (WAC'96), Monpellier, France, 1996.

[8] M. Sugeno, M.F. Griffin, A. Bastian, "Fuzzy hierarchical control of an unmanned Helicopter", *5th IFSA World Congress*, Seoul, 1993. pp. 1262-1265.

[9] Y. Yam "Fuzzy approximation via grid point sampling and singular value decomposition", *IEEE Trans. on System Man and Cybernetics,* Vol. 27, No. 6, 1997, pp. 933-951.

[10] Y. Yam, P. Baranyi and C.T. Yang "Reduction of fuzzy rule base via singular value decomposition", *IEEE Trans. on Fuzzy Systems*. Vol. 7, No. 2, ISSN 1063-6706, 1999, pp. 120-131.

[11] P. Baranyi, Y. Yam and C.T. Yang "Complexity reduction of the rational general form", *8th IEEE Int. Conf. on Fuzzy Systems* (FUZZ-IEEE'99), Seoul, Korea, 1999, pp. 366-371.

[12] P. Baranyi, A. Martinovics, Sz. Kovács, D.Tikk and Y. Yam "A general extension of the fuzzy SVD rule base reduction using arbitrary inference algorithm", *IEEE Int. Conf. System Man and Cybernetics* (IEEE SMC'98), 1998, San Diego, California, USA, pp 2785-2790.

[13] P. Baranyi, Y. Yam and C.T. Yang "Singular value decomposition of linguistic symbol-array", *IEEE Conf. on Systems Man and Cybernetics* (IEEE SMC'99), 1999, Tokyo, Japan, pp.:III/822-III/826.

[14] P. Baranyi, Y. Yam, C.T. Yang and A. Várkonyi-Kóczy "Practical extension of the SVD based reduction technique for extremely large fuzzy rule bases", *IEEE Int. Workshop on Intelligent Signal Processing*, (WISP'99), 1999, Budapest, Hungary, pp. 29-33.

[15] C.T. Yang, P. Baranyi, Y. Yam and Sz. Kovács "SVD reduction of a fuzzy controller in an AGV steering system", EFDAN'99, Dortmund, Germany, 1999, pp 118-124

[16] P. Baranyi, I. Mihálcz, P. Korondi, Z. Gubinyi and H. Hashimoto "Fuzzy rule base reduction for robot finger furnished with shape memory alloy", *IEEE 24th Industrial Electronics Society Conference* (IEEE IECON'98), 1998, pp 6-11.

[17] P. Baranyi and Y. Yam "Singular value-based approximation with non-singleton fuzzy rule base", *7th Int. Fuzzy Systems Association World Congress* (IFSA'97), Prague, 1997, pp. 127-132.

[18] P. Baranyi and Y. Yam "Singular value-based approximation with *Takagi-Sugeno* type fuzzy rule base", *6th IEEE Int. Conf. on Fuzzy Systems* (FUZZ-IEEE'97), Barcelona, Spain, 1997, pp. 265-270.

[19] P. Baranyi, Y. Yam and L.T. Kóczy "Multi variables singular value based rule interpolation", *IEEE Int. Conf. System Man and Cybernetics* USA, 1997, pp. 1598-1603.

[20] L. Wang, R. Langari, and J. Yen "Principal components, B-splines, and fuzzy systems reduction", in Fuzzy Logic for the applications to Complex systems, W. Chiang and J. Lee, Eds. Singapure: World Scientific, 1996, pp. 255-259.

[21] Sz. Kovács and L.T. Kóczy "The use of the concept of vague environment in approximate fuzzy reasoning", *Fuzzy Set Theory and Applications, Tatra Mountains Mathematical Publications*, Math. Inst. Slovak Academy of S. 1997, vol.12, pp.169-181.

[22] P. Baranyi, T.D. Gedeon and L.T. Kóczy "A general interpolation technique in fuzzy rule bases with arbitrary membership functions", *IEEE Int. Conf. S.M.C.*, Beijing, China, 1996, pp. 510-515.

[23] P. Baranyi, D. Tikk, Y. Yam, L.T. Kóczy and L. Nádai "A new method for avoiding abnormal conclusion for α-cut based rule interpolation", *8th IEEE Int. Conf. on Fuzzy Systems* (FUZZ-IEEE'99), Seoul, Korea, 1999,

[24] M.F. Kawaguchi and M. Miyakoshi "Fuzzy spline interpolation in sparse fuzzy rule bases", *Proc. of 5th Int. Conf. on Soft Comp. and Inf./Int. Systems* IIZUKA 98, Iizuka, Japan, 1998, pp 664-667.

[25] G. Farin "Curves and surfaces for computer aided geometric design", Academic press 1997
[26] M. Mizumoto "Fuzzy controls by product-sum-gravity method", Advancement of Fuzzy Thory and Systems in China and Japan, Eds. Liu and Mizumoto, International Academic Publishers, c1.1.-c.1.4. 1990.
[27] T. Takagi and M. Sugeno "Fuzzy identification of systems and its applications to modelling and control", *IEEE Trans. On System Mans and Cybernetics* Vol.15, 1985, 116-132.

Chapter 8

REASONING WITH COGNITIVE STRUCTURES OF AGENTS I: ACQUISITION OF RULES FOR COMPUTATIONAL THEORY OF PERCEPTIONS BY FUZZY RELATIONAL METHODS

Ladislav J. Kohout and Eunjin Kim
Dept. of Computer Science,
Florida State University,
Tallahassee, Florida 32306-4530, USA
{kohout, ejkim}@cs.fsu.edu

Abstract Generalized fuzzy rules provide the essential means for computational representation of equational and inequality constraints over fuzzy granular structures. In this chapter, we are concerned with the ways the rules can be acquired from the real world data. We also provide a method for elicitation of knowledge of experts and link this to Zadeh's Computational Theory of Perceptions. The algorithms for extracting and consistency checking of the rules explore the BK-products of relations. This makes it possible to treat the numerical and linguistic computations in a unified framework with the methods for extrapolation of knowledge and data.

Keywords: Knowledge elicitation, BK-products of relations, computational theory of perceptions, non-associative products of relations, preorder, generalized morphisms, Hasse diagram, relational properties, repertory grid, manufacturing, industrial engineering, cost drivers, consensus of experts, 1-tautology, BL-logic, triangle subproducts, t-norm.

1. INTRODUCTION: FUZZY RULES AS THE MEANS OF EXTENDING THE SCOPE OF COMPUTATIONS

What computers are used for includes: mathematical modeling of the real world, information retrieval, controlling technological artifacts, simulation of dynamics of processes of various kinds, recognizing and generating speech, generating the virtual reality environments, constructing AI and other symbolic computing systems. In all these tasks, rules play an important albeit not completely exclusive role. Without rules applied to data and knowledge at various levels of abstraction, quality of functioning and performance of such computing systems would be considerably impoverished. Rules can be applied at various levels of granularity and can handle not only numerical but also linguistic information. Indeed, transformational rules play important role in linguistic as well as in formal systems for symbolic computation and computer algebra systems. Generalized fuzzy rules provide essential vehicle for computational representation of equational and inequality constraints over fuzzy granular structures [45]. This makes it possible to treat in a unified framework numerical and linguistic computations and also embrace approximations and extrapolations of both numeric and linguistic/symbolic computations.

In this chapter we are concerned with the ways rules can be acquired from the real world data. Because our knowledge of the real world is never precise and complete, and sometimes it may be conflicting across various contexts and environments, we employ fuzzy relational methods in the acquisition and representation of rules.

The importance of our approach stems from the fact that our methodology for the acquisition of rules and generalized fuzzy rules can be applied not only to the situations where the data are obtained by measurement but also to eliciting, comparison and validation of *perceptions* and internal symbolic knowledge and experience of experts. Thus it is a methodology that is also applicable within a new paradigm of the Computational Theory of Perceptions recently introduced by Lotfi Zadeh [43]

In section 1 of our chapter we have introduced the reader into motivation for the work. In Sec. 2, the basic notions in the Computational Theory of Perceptions are briefly discussed. Sec. 3 describes the three basic steps of modeling used to build relational models of knowledge. Sec. 4 outlines the methodology of Activity Structures which supports the knowledge engineering of fuzzy relational computations. Sec. 4.1 and 4.2 present an outline of semiotic descriptors that capture the ontology and epistemology of the relational model into which the knowledge is captured. The basic ontological entities into which the semiotic descriptors are embedded are: *objects, qualities and relationships.*

Section 5 gives the necessary background in fuzzy relational methods introducing BK-products of relations and fuzzy relational algorithms used in our approach.

In section 6 we describe how relational models of cognitive structures of experts are elicited by means of repertory grids, analyzed and transformed into fuzzy rules.

In section 7 we present a case study from the domain of engineering.

Section 7.8 deals with semantic meaning of fuzzy rules acquired by transformation of relational models of expert knowledge.

2. RULES IN COMPUTATIONAL THEORY OF PERCEPTIONS

Human models and conceptualizations of the world are either based on perception (direct or with the help of some tools that enhance these perceptions), or on quantitative measurements. In the 19th and first half of the 20th century descriptive theories based on perception and qualitative observation have been mostly eliminated from science, where the precision of measurement was the main criterion for the scientific respectability of scientific models. This lead to elimination of the representations of the world that are not obtained by precise measurement and quantization, regarding these as "unscientific". Yet such non-measurable qualitative notions may be crucial for healthy functioning of human society. Introducing automation and information technology that cannot capture and assess such notions has lead to the impoverishment of quality of human life in some aspects, despite the vast successes in other domains.

Zadeh has recently suggested [43] that computing with words paves the way to a paradigm shift:

> *One of the most deep-seated traditions in science has been and continues to be that of according much more respect to numbers than to words. ...*
>
> *It is a truism that the quest for precision has led to brilliant successes. .... But alongside the brilliant successes we see many problem-areas where progress has been slow and many problems which cannot be solved by any prolongation of existing theories, methodologies and technologies.*
> ...
>
> *With the vast computational power at our command, what is becoming feasible is a counter-traditional move from measurements to perceptions. What this implies is a major enlargement of the role of natural languages in scientific theories. This is the essence of the paradigm shift which, in my view, is likely to take place in coming years.*

According to Zadeh, the principal aim of the computational theory of perceptions (CTP) is to develop an automated capability to reason with perception-based information. Existing theories do not have this capability and rely instead on conversion of perceptions into measurements – a process which in many cases is "infeasible, unrealistic or counterproductive" [43]. It is plausible to expect that CTP will have impact because it is based on the newly emerging methodology of computing with words (CW) [39],[44]. In CW, the objects of computation are words and propositions drawn from a natural language. In his talk entitled "From Computation with Measurements to Computation with Perceptions: A Paradigm Shift" given on September 16th, 1999 Zadeh introduced the computational theory of perceptions – a theory which comprises a conceptual framework and a methodology for computing and reasoning with perceptions. The point of departure in the computational theory of perceptions is the assumption that perceptions are described as propositions in a natural language, e.g., "Jan is tall", "it is likely to rain tomorrow," etc. Hence, computing and reasoning with perceptions is reduced to computing and reasoning with words.

In the computational theory of perceptions, representation of meaning is a preliminary to reasoning with perceptions – a process which starts with a collection of perceptions which constitute the initial data set (IDS) and terminates in a proposition or a collection of propositions which play the role of an answer to a query, that is, the terminal data set (TDS).

3. MODELS

Rules are acquired from the real world situations by selection of features relevant to the purpose in hand and rejection of those which are irrelevant [34]. The knowledge to be embedded into rules may serve either analytic or synthetic modeling process.

3.1 THE THREE STAGES OF RELATIONAL MODELING

There are broadly two main kinds of modeling[1]. One takes an existing situation or system or organism or factory or organization in the real world, and constructs a model of it. The other builds a model first, with a view to setting up the large-scale real-world version later. These two kinds could be called respectively *analytic* and *synthetic*. They are not as different as might at first appear. In both these approaches rewriting rules may serve as the means for representation and transformation of various kinds of knowledge.

Here are the three steps by which anyone must begin in acquisition of knowledge [28] faithfully describing a real-world system [34]:

1. *Selection* of features judged to be significant, and *discarding* of irrelevant ones.
2. Further *reduction* of details by grouping similar cases together.
3. *Representation* of the results of Step 2 in a (visual or verbal or analogue or digital or ...) model.

Steps 1 and 2 are *simplification* of the unwieldy mass of detail. Step 3 is the *translation* of the system "from the field to the laboratory", that is, from where it functions in the real world in all its inscrutability to a *copy* which the investigator can handle. These steps are all spontaneous human activities. Mathematics has formalized each of them, as follows:

1. This is the passage to a *partial substructure* of the original.
2. This is the passage to a *partition* of the structures into *equivalence or congruence classes* or *blocks*.
3. This is an *isomorphism* between the simplified "outside" system and the model of "inside" system. Each *class* resulting from Step 2 is *mapped* into its representative in the model.

In its most general form, all three steps employ mathematical relations as the tools. In general, step 3 yields the representation of a relational model. When special restrictions are imposed on the transformation from step 2 to step 3 so that the only subrelations that have the relational property of the preorder are represented, this yields a representation by a rule system.

We require that our models are computational. Indeed, with vast amounts of data and knowledge rules, such models are useful only if they can be manipulated by a computer and produce computable solutions. It would be a mistake to think that giving a relational foundation by means of fuzzy relations to our modeling process means that we restrict ourselves to quantization and measurement. The three steps of modeling can also be performed on linguistic statements represented relationally.

4. ACTIVITY STRUCTURES: KNOWLEDGE ENGINEERING METHODOLOGY FOR DESIGN OF FUZZY COMPUTATIONS

Relational modeling and computing with words cannot be done without appropriate computer tools. Knowledge elicitation also requires support by adequate computer environment. For the purpose constructing it we have used the methodology of Activity Structures. This methodology cuts across the field boundaries of Software Engineering, Knowledge Engineering, Artificial Intelligence and formal methods. Activity Structures [3],[19],[29] provide a connection between the fundamental concepts and techniques of the above fields and also make it possible to integrate the description of the activities within an Information Processing System with those outside it, in its environment.

4.1 SEMIOTIC DESCRIPTORS

4.1.1 Basic Notions. The essential facet of the fuzzy approach is the fact that fuzzy relations can manipulate semantic information that is carried by linguistic labels. Conditions imposed on logic operations applied to fuzzy relational systems are, however, strictly mathematically defined. Such mathematically defined semantics of the logic of relations is necessary, but it is not sufficient to deal satisfactorily with the meaning of linguistic labels that carry the conceptual meaning of intended applications.

To be able to carry out the three steps of relational modeling requires adopting some minimal ontology. The *ontology* tells us what kinds of things, elements or entities we shall be using in our computational models. This is always required, whether or not we use linguistic statements or quantized numerical statements in our computational models. In order to capture the ontology in linguistic terms we use semiotic descriptors as defined in our methodology of Activity Structures [27],[28],[29], which has been motivated by the approaches used General Systems studies [3], [26],[37].

The purpose of semiotic descriptors is to capture the ontology and epistemology of the relational model in which the knowledge is represented. The basic ontological categories of semiotic descriptors are: *objects, qualities and relationships.* Each category may contain various kinds of semiotic descriptors as we shall see from the Glossary given in the sequel.

The essential facet of the fuzzy approach is the fact that fuzzy relations can manipulate semantic information which is carried by linguistic labels. The notion of semiotic descriptor helps in systematizing the kinds of linguistic variables used in fuzzy sets approaches to modeling.

One may ask why do we need semiotic descriptors in relational models. Indeed, conditions imposed on logic operations applied to fuzzy relational systems are strictly mathematically defined. Such mathematically given semantics of the logic of relations is necessary, but it is not sufficient to deal satisfactorily with the meaning of linguistic labels which carry the conceptual meaning of applications. It has to be supplemented by some semiotic notions that can be expressed as algebraic restrictions over the basic fuzzy relational system. The carrier of these semiotic notions is the system of semiotic descriptors. This leads us to the following definition.

Semiotic Fuzzy Knowledge Representation Structure consists of the pair of structures, namely < FRS, SD >. FRS is Fuzzy Relational Structure consisting of a family of fuzzy relations, and SD is collection of semiotic descriptors [28],[29]. Semiotic descriptors can be classified into the basic kinds. These kinds are briefly characterized in the Glossary that follows.

4.1.2 Glossary: Basic Kinds of Semiotic Descriptors.

Element: Any thing that forms a coherent whole capable of potential or actual being. We shall recognize four important kinds of elements: object, agent, property and situation.

Presentation: Element can appear as a *presentation* in form of a symbol in a text (as used in XML), or a cognitive image by another kind of element (agent) as underflow of stream of experience [21].

Object: An abstract or concrete thing (element) that is characterized by properties.

Composed object: An object composed of other elements mutually related (organized in some structure). Computer science object as it appears in object-oriented programming is a special kind of composed object in our terminology.

Atomic object: Object that is not composed, when viewed at a particular resolution level (level of granularity).

Property: Basic characteristic, a relationship between properties. Defines objects by intension. Composed property is defined by a logic expression that takes as its arguments atomic properties.

Attribute: Property (or a composed property) that is chosen according to a specific selection criterion.

Agent: A thing (element) that can take actions on objects and properties and is a meta-object with respect to objects and properties [30].

Subject: A thing (element) of kind *object* that can take actions on other objects [40].

Situation: It is an n-ary relation that connects a class of objects with some other kinds of elements.

Scenario: A sequence or aggregation of situations viewed from the perspective of a meta-agent.

Basic kinds of semiotic descriptors: Object, property, agent, situation are the basic kinds of semiotic descriptors. They can be used as elements in texts.
Activity: An activity is composed of actions. An activity is determined by the *repertory of actions*, and the classes of participants. Participants may be objects and/or agents.
Meta-Agent: A meta-agent forms a *scenario* using four basic kinds of semiotic descriptors: object, property, agent and situation. Meta-Activities are activities of meta-agents.

4.1.3 Forming a Scenario. In order to be able to form a scenario, a meta-agent has to be given a leading activity. There are several leading activities that can be applied to the quadruple of basic semiotic kinds. Each leading activity leads to a different scenario defining a different computational model.

The following are three important alternatives:

1. Select an object and investigate the interrelationship of its properties with respect to the views of individual agents observing the object.

2. Select a property (or a class of properties) and investigate the interaction of the interrelationship of views of individual agents with respect to the interrelationship of objects.

3. Select an agent and investigate the interrelationship of objects with respect to properties as viewed by that agent.

4.2 SEMIOTIC DESCRIPTORS IN COMPUTING WORDS

In order to see how the Activity Structures methodology is to be used we shall use some of its concepts to analyze an example from the domain of Computing with Words (CW).

Zadeh gives an example which shows how CW works. [43]:

> *A typical problem in CW is the following. Assume that a function f, Y=f(X), is described in words as: if X is small then Y is small; if X is medium then Y is large; if X is large then Y is small, where small, medium and large are labels of fuzzy sets. The question is: What are the maximum and maximizing values of Y and X respectively?*

In order to start reasoning with perceptions we need the initial premises, in Zadeh's terminology *collection of perceptions which constitute the initial data set (IDS)*. In case of the above example, the perceptions that characterize the function f are presented in words such as *if X is small then Y is small; if X is medium then Y is large; ...* etc. The constraint process that produces a solution terminates in a proposition or a collection of propositions which according to Zadeh *play the role of an answer to a query, that is, the terminal data set (TDS).*

This process was described in more general terms in the abstract of Zadeh's lecture [43]:

> *Canonical forms of propositions in TDS constitute the initial constraint set (ICS). The key part of the reasoning process is goal-directed propagation of generalized constraints from ICS to a terminal constraint set (TCS) which plays the role of the canonical form of TDS.*

Let us analyze Zadeh's example in terms of ontology of Activity Structures, investigating which kinds of semiotic descriptors are involved. Function f is an object, the labels of fuzzy sets *small, medium, large* denote properties. The question *What are the maximum and maximizing values of Y and X respectively?* imposes further constraints of the aggregation of properties. These constraints filter out some potential candidates for solution defining the required properties of the computation that is the acceptable solution. This restriction is an *intension* in logic terms. But we require that an acceptable solution is an object formed by the restriction of the values of the initial object, namely the function f. This may or may not exist, but even if it does not exist, an acceptable approximation, close enough in terms of its properties to the required solution may be available.

4.3 SEMIOTIC DESCRIPTORS, GRANULARITY AND MEANING

The main goal of semiosis is interpretation of signs and symbols – recognition of their meaning. The meaning interpretation unit is a very complicated one. The meaning of something can be different depending on the scale which is used for representation. This makes the notion of granularity a very important one. Levels of granularity also referred to as levels of resolution are closely related to generalization and abstraction.

Interpretation and meaning depend on context. A process considered with all its details cannot be properly understood unless the details irrelevant within a particular context or perspective are brushed away. But this interpretation will often be incomplete. With incompleteness at work, there may be more than one interpretation. Indeed, a whole family of interpretations may be possible, some of which may be conflicting. Fuzzy sets, relations and logic can play an important role here: they allow us, through the theory of potentiality (or virtual plurality) to deal with the whole family of virtual outcomes, and also to measure the degree of conflict of individual members of some possibilistic family of outcomes produced by the meta-process of interpretation.

5. A BRIEF OVERVIEW OF BK-PRODUCTS

5.1 GENERAL CHARACTERISTICS OF BK-PRODUCTS OF RELATIONS

In our approach, a significant emphasis is put on relational representation of computational structures and simulation models, because this approach makes it possible to perform all the computations and decision making in a uniform relational way [35]. The indispensable unifying constructs are provided by means of *special relational compositions* called triangle and square products. These were first introduced by Bandler and Kohout in 1977, and are referred to as the BK-products in the literature [13],[14],[16]. See the survey in [35] with a list of 50 selected references on the theory and applications. The most recent overview is [33].

There are several types of product used to produce product-relations [5],[9], [35].

Definition 1 *For arbitrary fuzzy relations in [0, 1], R from the set X to Y, S from Y to Z define:*
1. $R \circ S = (\forall x)(\forall z)(\exists y)(xRy \ \& \ ySz)$;
2. $R \lhd S = (\forall x)(\forall z)(\forall y)(xRy \to ySz)$;
3. $R \rhd S = (\forall x)(\forall z)(\forall y)(xRy \leftarrow ySz)$;
4. $R \Box S = (\forall x)(\forall z)(\forall y)(xRy \equiv ySz)$
Only the conventional $\circ$ is associative. $\Box$ is not associative at all, and the following pseudo-associativities hold: [6]:
1. $Q \lhd (R \rhd S) = (Q \lhd R) \rhd S$,
2. $Q \lhd (R \lhd S) = (Q \circ R) \lhd S$
3. $Q \rhd (R \rhd S) = Q \rhd (R \circ S)$.

On the abstract side of relational algebras (RA), Tarski and his school have investigated the interrelationship of various RAs. Namely, representable (RRA), semiassociative (SA), weakly associative (WA) and non-associative (NA) relational algebras. Maddux [41] gives the following result: RRA $\subset$ RA $\subset$ SA $\subset$ WA $\subset$ NA. These results do not say anything about representations of non-associative products in these extended relational algebras.

The BK-products defined over relational calculi give the constructive realization of such non-associative products. For further details see papers of Bandler and Kohout [1],[2],[4] which predate the papers by Maddux. Hence, non-associative products have representations in crisp as well as fuzzy relational calculi. Applications of BK-products demonstrate that these non-associative products offer various computational advantages. For example, the following **universal representation of preorders** is given for all the crisp relations that are in $\mathcal{B}(X) = \mathcal{B}(X \rightsquigarrow X)$:

Theorem 2 *[11],[32] (a) R is a preorder if and only if* $R = R \triangleright R^{-1}$.
(b) Every preorder on relations can be expresses that way.
(c) $R = R \square R^{-1}$ *if and only if R is an equivalence.*

$\triangleleft, \triangleright, \square$ products add the expressive power to the mathematics of relations. Very important for distributed knowledge networking is a generalization of conventional homomorphisms defined constructively by BK-products:

Definition 3 *Let* F, R, G, S *be the relations between the sets* A, B, C, D *such that* $R \in \mathcal{R}(A \rightsquigarrow B)$. *The conditions that (for all* $a \in A, b \in B, c \in C, d \in D$*)* aFc *and* aRb *and* bGd *imply* cSd*, will be expressed in any of the following ways: (i)* $FRG; S$ *are* **forward compatible** *(ii)* F, G *are* **generalized homomorphisms** *from R to S.*

Theorem 4 *Compatibility [6] 1.* $FRG; S$ *are forward compatible if and only if* $F^T \circ R \circ G \sqsubseteq S$.
2. Formulas for computing the explicit compatibility criteria for F and G are: $FRG; S$ *are forward-compatible iff* $F \sqsubseteq R \triangleleft (G \triangleleft S^T)$

Similarly, the *backward compatibility* is defined and constructive conditions for relations both-way compatible (i.e. forward *and* backward) are given in [6]. *Both-ways* compatibility subsumes the conventional homomorphisms.

5.1.1 Many-Valued Logic Based Extensions of Relations.

The BK-products can be expressed in many-valued logics in the following way:

PRODUCT TYPE	MANY-VALUED LOGIC FORMULA
Circle product:	$(R \circ S)_{ik} = \bigvee_j (R_{ij} \bigwedge S_{jk})$
Triangle Subproduct:	$(R \triangleleft S)_{ik} = \bigwedge_j (R_{ij} \rightarrow S_{jk})$
Triangle Superproduct:	$(R \triangleright S)_{ik} = \bigwedge_j (R_{ij} \leftarrow S_{jk})$
Square product:	$(R \square S)_{ik} = \bigwedge_j (R_{ij} \equiv S_{jk})$

By choosing appropriate many-valued logic (MVL) operations for the logic connectives in the above definitions of products, the Boolean (crisp) case extends to a wider variety of many-valued logic based (fuzzy) relational systems. The details of choice of the appropriate many-valued connectives are discussed in [5], [7],[8],[35],[38].

The above definitions and theorems hold in crisp relation calculi as well as in fuzzy systems based on t-norms and residuated implication operators. Thus the unification and extension of the technical apparatus of relations from crisp to fuzzy, including a substantial generalization of BK-products can be fully developed in predicate logic BL (Basic Logic) of Hájek [18]. The BL axiomatization puts on a firm logical footing the whole family of t-norm [25] based fuzzy logics and provides a rigorous basis for extending the theory of generalized morphisms using BK-products to the infinite family of t-norm fuzzy logics that can be axiomatized by BL systems [31]. Furthermore, Kohout has recently demonstrated [32] that the BK-product based theorems hold even in the general monoidal fuzzy logics of Höhle [20], which are more general than BL systems of Hájek.

5.2 USE OF BK-PRODUCTS FOR ELICITING, REPRESENTING AND ANALYSIS OF COGNITIVE STRUCTURES OF HUMAN EXPERTS

When we want to employ the Computational Theory of Perceptions in solving practical problems we have to elicit chains-of-thought of experts [12],[22] and represent these relationally. First we perform exploratory knowledge elicitation [34] to select the relevant semiotic descriptors. We use repertory grids to elicit the meaning as used by human experts [23], [24] Once the relevant semiotic descriptors are identified, relationships between them can be captured by repertory grids using bi-polar constructs, each construct consisting of a pair of semiotic descriptors [5]. After the repertory grids are applied to a particular problem, and the semantic relationship captured by the grids, the grids are transformed into fuzzy relations which relate the semiotic descriptors.

Relational methods of analysis are then applied in order to discover meaningful conceptual structures implicit in these fuzzy relations. This is done by forming fuzzy relational products and further processing thus obtained composed relations by algorithms computing closures and interiors of fuzzy relations [10].

Particularly useful structural relationships are equivalences, similarities and preorders on relations holding between individual semiotic descriptors. These structural properties relating the meaning of concepts intrinsically contained in the data captured by the repertory grids must be made explicit by appropriate relational computations. The computational algorithms for this purpose are based on BK-Products of fuzzy relations and Fast Fuzzy Relational Algorithms [10],[36].

6. A METHODOLOGY FOR ELICITATION OF RELATIONAL KNOWLEDGE STRUCTURES

6.1 DESCRIPTION OF THE GOALS AND TOOLS

6.1.1 Knowledge Elicitation. *Knowledge Elicitation* is extraction of expert's declarative and procedural knowledge. The *knowledge elicitation* process is concerned with identification and formation of domain concepts and structural domain knowledge. Various methods can be used to abstract internal and causal knowledge possessed by experts.

In our study, we use a method of *Repertory Grid* techniques to elicit the initial information that is further processed by relational computations [5],[23].

6.1.2 Repertory Grid (RG). The *Repertory Grid* is a tool for eliciting a relation between some entities. It refers to three major components: *objects*, *attributes* and *respondents.* **Objects** consist of a name and a physical/functional description of items or elements, or things.
Attributes (properties, or discriminants) characterize objects in some way. For example, in Urban Studies models [42], objects are the places in their urban environment, where the inhabitants live, go, eat, etc. and attributes are some psychological constructs attached to places. In a medical system, objects may be diseases and attributes may be signs or symptoms [7]. In a propulsion system, its components may be objects; factors which impact the cost in each process may be represented as attributes [23] Attributes are made up of two opposing concepts called poles; the emphasized, positive or favored pole is alleged to be the primary pole.
Respondents (or participants) are, for example, the set of individual inhabitants of the urban area whose opinion is elicited by the repertory grid, medical doctors assessing patients or technologists involved in a high technology design.
An individual respondent is presented with a list of objects (elements) and a list of attributes (e.g.,constructs or properties) in a repertory grid form. Then, the person is asked to apply those properties to the objects using a graded scale expressing the degree to which the relation between objects and attributes holds (semantic hedges). Such degree not only describes the level of emphasis placed on a pole but also which pole is being stressed. Thus, the objects are evaluated and an opinion is formed.

6.1.3 Structure Formation. Given a set of attributes (or objects), we seek to identify their relationships implicit on data, pertinent to a particular perspective of our concerns. It falls upon fuzzy relational theory to approximate the connection of attributes of objects by extracting them out of the knowledge of how the respondents apply the attributes to the objects. The attributes and their connection form a relational structure together.

Structure formation consists of two major steps:

- formation of an attribute-attribute (or object-object) composed relation from the summary table of a repertory grid by a BK-product and

- extraction of an ordering from the relation, forming a graphical hierarchy.

6.1.4 Forming the Composite Relation by the $\triangleleft$ BK-product. The values in the repertory grids are normalized over the values [0, 1] in the many-valued binary relation. The normalized grid yields the fuzzy relational matrix, say R, to be further processed.

Let R be a relation from X to Y where X is the set of *Properties* and Y is the set of *Objects*. Then, R_{xy} is the degree to which a respondent assigns property x to object y. On the other hand, R^T is a relation Y to X where $R^T_{yx} = R_{xy}$ is the degree to which a respondent assigns to object y property x. By composing the normalized matrix R with its transpose R^T, the fuzzy relational triangle subproduct $R \triangleleft R^T$ yields a property-property relation over objects by applying a fuzzy implication operator. $(R \triangleleft R^T)_{x_i x_j}$ gives the degree to which property x_i implies property x_j based on how a respondent applied both properties to the objects. Analogically, the subproduct $R^T \triangleleft R$ yields the object-object relation.

6.1.5 Extracting an Oder. Fast Fuzzy Relational algorithms [10] are used to test relational properties and extract local (and global) equivalences, preorders and other mutual dependencies between objects or properties.

It is essential for a relation to have the properties of reflexivity, transitivity, antisymmmetry to be an ordering. We compute local preorder closure of a relation, say R, which is the least inclusive relation which has the property of reflexivity and transitivity, and includes R. And, we compute the degree of approximation of the original relation to the closure. When the degree is sufficiently high, the closure is taken as the the relation used in further analysis.

Alpha(α)-cuts By taking proper α-cuts, crisp structures are extracted from fuzzy relations. These α-cuts are used for additional analysis and interpretation in a variety of crisp perspectives. So an α-cut may be viewed as a filtering process and crisp examination of a fuzzy relation.

Hasse Diagram (HD) The order relationships of properties are represented in the use of HD and equivalence classes. The graphical representation of the implicit hierarchy on the crisp matrix is the HD which shows ordering. The properties hierarchy gives us insight into the opinions of a respondent's view of her/his modeling system. See Fig. 1 for an example of Hasse Diagram.

7. CASE STUDY

The relational model presented in this case study is intended to capture and represent the features relevant to the affordability analysis of engineering design and cost prediction. Each kind of semiotic descriptor has a specific use in our case study taken from the domain of industrial engineering. The general framework for defining semiotic descriptors consists of the conceptual categories described in the next section. This is followed in Sec. 7.2 and 7.3 by the description of the specific details of the case study.

7.1 CONCEPTUAL CATEGORIES OF SEMIOTIC DESCRIPTORS USED IN THE CASE STUDY

Here we explain the general meaning of each kind of semiotic descriptor. More specific meaning is induced by the context of the engineering application — a case study described in the sections that follow.

Each conceptual category listed below represents a specific level of granularity. Each level of granularity may possess conceptual refinements that are interlinked with other granular structures. Furthermore, each conceptual category has specific meaning that can be understood and interpreted by a domain expert.

Objects are either parts of a manufactured product, entire manufactured products or whole technologies, depending on the resolution level.

Attributes are characterized by *linguistic descriptors.* Examples of these are: low_raw_material_cost, small_processing_windows, high_temperature, good_lubricity, etc. Attributes, however, can also be characterized by *measurable* physical or fiscal parameters, such as: temperature, lubricity, potential_investment, cost, etc.

Agents. In this case study, agents are the observers (e.g.,engineers or accountants) assessing the degree to which an attribute is possessed by an object [23].

The repertory grid has been used to capture the evaluation of an LPT cover plate by engineers assessing to what degree various attributes listed in the grid can be assigned to the LPT plate.

Perspectives. An object or a family of objects can be evaluated within different perspectives. For example, an LPT cover plate can be evaluated from the perspective of an engineer, or from the perspective of a business analyst performing value analysis of the part, or from the perspective of an accountant. Each perspective may employ attributes that are different from the attributes of a different perspective for the same object. Some attributes may, however, be shared by different perspectives.

Contexts. Each object or family of objects can appear in several different contexts. For example an LPT cover plate may appear e.g.,in context of ingot process, forging process, extrusion process, or other processes. Here we present the results obtained in the context of the forging manufacturing process.

Views. Even in one particular perspective or context, different experts may assess the objects and situations in which objects appear differently. The views, although different, may have some features in common. These features that are shared by all or the majority of respondents are captured structurally by the composed relation over constructs (e.g.,cost drivers) contained in the repertory grid. They represent the consensus of the respondents. In our case study described here, we capture the consensus of four experts that is expressed by the 1-tautologies of Basic Logic [17] of t-norms and transform it into a set of rules.

These differences and agreements of views of different experts are captured by repertory grids and compared by relational methods using algorithms provided by TRISYS. TRISYS computations yield the subrelations that satisfy the prescribed relational properties. We seek to extract from the "train-of-thought" [23] of engineers the subrelations that have the property of *local preorder*. This follows from the minimal logic requirements imposed on the rules of inference in t-norm logics: the rules of inference have to possess fuzzy transitivity and (local) reflexivity. Because the quantifiers $\forall, \exists$ of Basic Logic BL are semantically defined as the *infimum* and the *supremum* respectively, we extract the subrelations that are *max-min* transitive.

7.2 DESCRIPTION OF DATA

The data used in our case study are one of the samples used in the development of new propulsion technologies. In the aeronautic industry, *early cost-driven decisions* are critical to low cost development and production of jet engines. For firms that are at the leading edge of high technology, it often involves determining the essential cost factors of a *design that has never before been manufactured.* This creates a formidable problem, because technological and business decisions are based on incomplete, uncertain information about the product that is to be designed and manufactured [15],[24].

7.2.1 Engineering Scenario: A Mechanical Component to be Evaluated. The structure has been elicited within the following scenario: A *Low Pressure Turbine (LPT) Cover Plate* is to be manufactured, using new material, namely, gamma titanium. Prior to any production characterization, the part is to be costed out, using the expert knowledge concerning manufacturing processes and available cost estimation that is available for other small gamma titanium parts. There are five main processing steps that will be involved in manufacturing the Low Pressure Turbine Cover Plate:

1. Initial ingot production
2. Extrusion
3. Forging
4. Heat treatment
5. Machining

In our study described in a series of papers (see [23] for list of references) we are using this LPT cover plate as the evaluated object and the above five steps provide the contexts. The forging process is better known than the extrusion or any other process listed above. However, there are some unknowns: the size is bigger than any similar component produced previously, and there is no production knowledge available for the component. If we ask several experts/engineers with prior experience in this matter to assess the situation by means of repertory grid, can we extract from this assessment some meaningful extrapolation? The experience of each expert is different due to different nature of their work, but partially overlapping. The results of the following section, however, show that there is a remarkable consensus on the dependencies of some constructs, despite different background experience of each respondent.

7.3 SEMANTICS OF RELATIONSHIPS OF THE MODEL OF LPT COVER PLATE

Our model relation M to be used in the sequel $M \in \mathcal{R}(C \times (O \times G) \times P)$, relates the following three lists (sets) of entities:

- the set C of process attributes;
- The Cartesian product $O \times G$ of the set O of observers with the set G of process identifiers;
- the set of Parts or Components.

Applying the usual selection and projection operators the ternary relation is decomposed into a family of 2-ary relations in $\mathcal{R}(C \times (O \times G))$, indexed by the set P. The relation R from the set of process attributes C to the set $(O \times G)$ is composed with its transpose R^T by means of *triangle subproduct* and the local preorder closure computed by the TRISYS system [10]. The result of this computation is a relation from the process attributes C to C. This relation shows the dependencies of process attributes represented as a preorder relation.

7.4 RULE EXTRACTION FROM THE HASSE DIAGRAM STRUCTURE OF FUZZY RELATIONS

7.4.1 Hasse Diagram of the Fuzzy Consensus of Experts.

TRISYS computations have identified subrelations that satisfy the property of reflexivity and max-min transitivity. That means that the resulting relation is a preorder as pictured by the Hasse diagram (HD) of Fig. 1. This Hasse Diagram shows the natural dependences of properties (constructs) of the forging process that capture the fuzzy consensus of the respondents.

The implication operators used to compute the Hasse diagram are indicated directly in the figure. Numerous fuzzy implication operators have been applied to the computation of fuzzy relational products, such as Gödel, Goguen, Łukasiewicz, Reichenbach, Kleene-Dienes, Early Zadeh, Willmott, etc., and the Hasse Diagram structures have been generated at the different levels of α-cuts.

From Fig. 1 we can see that the Hasse diagram is parameterized by fuzzy implication operator used for computation of triangle subproducts, α-values and by a label that indicates if the computed product is of the mean or harsh type. We can also see that some attributes form equivalence classes. For example, the process attributes {*C2* (Large Processing

Windows), *C5* (Air Furnace Atmosphere)} or {*C10* (Conservative L/D), *C19* (Acceptable Press Capability/Load Limits)} are in the same equivalence class. This indicates that these have equivalent (or closely similar) effect in characterizing the LPT cover plate.

The attributes which construct the Hasse Diagram structure are arranged in several levels through the structure. In Fig. 1, the attributes are arranged from the highest level 1 at the top to the lowest level 7 at the bottom.

From this HD structure, we may extract the rules of the inference. Since the HD structure is a preorder, the direct link between the attribute (C_j) which goes directly from the attribute(C_i) yields a rule $C_i \vdash C_j$, where C_j is a consequent which may be derived from an antecedent C_i when used for inference. Every rule may be represented as a simple rule of the form $C_i \vdash C_j$. An equivalent representation in a composed form reduces the number of rules. There are two different ways of representing rules:

- in the form having a single consequent and the disjunction of several antecedents: $C_{i_1} \vee C_{i_2} \vee \cdots \vee C_{i_m} \vdash C_j$;
- having a single antecedent and the conjunction of several consequents: $C_i \vdash C_{j_1} \wedge C_{j_2} \wedge \cdots \wedge C_{j_n}$.

We can also further combine these two forms in order to obtain rules with multiple antecedents and multiple consequents.

A remarkable feature of our results is the fact that despite of differences in the nature experience of individual respondents, and differences in the actual values of ratings entered by each expert into the repertory grid, there is absolute consensus of all experts with respect of some dependencies of the constructs. These are captured by the Hasse diagram at the α-cut at the value 1. This structure represents the *1-tautologies* of our relational model capturing the relationship of cost-drivers (constructs) in the forging process. These 1-tautologies of t-norm general logic BL have captured the total consensus of four engineers-experts that was contained in their responses elicited by the repertory grids.

7.5 EXAMPLES OF THE RULES

In this section, we show the rules extracted from the HD structure in Fig. 1. We show three different forms of rules discussed above.

Representation of Equivalence classes:

1. $\{C2, C5\} \Leftrightarrow (C2 \equiv C5)$
   $\Leftrightarrow (C2 \rightarrow C5) \wedge (C2 \leftarrow C5) \wedge (C2 \rightarrow C2) \wedge (C5 \rightarrow C5)$

Part : Gamma_Titanium based LPT Cover Plate
Process : Forging
Respondents : {R1, R2, R3, R5}
Grids : Forging.All.R1235.Arrange
Representation : (Fuzzy Logic Operator, Alpha-Cut, Criteria)

(S.H_only(=1).mean) = (S.H_only(=1).harsh)
= (S*.(H(=1),HU(=.94)).mean) = (S*.(H(=1),HU(=.75).harsh)
= (G43.(H(=1),HU(=.97)).mean) = (G43.(H(=1),HU(=.90)).harsh)
= (G43'.(H(=1),HU(=.97)).mean) = (G43'.(H(=1),HU(=.90)).harsh)
= (L.(H(=1),HU(=.99)).mean) = (L.(H(=1),HU(=.96),M(=.91)).harsh)

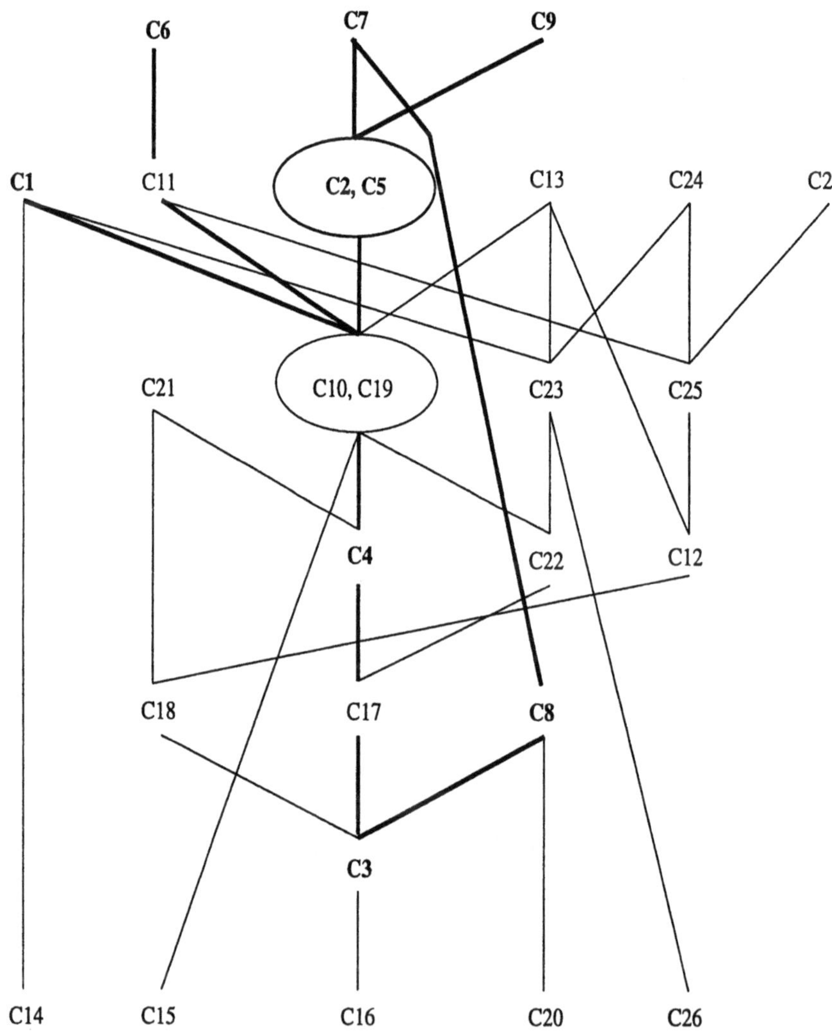

Figure 1.1

2. $\{C10, C19\} \Leftrightarrow (C10 \equiv C19) \Leftrightarrow$
$(C10 \rightarrow C19) \wedge (C10 \leftarrow C19) \wedge (C10 \rightarrow C10) \wedge (C19 \rightarrow C19)$

Representation of rules in the form of *disjunctions* of multi antecedents and a single consequent:
At the level i, $(Antecedent_1^{i+} \vee Ant_2^{i+} \vee \cdots \vee Ant_n^{i+}) \vdash Consequent^i$

1-3: $((C2 \equiv C5) \vee C8) \vdash C7$, $C11 \vdash C6$, $(C2 \equiv C5) \vdash C9$

4-9: $((C10 \equiv C19) \vee C23 \vee C12) \vdash C13$, $(C23 \vee C25) \vdash C24$,
$C25 \vdash C27$, $((C10 \equiv C19) \vee C25) \vdash C11$,
$((C10 \equiv C19) \vee C14 \vee C23) \vdash C1$, $(C10 \equiv C19) \vdash (C2 \equiv C5)$

10-12: $(C15 \vee C4 \vee C22) \vdash (C10 \equiv C19)$, $(C18 \vee C4) \vdash C21$, $(C22 \vee C26) \vdash C23$

13-15: simple rules
$C17 \vdash C4$, $C17 \vdash C22$, $C18 \vdash C12$

16-18: simple rules
$C3 \vdash C18$, $C3 \vdash C17$, $C3 \vdash C8$

19: $(C3 \vee C20) \vdash C8$

20: a simple rule: $C16 \vdash C3$

Representation of rules in the form of *conjunctions* of multi consequents and a single antecedent:
At the level i, $Antecedent^i \vdash (Consequent_1^{i-} \wedge Cons_2^{i-} \wedge \ldots \wedge Cons_m^{i-})$

1-2: $(C2 \equiv C5) \vdash (C7 \wedge C9)$, $C11 \vdash C6$

3-5: $(C10 \equiv C19) \vdash (C1 \wedge C11 \wedge (C2 \equiv C5) \wedge C13)$,
$C23 \vdash (C1 \wedge C13 \wedge C24)$, $C25 \vdash (C11 \wedge C24 \wedge C27)$

6-8: $C4 \vdash (C21 \wedge (C10 \equiv C19))$, $C12 \vdash (C13 \wedge C25)$,
$C22 \vdash ((C10 \equiv C19) \wedge C23)$

9-11: $C17 \vdash (C4 \wedge C22)$, $C18 \vdash (C21 \wedge C22)$, $C8 \vdash C7$

12: $C3 \vdash (C18 \wedge C17 \wedge C8)$

13-17: simple rules:
$C14 \vdash C1$, $C15 \vdash (C10 \equiv C19)$, $C16 \vdash C3$, $C20 \vdash C8$,
$C26 \vdash C23$,

In the next step, we try to reduce simple rule representation from the above two sets of rules, and to refine them in the combined forms of disjunction and of conjunction for the compactness of the number of rules.

The combined form of rules using multiple antecedents and consequents is presented below. At the level i, $(Antecedent_1^{i+} \vee Ant_2^{i+} \vee \ldots \vee Ant_n^{i+}) \vdash Consequent^i$ or $Antecedent^i \vdash (Consequent_1^{i-} \wedge Cons_2^{i-} \wedge \ldots \wedge Cons_m^{i-})$

1. $(C2 \equiv C5) \vdash (C7 \wedge C9)$
2. $(C10 \equiv C19) \vdash (C1 \wedge C11 \wedge (C2 \equiv C5) \wedge C13)$
3. $C23 \vdash (C1 \wedge C13 \wedge C24)$
4. $C25 \vdash (C11 \wedge C24 \wedge C27)$
5. $C4 \vdash (C21 \wedge (C10 \equiv C19))$
6. $C12 \vdash (C13 \wedge C25)$
7. $C22 \vdash ((C10 \equiv C19) \wedge C23)$
8. $C17 \vdash (C4 \wedge C22)$
9. $C18 \vdash (C21 \wedge C22)$
10. $C3 \vdash (C18 \wedge C17 \wedge C8)$
11. $((C2 \equiv C5) \vee C8) \vdash C7$
12. $((C10 \equiv C19) \vee C14 \vee C23) \vdash C1$
13. $(C15 \vee C4 \vee C22) \vdash (C10 \equiv C19)$
14. $(C22 \vee C26) \vdash C23$
15. $(C3 \vee C20) \vdash C8$
16. the simple rules which can not be reduced:
    $C11 \vdash C6, \quad C11 \vdash C6$

7.6 THE SEMANTIC MEANING AND KNOWLEDGE DOMAIN INTERPRETATION OF THE ACQUIRED FUZZY RULES

7.6.1 The Surface and Deep Structure of the Rules. In our application, the rule is a relation between constructs.

Some information has been eliminated during the formation process, in the reduction step. This information, however, provides the semantic context for the interpretation of the rules.

The rule is obtained from the formula for harsh BK-subproduct

$$c_i(R \lhd R^T)c_k \Leftrightarrow (c_i R \subseteq R^T c_k) \Leftrightarrow (\forall o_j)(c_i R o_j \rightarrow o_j R^T c_k)$$

The Hasse diagram at $\alpha = 1$ yields all 1-tautologies of the structure. This follows from the evaluation of the $\lhd$ that is performed in two steps:

1. Compare the construct c_i over the element o_j of the reduction set A (agents/observers/engineers in this case study) by means of an implication operator, obtaining the implicant of the form $c_i R o_j \rightarrow o_j R^T c_k$.

2. Quantify over all the implicants by $(\forall o_j)$.

For any t-norm, its adjoint residual $\rightarrow$ (1) gives the truth value with respect to the ordering values of the scale on which the constructs are assessed. The truth values are assigned according to the following criteria:

1.1 $x \rightarrow y = 1$ if $x \leq y$,

1.2 $x \rightarrow y = f(x, y)$ if $x > y$.

The value $f(x, y) \in [0, 1]$ depends on the selected implication operator. This value represents the fuzzy degree of truth reflecting the contribution of the arguments of f when **1.2** above holds. The larger f is, the smaller the difference $1-(x-y)$ is. When $x = y$, the value of f is 1 (crisp truth). When $x = 1$ and $y = 0$, f has the value 0 (crisp false). Hence, the degree of truth of [1.2] is anti-monotonic with respect to the difference $x - y$ when $x > y$.

It is obvious that the relation R represented by the Hasse diagram, when taken at the alpha-cut $\alpha = 1$ represents the 1-tautologies of BL (Basic Logic system of Hájek [17]). This is important because BL is the axiomatic core of all of the t-norm based residuated fuzzy logics.

The tautologies extracted from the repertory grids capture the consensus of all observers-engineers about the interrelationship of the values of constructs C characterizing the observed object — the LPT cover plate. Because the constructs relevant to the affordability studies are cost drivers, the relational model viewed from the engineering perspective represents the inequality constraints on the values of the cost drivers (constructs).

7.6.2 Minimal Representation. Computing the Hasse diagram from the relation $Q = \text{preclo}(R \lhd R^T)$ reduces the number of rules that have to be stored in the computer memory. It is the minimal representation of the preorder as far as the space needed for storing of the rules is concerned. All other rules implicitly contained in the Hasse structure can be generated from this minimal representation containing just simple rules.

A *simple rule* (or atomic rule) is of the form $c_i \vdash c_j$. The Hasse diagram of Fig. 1 contains 31 simple rules, but the full directed graph of the preorder closure relation Q contains 164 simple rules. Let us call the set of simple rules obtained form the Hasse diagram of the preorder closure relation Q the *Hasse base* of the rules determined by Q, or briefly the Hasse base of Q. If we use only the Hasse base of Q, we can generate the remaining simple rules by the meta-rule

$$\frac{c_i \vdash c_j, c_j \vdash c_k}{c_i \vdash c_k}$$

This meta-rule obviously follows from the preorder properties of R.

Acknowledgments

This work has been partially supported by the NSF grants DMI-972 6027, *"Decision-Making with Incomplete Information in an Integrated Product and Process Development Enterprise – A Management Decision Tool for Cost Modeling and Affordability Analysis"*

Notes

1. Section 3 is based on excerpts from Bandler, W. and Kohout, L.J., *Modeling for Computer Scientists*, Lecture Notes, Brunel University, U.K., 1983

References

[1] W. Bandler and L.J. Kohout. *Mathematical Relations, their Products and Generalized Morphisms.* Tech. report, Man-Machine Systems Laboratory, EES-MMS-REL 77-3, Dept. of Electrical Eng., University of Essex, Colchester, Essex, U.K., 1977. Reprinted as Ch. 2 in: Kohout, L.J. & Bandler, W., Survey of Fuzzy and Crisp Relations, Lect. Notes in Fuzzy Mathematics and Computer Sci., Creighton Univ. Omaha (to appear in 2000).

[2] W. Bandler and L.J. Kohout. Fuzzy relational products and fuzzy implication operators. In *International Workshop on Fuzzy Reasoning Theory and Applications*, London, September 1978. Queen Mary College, University of London.

[3] W. Bandler and L.J. Kohout. Activity structures and their protection. In R.F. Ericson, editor, *Improving the Human Condition: Quality and Stability in Social Systems (Proc. Silver Anniversary 1979 International Meeting of the Society for General Systems Research).*, pages 239–244, Louisville KY, 1979. Society for General Systems Research, Distributed by Springer Verlag, New York.

[4] W. Bandler and L.J. Kohout. Fuzzy relational products as a tool for analysis and synthesis of the behavior of complex natural and artificial systems. In P.P. Wang and S.K. Chang, editors, *Fuzzy Sets: Theory and Applications to Policy Analysis and Information Systems*, pages 341–367. Plenum press, New York and London, 1980.

[5] W. Bandler and L.J. Kohout. Semantics of implication operators and fuzzy relational products. *Internat. Journal of Man-Machine Studies*, 12:89–116, 1980. Reprinted in Mamdani, E.H. and Gaines, B.R. eds. *Fuzzy Reasoning and its Applications*. Academic Press, London, 1981, pages 219-246.

[6] W. Bandler and L.J. Kohout. On the general theory of relational morphisms. *International Journal of General Systems*, 13:47–66, 1986.

[7] W. Bandler and L.J. Kohout. A survey of fuzzy relational products in their applicability to medicine and clinical psychology. In L.J. Kohout and W. Bandler, editors, *Knowledge Representation in Medicine and Clinical Behavioral Science*, pages 107–118. an Abacus Book, Gordon and Breach Publ., London and New York, 1986.

[8] W. Bandler and L.J. Kohout. Fuzzy implication operators. In M.G. Singh, editor, *Systems and Control Encyclopedia*, pages 1806–1810. Pergamon Press, Oxford, 1987.

[9] W. Bandler and L.J. Kohout. Relations, mathematical. In M.G. Singh, editor, *Systems and Control Encyclopedia*, pages 4000 – 4008. Pergamon Press, Oxford, 1987.

[10] W. Bandler and L.J. Kohout. Special properties, closures and interiors of crisp and fuzzy relations. *Fuzzy Sets and Systems*, 26(3):317–332, 1988.

[11] W. Bandler and L.J. Kohout. On the universality of the triangle superproduct and the square product of relations. *Internat. Journal of General Systems*, 25(4):399–403, 1997.

[12] W. Bandler, V. Mancini, and L.J. Kohout. Investigating the structures of knowledge and thought. Presented at ICSRIC'88 (4th International Conference on Systems Research, Informatics and Cyber-

netics, Baden Baden, Germany, 15-21 August 1988). In: *Proceedings* thereof.

[13] B. De Baets and E. Kerre. Fuzzy relational compositions. *Fuzzy Sets and Systems*, 60(1):109–120, 1993.

[14] B. De Baets and E. Kerre. A revision of Bandler-Kohout composition of relations. *Mathematica Pannonica*, 4:59–78, 1993.

[15] B. Dubrosky, L.J. Kohout, R.M. Walker, E. Kim, and H.P. Wang. Use of fuzzy relations for advanced technology cost modeling and affordability decisions. In *35th AIAA Aerospace Sciences Meeting and Exhibit (Reno, Nevada, January 6-9,1997)*, pages 1–12, Reno, NV, January 1997. AIAA, the American Institute of Aeronautics and Astronautics. Paper AIAA 97-0079.

[16] P. Hájek. A remark on Bandler-Kohout products of relations. *Internat. Journal of General Systems*, 25(2):165–166, 1996.

[17] P. Hájek. *Metamathematics of Fuzzy Logics.* Kluwer, Dordrecht, 1998.

[18] P. Hájek and L. Godo. Deductive systems of fuzzy logic. In R. Mesiar and B. Riečan, editors, *Tatra Mountains, Mathematical Publications (Special Issue: Fuzzy Structures - Current Trends)*, pages 35–68. Mathematical Institute, Slovak Academy of Sciences, Bratislava, 1997.

[19] L.O. Hall. Some comments on, and an extension to Activity Structures for intelligent systems. *Journal of Fuzzy Logic and Intelligent Systems*, 3(1):23–28, 1993. A Special Issue in Honor of Wyllis Bandler, edited by L.A. Zadeh et al.

[20] U. Höhle and E.P. Klement. *Non-Classical Logics and their Applications to Fuzzy Subsets: A Handbook of Mathematical Foundations of Fuzzy Sets.* Kluwer, Boston, 1995.

[21] E. Husserl. *Ideas.* Collier Macmillan, London, 1962.

[22] B. Juliano and W. Bandler. *Tracing Chains-of-Thought: Fuzzy Methods in Cognitive Diagnosis.* Physica Verlag. A Springer-Verlag Company, Heidelberg, 1996.

[23] E. Kim and L.J. Kohout. A new tool for comparative evaluation of fuzzy implications, t-norms and co-norms based on generalized morphisms. *Fuzzy Sets and Systems*, 1999. In press.

[24] E. Kim, L.J. Kohout, B. Dubrosky, and W. Bandler. Use of fuzzy relations for affordability decisions in high technology. In R.A. Adey, G. Rzevski, and A.K. Sunol, editors, *Applications of Artificial Intelligence in Engineering XI.* Computational Mechanics Publications, Southampton, UK - Bilerica, MA, USA, 1996.

[25] E.P. Klement and R. Mesiar. Triangular norms. In R. Mesiar and B. Riečan, editors, *Tatra Mountains, Mathematical Publications (Special Issue: Fuzzy Structures - Current Trends)*, pages 169–194. Mathematical institute Slovak academy of sciences, Bratislava, 1997.

[26] L.J. Kohout. The functional hierarchies of the brain. In G.J. Klir, editor, *Applied General Systems Research: Recent Developments and Trends*, pages 531–544. Plenum Press, New York, 1978.

[27] L.J. Kohout. Activity structures as a tool for design of technological artifacts. *Systems and Cybernetics: An International Journal*, 18(1):27–34, 1987.

[28] L.J. Kohout. *A Perspective on Intelligent Systems: A Framework for Analysis and Design.* Chapman and Hall & Van Nostrand, London & New York, 1990. In 1991, received a prize from The International Institute for Advanced Studies in Systems Research: *"The best book of the year in AI Systems"*.

[29] L.J. Kohout. Activity Structures: A methodology for design of multi-environment and multi-context knowledge-based systems. In L.J. Kohout, J. Anderson, and W. Bandler, editors, *Knowledge-Based Systems for Multiple Environments*, chapter 5. Ashgate Publ. (Gower), Aldershot, U.K., 1992.

[30] L.J. Kohout. Dynamic sharing of cognitive structures by means of generalized morphisms. In P.P. Wang, editor, *Proc. of the 2nd Joint Conf. on Information Sciences (Sept. 28 - Oct. 1, Wrightsville Beach, NC, USA)*, pages 597–600. Duke University, 1995.

[31] L.J. Kohout. Generalized morphisms in BL-logics. In *Logic Colloquium 1998(The 1998 ASL European Summer Meeting, Prague).* Association for Symbolic Logic, August 9-15 1998. (Extended abstract presenting the main mathematical theorems, to appear in the Bulletin of ASL, 1999).

[32] L.J. Kohout. Theory and applications of non-associative products of relations. In T. Childers, editor, *The Logica Yearbook 1997*, chapter 16, pages 152–164. Filosofia (Publisher of the Institute of Philosophy, Czech Academy of Sciences), Prague, 1998. (ISBN 80-7007-109-5).

[33] L.J. Kohout. Boolean and fuzzy relations. In P.M. Pardalos and C.A. Floudas, editors, *The Encyclopedia of Optimization.* Kluwer, Boston, 1999. In press.

[34] L.J. Kohout, J. Anderson, and W. et al. Bandler. *Knowledge-Based Systems for Multiple Environments.* Ashgate Publ. (Gower), Aldershot, U.K., 1992. Awarded *"Outstanding Scholarly Contribution Award"* by the Systems Research Foundation in 1993.

[35] L.J. Kohout and W. Bandler. Fuzzy relational products in knowledge engineering. In V. Novák et al., editor, *Fuzzy Approach to Reasoning and Decision Making*, pages 51–66. Academia and Kluwer, Prague and Dordrecht, 1992.

[36] L.J. Kohout and W. Bandler. Use of fuzzy relations in knowledge representation, acquisition and processing. In L.A. Zadeh and J. Kacprzyk, editors, *Fuzzy Logic for the Management of Uncertainty*, pages 415–435. John Wiley, New York, 1992.

[37] L.J. Kohout and B.R. Gaines. Protection as a general systems problem. *Internat. Journal of General Systems*, 3:1–21, 1976.

[38] L.J. Kohout and M. Kallala. Choice of fuzzy optimal logics for pattern classifiers by means of measure analysis. In *Proc. of 7th International Congress on Cybernetics and Systems*, Imperial College, London, September 1987.

[39] L.J. Kohout and E. Kim. Granular relational computing with semiotic descriptors using BK-products of fuzzy relations. In P. Wang, Paul, editor, *Computing with Words*, chapter 4. John Wiley, New York, 1999. In press.

[40] L.J. Kohout and I Stabile. Logic and methodology of galen: Its relevance to contemporary systems science. In G.E. Lasker and R.R. Hough, editors, *Advances in Support Systems Research (Proc. of 2nd Internatinal Congress on Systems Research, Informatics and Cybernetics 1989)*, pages 271–276, 1990. ISBN 0-921836-06-6.

[41] R.D. Maddux. Some varieties containing relation algebras. *Transact. of the American Mathematical Society*, 272(2):501–526, 1982.

[42] V. Mancini, W. Bandler, and L.J. Kohout. Overview of the urbs project. In P.P. Wang, editor, *Proc. of the 2nd Joint Conf. on Information Sciences (Sept. 28 - Oct. 1, Wrightsville Beach, NC, USA)*, pages 131–133. Duke University, 1995. (In: Invited Session on Fuzzy Relational Modeling as a Tool for Improving Human Sustainable Environment - URBS Project).

[43] L.A. Zadeh. From computation with measurements to computation with perceptions: A paradigm shift. *Abstract of a lecture given at BISC, University of California, Berkeley on September 16th, 1999* ,

[44] L.A. Zadeh. Fuzzy Logic = Computing with Words. *IEEE Trans. on Fuzzy Systems*, 4(2):103–111, 1996.

[45] L.A. Zadeh. Toward a theory of fuzzy information granulation and its centrality in human reasoning and fuzzy logic. *Fuzzy Sets and Systems*, 90(2):111–127, 1997.

Chapter 9

DIFFERENT PROPOSALS TO IMPROVE THE ACCURACY OF FUZZY LINGUISTIC MODELING

Oscar Cordón, Francisco Herrera
Dept. of Computer Science and A. I. - University of Granada. Granada. Spain
{ocordon,herrera}@decsai.ugr.es

María José del Jesus
Dept. of Computer Science - University of Jaen. Jaen. Spain
mjjesus@ujaen.es

Pedro Villar
Dept. of Computer Languages and Systems - University of Vigo. Ourense. Spain
pvillar@uvigo.es

Igor Zwir
Dept. of Computer Science - University of Buenos Aires. Buenos Aires. Argentina
zwir@dc.uba.ar

Abstract

Nowadays, Linguistic Modeling is considered as one of the most important applications of Fuzzy Set Theory, along with Fuzzy Control. Linguistic models have the advantage of providing a human-readable description of the system modeled in the form of a set of linguistic rules. In this contribution, we will analyze several approaches to improve the accuracy of linguistic models while maintaining their descriptive power. All these approaches will share the common idea of improving the way in which the Fuzzy Rule-Based System performs interpolative reasoning by improving the cooperation between the rules in the linguistic model Knowledge Base.

Introduction

Nowadays, Linguistic Modeling (LM) is considered as one of the most important applications of Fuzzy Set Theory, along with Fuzzy Control. Linguistic models have the advantage of providing a human-readable description of the system modeled in the form of a set of linguistic rules [29], which is a desirable characteristic in many modeling problems. Unfortunately, their accuracy is sometimes not as high as desired when dealing with complex modeling problems, thus causing the designer to discard them and replace them by other kinds of more accurate but less interpretable models. This drawback is due to some problems related to the inflexibility of the concept of the linguistic variable, which is the one involved in the fuzzy rule structure.

In this contribution, we review several approaches to improve the accuracy of linguistic models while maintaining their descriptive power. All these approaches will share the common idea of improving the way in which the Fuzzy Rule-Based System (FRBS) performs interpolative reasoning by improving the cooperation between the rules in the linguistic model Knowledge Base (KB).

The rule cooperation may be induced in four different FRBS components, namely the Inference System (IS), the KB as a whole and both KB components in isolation, the Data Base (DB) and the Rule Base (RB). All of them will be analyzed. To be precise, we will deal with the following aspects:

- Genetic tuning of the membership functions.
- Simulated Annealing-based Learning of the DB from examples.
- Genetic selection of fuzzy rules.
- The Accurate Linguistic Modeling paradigm, based on a double-consequent linguistic rule generation and selection.
- The Hierarchical Accurate Linguistic Modeling paradigm, based on a hierarchical linguistic rule generation and selection.
- Cooperative Fuzzy Reasoning Methods for classification problems.

The behaviour of the first five methods in solving the real-world Spanish electrical distribution problem shown in the Appendix will be analyzed. On the other hand, the performance of the last one, the only dealing with classification problems, will be tested with the IRIS and

PIMA data sets. In every experiment, the same basic rule generation process will be considered, the Wang and Mendel's one (WM-method) [30]. Two variants of this method to deal with modeling and classification problems are also introduced in the Appendix.

In order to put this into effect, this chapter is set up as follows. In Section 1., the framework is presented, i.e., System Modeling with FRBSs, structure of linguistic models and problems of LM. Section 2. describes our proposals to improve the accuracy of LM, by presenting a short study of rule cooperation in FRBSs and a brief description of the different approaches. Sections 3., 4. and 5. present the specific proposals to induce cooperation from the DB, the RB and the KB respectively. On the other hand, our proposals to induce cooperation from the IS, including their own experiments, are introduced in Section 6. Finally, a summary of the contribution is presented in Section 7., and an Appendix describing the WM-method and the electrical problem used as benchmark is included.

1. FRAMEWORK

In this section, some preliminary concepts will be presented. First, System Modeling with FRBSs will be introduced and the two different existing approaches will be reviewed. The section will focus then on LM and the basic structure of two different kinds of linguistic models, for regression and classification problems, will be described. Finally, the problems of LM will be analyzed.

1.1 SYSTEM MODELING WITH FRBSS

One of the most important applications of FRBSs is *system modeling* [5, 28], which in this field may be considered as an approach used to model a system making use of a descriptive language based on Fuzzy Logic with fuzzy predicates [29]. In this kind of modeling we may usually find two contradictory requirements, the accuracy and the interpretability of the model obtained.

It is possible to distinguish between two types of modeling when working with FRBSs: *Linguistic Modeling* and *Fuzzy Modeling*, according to the fact that the main requirement is the interpretability or the accuracy of the model, respectively. The former is developed by means of descriptive Mamdani-type FRBSs, which use fuzzy rules composed of linguistic variables [34] that take values in a term set with a real-world meaning, thus the linguistic model consists of a set of linguistic descriptions regarding the behaviour of the system being modeled [29]. On the other hand, Fuzzy Modeling is put into effect by means of approximate Mamdani-type FRBSs [3, 5, 10], systems in which the fuzzy rules are composed of fuzzy predicates without a linguistic meaning, i.e., the variables forming the rules do not take as a value a linguistic term with a fuzzy set associated defining their meaning, but a fuzzy set directly.

Therefore, a *linguistic model* is a system description in the form of a linguistic rule set interpretable by human beings, which is a desirable characteristic in some problems.

1.2 STRUCTURE OF A LINGUISTIC MODEL

The basic structure of a linguistic model [31] is showed in Fig. 1.1.

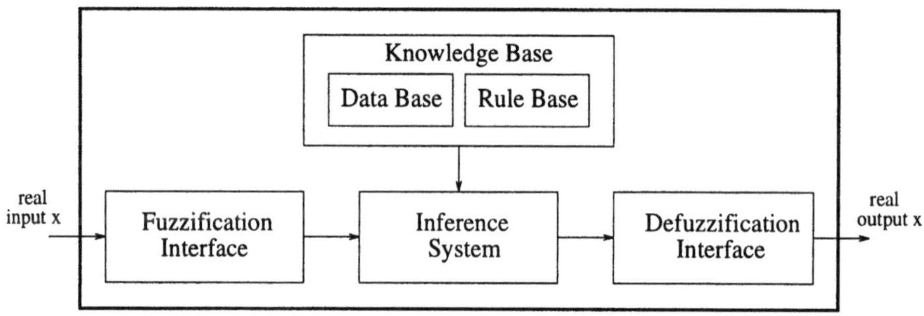

Figure 1.1 Structure of a linguistic model

The *Knowledge Base* (KB) is the component containing the knowledge about the system modeled in the form of linguistic rules. It is composed of two components:

- *Rule Base* (RB): Collection of linguistic rules:

$$R_i : \text{IF } x_1 \text{ is } A_{i1} \text{ and } \ldots \text{and } x_n \text{ is } A_{in} \text{ THEN } y \text{ is } B_i$$

  with x_j and y being linguistic system variables, and with A_{ij} and B_i being the linguistic labels associated with fuzzy sets specifying their meaning.

- *Data Base* (DB): Semantics of the linguistic labels (Fig. 1.2).

The *Fuzzification Interface* has the function of computing the firing degree of each single rule in the KB with respect to the current system input. This is done by computing the matching degree between the input and the rule antecedents, considering a conjunctive operator (a t-norm) when there is more than one input variable. The *Inference System* (IS) performs then the fuzzy reasoning process by applying the Compositional Rule of Inference [33] on each individual rule in the KB.

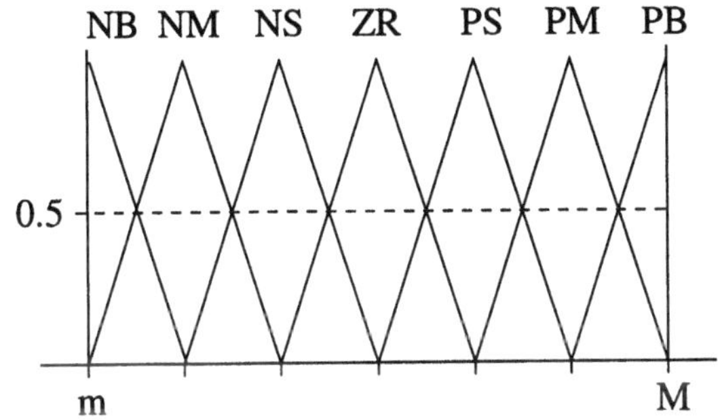

Figure 1.2 Example of semantics of the linguistic labels

The output obtained from the latter process is a number of fuzzy sets equal to the number of rules fired. The *Defuzzification Interface* works by aggregating these individual fuzzy sets in a single one and transforming it into a real number, the final output of the FRBS. For more information on the reasoning process, refer to [9].

In this contribution, the membership functions considered in the DB will always be triangular-shaped. The minimum t-norm will be used as conjuctive and implication operators, while the Center of Gravity weighted by the matching will be the defuzzification method considered [9].

We use a global error measure, the mean square error (MSE), as evaluation measure for our proposals. The MSE will allow us to determine the accuracy of the linguistic model obtained, which directly depends on the cooperation levels of the rules existing in the KB. The MSE over a training data set, E_p, is represented by the following expression:

$$F(C_j) = \frac{1}{2|E_p|} \sum_{e_l \in E_p} (ey^l - S(ex^l))^2$$

where $S(ex^l)$ is the output value obtained from the FRBS when the input variable values are $ex^l = (ex_1^l, \ldots, ex_n^l)$, and ey^l is the known desired value.

1.3 FUZZY RULE-BASED CLASSIFICATION SYSTEMS

In this section, a special type of linguistic models is introduced: Fuzzy Rule-Based Classification Systems (FRBCSs), the FRBSs used for classification problems.

The structure of an FRBCS is very similar to the one of an FRBS. In an FRBCS, two components are distinguished: 1) The KB, composed of

RB and DB as in every linguistic model, and 2) an a Fuzzy Reasoning Method (FRM), an inference procedure which derives conclusions from a fuzzy rule set and an example.

The FRBCS design implies finding both components, and this process is carried out through a supervised learning process, which starts with a set of correctly classified examples (training examples) and whose ultimate objective is to design a Classification System, assigning class labels to new examples with a minimum error. Finally, the system performance on the test data is computed, to gain an estimate about the FRBCS real error.

The composition of the DB is the usual one in LM. The main difference between an usual linguistic model and an FRBCS lies on the structure of the linguistic rule considered in the latter and, more concretely, on the form of the consequent of the fuzzy classification rule. Three different fuzzy classification rules have been proposed in the specialised literature with the consequent being: a class [2, 21], a class and a certainty degree associated to the classification of that class [24], and the certainty degree associated to the classification of each one of the possible classes [26].

In this work, we will consider FRBCSs composed of RBs of the former two types:

$$
\begin{aligned}
R_k &: \text{ If } x_1 \text{ is } A_1^k \text{ and } \ldots \text{ and } x_N \text{ is } A_N^k \text{ then } Y \text{ is } C_j \\
R_k &: \text{ If } x_1 \text{ is } A_1^k \text{ and } \ldots \text{ and } x_N \text{ is } A_N^k \text{ then } Y \text{ is } C_j \text{ with } r^k
\end{aligned}
$$

where:

- $x_1, \ldots, x_N$ are the selected features for the classification problem,
- $A_1^k, \ldots, A_N^k$ are linguistic labels used to discretise the continuous variable domain,
- Y is the class $C_j \in \{C_1, \ldots, C_M\}$ to which the example belongs, and
- r^k is the classification certainty degree in the class C_j for an example belonging to the fuzzy subspace defined by the rule antecedent.

Focusing on the FRM, the classical approach, called *maximum matching*, considers the rule with the highest association degree to make the final decision. This FRM classifies the pattern with the class of this rule. Graphically, this method could be seen as depicted in Fig. 1.3, where the rule R_k would show the highest association degree.

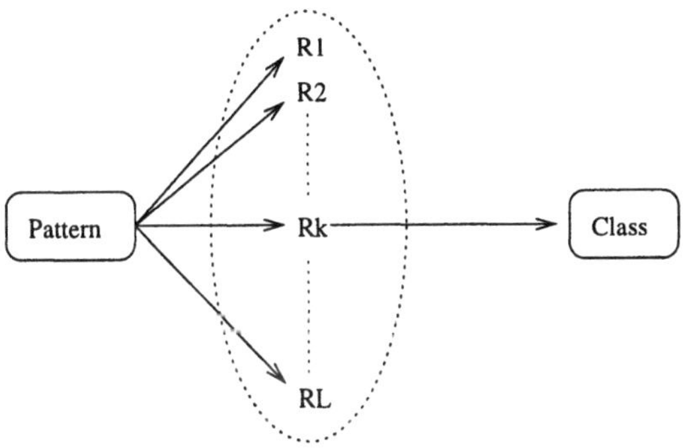

Figure 1.3 FRM that uses only the winner rule

1.4 PROBLEMS OF LINGUISTIC MODELING

As said, interpretability and accuracy are usually contradictory requeriments in System Modeling. Linguistic models present sometimes a lack of accuracy in complex modeling problems. As Zadeh pointed out in his *principle of incompatibility* [33], *"as the complexity of a system increases, our ability to make precise and yet significant statements about its behaviour diminishes ... "*. Thus, although the use of Fuzzy Logic-based techniques, specifically of FRBSs, allows us to deal with the modeling of systems in which a certain degree of imprecision is involved, building a linguistic model clearly interpretable by human beings, the accuracy obtained is not always as good as desired and we prefer a loss in the model description ability to obtain an improvement in the overall model performance. The choice between how interpretable and how accurate the model must be usually depends on the user's needs for the specific problem and will condition the kind of FRBS selected to generate it.

The lack of accuracy is due to some problems relating to the fuzzy rule structure considered which are a consequence of the inflexibility of the concept of linguistic variable. A brief summary of these problems is shown as follows [6, 7]:

- Lack of flexibility due to the rigid partitioning of the input and output spaces.
- The homogeneous partitioning of these spaces when the input-output mapping varies in complexity within the space is inefficient and does not scale to high-dimensional spaces.
- Dependent input variables are very hard to partition.

- Limitation on the size of the RB.

Hence, in many cases the linguistic model designed is not accurate to a sufficient degree and has to be discarded and replaced by other less interpretable but more accurate model. In this contribution, some proposals allowing us to improve the accuracy of linguistic models while maintaing their descriptive power will be introduced.

2. HOW TO IMPROVE THE ACCURACY OF LINGUISTIC MODELING

One of the most interesting features of an FRBS is the interpolative reasoning it develops. This characteristic plays a key role in the high performance of FRBSs and is a consequence of the *cooperation among the fuzzy rules composing the KB*. As is known, the output obtained from an FRBS is not usually due to a single fuzzy rule but to the cooperative action of several fuzzy rules that have been fired, because they match the input to the system to some degree (Fig. 1.4).

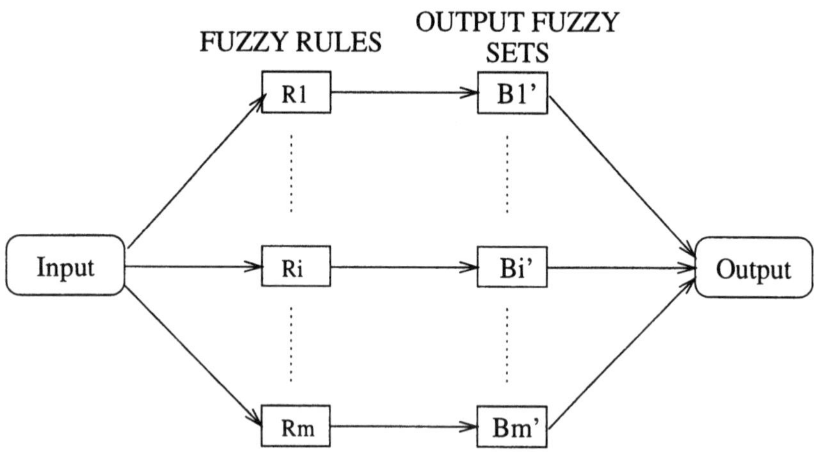

Figure 1.4 Cooperation among the fuzzy rules

Improving the cooperation among the fuzzy rules in the KB can be a good way to improve the accuracy of linguistic models. All our proposals will be based on this idea.

There are two components in an FRBS having a significant influence on the rule cooperation, the IS and the the KB. The accuracy of the FRBS can be increased, while its descriptive nature can be preserved, improving the cooperation among rules in the KB by dealing with four different aspects: the IS, the KB as a whole, and its two components, the DB and the RB, in isolation.

We propose six different approaches acting on these four different components, that are not isolated and can be combined among them:

- *Approaches to induce cooperation from the DB:*
  - Genetic tuning of the membership functions [10]
  - Simulated Annealing-based Learning of the fuzzy partition granularity [17]
- *Approaches to induce cooperation from the RB:*
  - Genetic selection of fuzzy rules [10, 19, 23]
  - Accurate Linguistic Modeling paradigm: Double-consequent fuzzy rule generation and selection [12, 13]
- *Approaches to induce cooperation from the RB:*
  - Hierarchical Accurate Linguisitc Modeling Paradigm [18]
- *Approaches to induce cooperation from the IS:*
  - Cooperative Fuzzy Reasoning Methods for Classification Problems [14]

In the next sections, all of these proposals will be analyzed in depth. The specific search procedures considered will not be introduced in the contribution with the aim of not extending it excessively. The reader can refer to [20, 27] and [1] for clear and wide descriptions on Genetic Algorithms (GAs) and Simulated Annealing (SA) respectively.

3. APPROACHES TO INDUCE COOPERATION FROM THE DATA BASE

In the last few years, many approaches have been presented to automatically learn the RB from numerical information (input-output data pairs representing the system behaviour). However, there is not much information about the way to derive the DB and most of these RB learning methods need of the existence of a previous definition for it.

A common way to proceed involves considering uniform fuzzy partitions with the same number of terms (usually an odd number between three and seven) for all the linguistic variables existing in the problem. Therefore, this operation mode makes the DB have a significant influence on the FRBS performance. This is why some approaches try to improve the preliminary DB definition considered once the RB have been derived. To put this into effect, a tuning process considering the whole KB obtained (the preliminary DB and the derived RB) is used a posteriori to adjust the membership function parameters. Our first proposal to improve the accuracy of LM is to do with this idea: given a complete KB, a *genetic tuning of the membership functions.* Nevertheless, the tuning process usually only adjusts the membership functions shapes and not the number of linguistic terms in each fuzzy partition, which remains fixed from the begining of the FRBS design process. Our second proposal has a different starting point, it is a DB learning method and tries to *learn an adequate fuzzy partition granularity for each linguistic variable* using Simulated Annealing.

3.1 GENETIC TUNING OF THE MEMBERSHIP FUNCTIONS

The genetic tuning process [10] is based on the existence of a previous complete KB, that is, an initial DB definition and an RB constituted by m fuzzy rules. The chromosomes only encode the primary fuzzy partitions constituting the DB in order to adjust the linguistic labels membership functions for all the fuzzy rules contained in the RB.

The GA designed presents a real coding issue that allows us to maintain the FRBS descriptive nature. Each chromosome encodes a different DB definition. A primary fuzzy partition is represented as an array composed by $3 \cdot N$ real values, with N being the number of terms forming the linguistic variable term set. The complete DB for a problem, in which m linguistic variables are involved, is encoded into a fixed length real coded chromosome C_r built by joining the partial representations of each one of the variable fuzzy partitions as is shown in the following:

$$(a_j, b_j, c_j) \rightarrow \text{ 3-tuple encoding of the fuzzy set } j$$
$$C_{ri} = (a_{i1}, b_{i1}, c_{i1}, \ldots, a_{iN_i}, b_{iN_i}, c_{iN_i}) \rightarrow \text{ fuzzy partition of the linguistic variable } i.$$
$$C_r = C_{r1}\ C_{r2}\ \ldots\ C_{rm} \rightarrow \text{whole DB definition.}$$

The initial gene pool is created making use of the initial DB definition. This one is encoded directly into a chromosome, denoted as C_1. The remaining individuals are generated by associating an interval of performance, $[c_h^l, c_h^r]$ to every gene c_h in C_1, $h = 1 \ldots \sum_{i=1}^{m} N_i \cdot 3$. Each interval of performance will be the interval of adjustment for the corresponding gene, $c_h \in [c_h^l, c_h^r]$.

If (t mod 3) $= 1$ then c_t is the left value of the support of a fuzzy number. The fuzzy number is defined by the three parameters (c_t, c_{t+1}, c_{t+2}) and the intervals of performance are the following:

$$c_t \in [c_t^l, c_t^r] = [c_t - \tfrac{c_{t+1} - c_t}{2}, c_t + \tfrac{c_{t+1} - c_t}{2}]$$

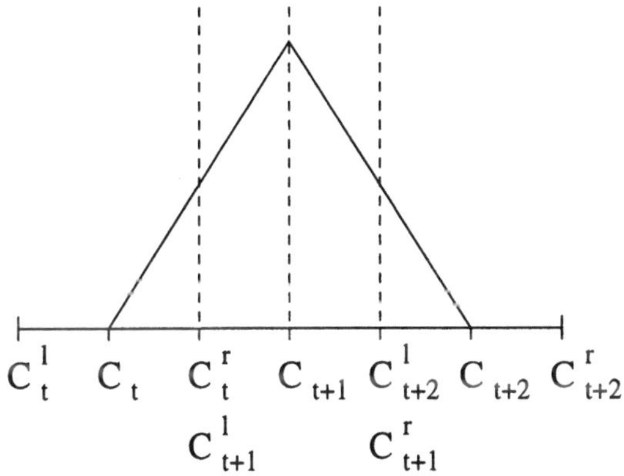

Figure 1.5 Membership function and intervals of performance for the tuning process

$$c_{t+1} \in [c^l_{t+1}, c^r_{t+1}] = [c_{t+1} - \frac{c_{t+1}-c_t}{2}, c_{t+1} + \frac{c_{t+2}-c_{t+1}}{2}]$$
$$c_{t+2} \in [c^l_{t+2}, c^r_{t+2}] = [c_{t+2} - \frac{c_{t+2}-c_{t+1}}{2}, c_{t+2} + \frac{c_{t+3}-c_{t+2}}{2}]$$

Fig. 1.5 shows these intervals. Therefore, we create a population of chromosomes containing C_1 as its first individual and the remaining ones initiated randomly, with each gene being in its respective interval of performance.

The GA designed uses the stochastic universal sampling as selection procedure together with an elitist scheme. The operators employed for performing the individual recombination and mutation are Michalewicz's non-uniform mutation [27] and the max-min-arithmetical crossover [22]. The MSE introduced in Section 1.2 plus a criterion penalyzing the lack of the completeness property compose the fitness function. Further information about this approach can be found in [10].

Table 1.1 collects the results of a brief experimentation where the KB obtained from the WM-method for the electrical problem shown in the Appendix is refined by means of the introduced tuning process. The large accuracy improvement can be clearly seen.

Table 1.1 Results obtained with the genetic tuning process

Method	*Granularity*	# Rules	$MSE_{training}$	MSE_{test}
WM-method	7 7 7	24	222622	240018
DB Tuning	7 7 7	24	144510	173167

3.2 LEARNING AN ADEQUATE FUZZY PARTITION GRANULARITY

As said, most of RB learning methods needs a previous definition of the DB, and the consideration of uniform fuzzy partitions with the same number of terms significatively affects the linguistic model accuracy. To solve this problem, we have developed a method to learn a good fuzzy partition granularity for a determinated problem [17]. We try to learn the number of linguistic terms for each variable, maintaining uniform fuzzy partitions. Since the exhaustive exploration of the search space is a very time consuming task, we consider the SA heuristic local search technique to perform the search.

In our case, given an RB generating method and an specific problem, each candidate solution is a concrete granularity level for each problem variable (number of labels), and the cost function is based on the MSE of the FRBS obtained with the WM-method using a DB with that granularity.

Three stopping criteria have been considered in order to reduce the run time of the procedure:

- The maximum number of iterations allowed without global improvement is reached.
- No solution was accepted in the last iteration.
- The maximum number of solutions have been generated.

It is interesting to point out that in all the runs done in [17] the procedure finished due to the first or second stopping criteria.

The basic operation mode of SA, adapted to our problem, is described in the next algorithm, with L being the number of possible values for the labels (seven in our case, $\{3, \ldots, 9\}$), with N being the number of problem variables, with α being the decreasing factor of the temperature, and with T_o, T being respectively the initial temperature, and the temperature in successive iterations.

SA (T_0, α, N, L):
  $T \leftarrow T_0$;
  $S_{act} \leftarrow Generate\_Initial\_Solution$;
  $S_{best} \leftarrow S_{act}$;
  while $(solutions \leq L^N)$ *and* $(iterations\_without\_improv. < N)$
   and *not*$(iteration\_without\_accepted\_solution)$ *do*
    begin
      $count \leftarrow 0$

$while\,(count < N^3)\ and\ (accepted\_solution\_number < N^2)\ do$
$begin$
$S_{cand} \leftarrow Generate\_Solution\_N(S_{act});$
$\delta \leftarrow cost(S_{cand}) - cost(S_{act});$
$if\ (U(0,1) < e^{(-\delta/T)})\ or\ (\delta < 0)$
$then\ S_{act} \leftarrow S_{cand};$
$if\ cost(S_{act}) < cost(S_{best})$
$then\ S_{best} \leftarrow S_{act};$
$count \leftarrow count + 1;$
$end;$
$T \leftarrow \alpha(T);$
$end;$
{Write as final solution, S_{best} }

The implementation of our SA procedure incorporates a taboo record of explored solutions, along with their cost, in order to eliminate the possibility of redundant executions of the RB generating method, with the consequent saving of run time. In fact, only 32 of the 59 solutions generated in the experiment developed in this contribution were evaluated. The results obtained are shown in Table 1.2, where the linguistic model generated by means of the granularity learning process can be compared with the one generated by the WM-method when considered the same number of labels (seven) for the three problem variables.

For more details about the SA procedure used and a wider experimentation, refer to [17].

Table 1.2 Results obtained with the SA-based fuzzy partition granularity learning

Method	*Granularity*	# Rules	$MSE_{training}$	MSE_{test}
WM-method	7 7 7	24	222622	240018
SA-based granularity learning	8 9 9	28	192980	230675

4. APPROACHES TO INDUCE COOPERATION FROM THE RULE BASE

In this section, two methods to improve the FRBS performance by increasing the cooperation among the rules belonging to the RB are proposed: a *genetic selection process of fuzzy rules* and the *Accurate Linguistic Modeling paradigm*, based on double-consequent fuzzy rules.

4.1 GENETIC SELECTION OF FUZZY RULES

The operation mode of many RB generation methods means that, in each input subspace, the rules are created individually from the examples in the input-output data set without taking into account the cooperation existing between them to give the final model output. That is, no information about the neighbour rules is considered in order to generate them. Because of this, the generated RB may present redundant or unnecessary rules making the model using this KB less accurate. In order to avoid this fact, a rule selection genetic process is proposed in [10, 23] with the aim of simplifying the initial linguistic rule set by removing the unnecessary rules from it and generating a KB with good cooperation.

The selection of the subset of linguistic rules best cooperating is a combinatorial optimization problem. Since the number of variables involved in it, i.e., the number of preliminary rules, may be very large, we consider an approximate algorithm to solve it, a GA. Another process solving the problem of selecting rules by means of the same technique is to be found in [25].

The rule selection genetic process is based on a binary coded GA, in which the selection of the individuals is performed using the stochastic universal sampling procedure together with an elitist selection scheme, and the generation of the offspring population is put into effect by using the classical binary multipoint crossover (performed at two points) and uniform mutation operators.

The coding scheme generates fixed-length chromosomes. Considering the rules contained in the initial linguistic rule set counted from 1 to m, an m-bit string $C = (c_1, ..., c_m)$ represents a subset of candidate rules to form the RB finally obtained, B^s, such that,

$$\textit{If } c_i = 1 \textit{ then } R_i \in B^s \textit{ else } R_i \notin B^s$$

The initial population is generated by introducing a chromosome representing the complete previously obtained rule set, i.e., with all $c_i = 1$. The remaining chromosomes are selected at random.

As regards the fitness funtion, $F(C_j)$, it is based on the MSE of the FRBS using the RB encoded in the chromosome over the training data set as well as a criterion penalyzing the lack of the completeness property of the said RB.

A possible improvement of this method is the *genetic multiselection process* [19], which obtains different simplified RBs for modeling and classification problems. It selects the rules cooperating best from the previous RB, by working as follows:

- The basic rule selection genetic process is run several times.
- Each time a simplified rule set is generated, the space zone where it is located is penalized by means of a genotypic sharing function (niching GAs [20]).
- The process ends when the desired number of simplified RBs is generated.

Results for this proposal will not be presented in this contribution, since the WM-method generates a small rule set which does not verify the completeness property. For some results obtained when applying the genetic selection process to RBs generated from other learning methods, refer to [10, 19, 23]. On the other hand, we will see that the genetic selection process is considered in the other approach proposed to induce cooperation from the RB (next subsection) and on the one presented to induce it from the whole KB (Section 5.).

4.2 THE ACCURATE LINGUISTIC MODELING PARADIGM

The *Accurate Linguistic Modeling* (ALM) [12, 13] is a methodology to obtain more cooperative RBs for linguistic models. It is based on the following two aspects:

- The usual linguistic model structure is extended allowing the RB to present rules where each combination of antecedents may have two consequents (the primary and secondary in importance) associated when it is necessary to improve the model accuracy. It is clear that this will improve the capability of the model to perform the interpolative reasoning and, thus, its performance.

  We should note that this operation mode does not constitute an inconsistency from the interpolative reasoning point of view but only a shift of the main labels making that the final output of the rule lie in an intermediate zone between them both. Hence, it may have the following linguistic interpretation. Let us suppose that a specific combination of antecedents, "x_1 is A_1 and ... and x_n is A_n," has two different consequents associated, B_1 and B_2. From a LM point of view, the resulting double-consequent rule may be interpreted as follows:

  IF x_1 is A_1 and ... and x_n is A_n THEN y is *between* B_1 *and* B_2

- The previous point deals with the improvement of the fuzzy reasoning in an input subspace defined by a specific combination of antecedents. On the other hand, the second aspect deals with the cooperation between the rules in the KB, i.e., with the overlapped space zones that are covered by different linguistic rules. Hence, it is considered an operation mode based on generating a preliminary fuzzy rule set in which single and double-consequent rules coexist and selecting the subset of them best cooperating. It is important to remark that each double-consequent rule is decomposed in two simple ones in the selection process. Thus, this stage will specify which double-consequent rules in the preliminary rule set will remain in the final RB, that is, those fuzzy input subspaces whose **two simple rules associated have been finally selected**.

On the other hand, it should be noted that the said operation mode gives more freedom to the RB generation process. As is known, the generation of the best fuzzy rule in each subspace does not ensure that the FRBS designed will perform well, due to the fact that the rules composing the KB may not cooperate suitably. The rule selection considered in ALM can make the final RB present single-consequent rules not being the best ones in their fuzzy input subspaces in order to improve the cooperation of the global RB.

In [12], two specific generation processes based on the ALM methodology are introduced. Both of them are based on two stages: double-consequent rule generation and rule selection. In the following, one of these processes is briefly described:

1. *A linguistic rule generation method* from examples based on a modification of the WM-method that involves generating the two most important consequents for each combination of antecedents (instead of only the most important one, as this method usually do). All the WM-method steps shown in the Appendix remains the same but the fourth one (*Obtain a final RB from the preliminary fuzzy rule set*). Whilst in that method the rule with the highest importance degree is the only one chosen for each combination of antecedents, in our case we allow two different rules, the two most important ones in each input subspace (if they exist), to form part of the RB, thus creating a double-consequent rule.

   Of course, a combination of antecedents may not have rules associated (if there are no examples in that input subspace) or only one rule (if all the examples in that subspace generated the same rule). Therefore, the generation of rules with double consequent is only addressed when the problem complexity, represented by the example set, shows that it is necessary.

2. *The rule selection genetic process*, introduced in Section 4.1, that selects the subset of rules in the preliminary linguistic set cooperating best working in the said way.

Another important characteristic of ALM is that it has no influence on the linguistic model inference system. The only restricition imposed is that the defuzzification method must consider the matching degree of the rules fired. In this contribution we work with the *Center of Gravity weighted by the matching degree* [9].

The inference mechanism designed will perform in the way shown next when it receives an input $x_0 = (x_1, \ldots, x_n)$:

1. *For each rule R_i, $i = 1, \ldots, T$, in the KB*:

   (a) Compute the matching degree, h_i, of the rule:

   $$h_i = Min(\mu_{A_{i1}}(x_1), \ldots, \mu_{A_{in}}(x_n))$$

   (b) Apply the Minimum t-norm in the role of implication operator to obtain the fuzzy set resulting from the application of the inference process on that rule, B_i':

   $$\mu_{B_i'}(y) = Min(h_i, \mu_{B_i}(y))$$

2. *Obtain the Center of Gravity for each individual fuzzy set B_i'*:

$$y_i = \frac{\int_{\mathbf{V}} y \cdot \mu_{B_i'}(y) \cdot dy}{\int_{\mathbf{V}} \mu_{B_i'}(y) \cdot dy}$$

3. *Compute the final output given by the system as output, y_0, by aggregating the partial actions obtained by means of the matching degree weighted average*:

$$y_0 = \frac{\sum_{i=1}^{T} h_i \cdot y_i}{\sum_{i=1}^{T} h_i}$$

The results obtained by the ALM-based process proposed in the solving of the electrical application tackled are showed in Table 1.3. In order to analyze the influence of the genetic selection process introduced in Section 4.1, two different rows will be associated to the ALM process in the table, each one collecting the results obtained after the application of each stage composing it. As can be seen, the linguistic model obtained is simpler and more accurate to a high degree than the WM-method one. We should note that the number of rules showed (20) stands for simple rules, i.e., the double-consequent rules existing in the RB have been counted twice for comparison purposes.

Table 1.3 Results obtained with ALM

Method	*Granularity*	# Rules	$MSE_{training}$	MSE_{test}
WM-method	7 7 7	24	222622	240018
ALM (generation)	7 7 7	34	231174	260067
ALM (selection)	7 7 7	20	155866	178601

5. APPROACHES TO INDUCE COOPERATION FROM THE KNOWLEDGE BASE

A single method will be introduced belonging to the group of approaches inducing cooperation from the whole KB (i.e., both from the DB and the RB), the *Hierarchical System of Linguistic Rules (HSLR) learning methodology*. In HSLRs, the linguistic variables involved in the fuzzy rules are defined in linguistic partitions with different granularity levels, thus making the rules belong to different hierarchical levels [18].

To do so, the KB structure of linguistic models is extended by introducing the concept of "*layers*". In this extension, which is also a generalization, the KB is composed of a set of layers where each one contains linguistic partitions with different granularity levels and linguistic rules whose linguistic variables take values in these partitions. This KB is called Hierarchical Knowledge Base (HKB), and it is formed by a Hierarchical Data Base (HDB) and a Hierarchical Rule Base (HRB), containing linguistic partitions of the said type and linguistic rules defined over them, respectively.

The description of the HKB and the relation between its components is studied next, and the methodology to automatically design an HSLR from generic linguistic rule generating methods is introduced later on. For more details about HSLR methodology, refer to [18].

5.1 HIERARCHICAL KNOWLEDGE BASE

This HKB is composed of a set of layers. We define a layer by its components in the following way:

$$layer(t, n) = DB(t, n) + RB(t, n)$$

with $DB(t, n)$ being the DB which contains the linguistic partitions with granularity level n of layer t, and with $RB(t, n)$ being the RB formed by those linguistic rules whose linguistic variables take values in the former partitions. From now on and for the sake of simplicity, we are going to refer to the components of a $DB(t,n)$ and $RB(t,n)$ as *n-linguistic partitions* and *n-linguistic rules*, respectively.

This set of layers is organized as a hierarchy, where the order is given by the granularity level of the linguistic partition defined in each layer. That is, given two succesive layers t and $t+1$, then the granularity level of the linguistic partitions of layer $t+1$ is greater than the ones of layer t. This causes a refinement of the previous layer linguistic partitions.

As a consequence of the previous definitions, we could now define the HDB as the union of the DBs of every layer t:

$$HDB = \cup_t DB(t,n)$$

and by the same token, the HRB is defined as:

$$HRB = \cup_t RB(t,n)$$

Focusing again on the HDB, we should note that, in this work, we are using *n-linguistic partitions* with the same number of linguistic terms for all input-output variables, composed of triangular-shaped, symetrical and uniformly distributed membership functions.

In order to build the HDB, we develop an strategy which satisfies two main requirements:

- To preserve all possible fuzzy set structures from one layer to the next in the hierarchy.
- To make smooth transitions between successive layers.

Hence, to build a new linguistic partition in the DB of the layer $t+1$ from a *n-linguistic partition* of the layer t with the minimum change between their granularity levels, we just add a new linguistic term between each two consecutive terms of the *n-linguistic partition*, after reducing the support of these linguistic terms in order to keep place for the new one, which is located in the middle of them. An example of the correspondence between a 3-*linguistic partition* and 5-*linguistic partition* is shown in Fig. 1.6.

Generically, we could say that a DB from a layer $t+1$ is obtained as:

$$DB(t,n) \rightarrow DB(t+1, 2\cdot n-1)$$

which means that an n-linguistic *partition* in $DB(t,n)$ with n linguistic terms becomes a *(2·n − 1)-linguistic partition* in $DB(t+1, 2\cdot n-1)$.

As regards the HRB, the *n-linguistic rules* contained in $RB(t,n)$ are those rules whose linguistic variables take values from the *n-linguistic partitions* contained in $DB(t,n)$.

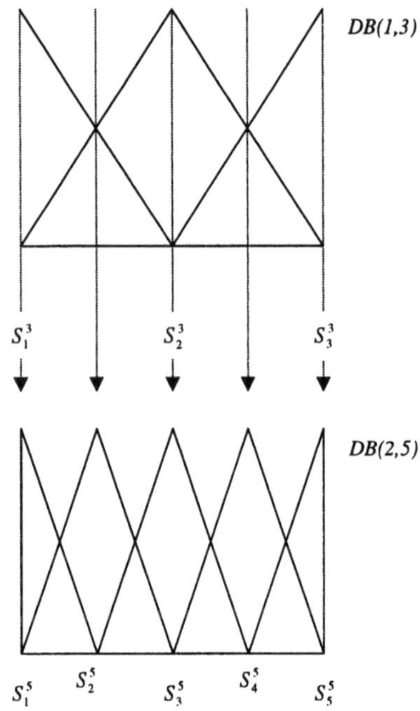

Figure 1.6 Transition from a partition in $DB(1,3)$ to another one in $DB(2,5)$

The main purpose of developing an HRB is to model the space of the problem in a more accurate way. To do so, those *n-linguistic rules* that model a subspace with bad performance are expanded in a set of *(2·n−1)-linguistic rules*, which become their image in $RB(t+1, 2\cdot n-1)$. This set of rules models the same subspace that the former one and replaces it.

We should note that not all *n-linguistic rules* are to be expanded. Only those *n-linguistic rules* which model a subspace of the problem with a significant error become the ones that are involved in this rule expansion process to build the $RB(t+1, 2\cdot n-1)$. The remaining rules preserve their location in $RB(t, n)$.

An explanation for this behaviour could be found in the fact that it is not always true that a set of rules with a higher granularity level perform a better modeling of a problem than other set composed of linguistic rules with a lower granularity level. Moreover, this is not true for all kinds of problems, and what is more, it is also not true for all linguistic rules that model a problem [16]. In an attempt to put this idea into effect, we consider a three-stage process to perform the mentioned rule expansion:

- Selection of those bad performance *n-linguistic rules* from $RB(t,n)$ that are going to be expanded in $RB(t+1, 2 \cdot n - 1)$.
- Selcction of those terms from $DB(t+1, 2 \cdot n - 1)$ that are going to be contained in the *(2·n − 1)-linguistic rules*, considcred as an image of the bad rules.
- Accomplishment of the *(2·n − 1)-linguistic rule* generation process, based on the previously selected term sets.

5.2 STRUCTURE OF THE HSLR LEARNING METHODOLOGY

Our HSLR learning methodology is composed of three main processes which will be described in detail in the following subsections [18]:

- The first process generates the HKB following the descriptions given previously.
- The second process performs a genetic rule selection task that removes the redundant or unnecessary rules from the HRB in order to select a subset of rules that cooperate better.
- In the third process, a user evaluation process extends this approach to an iterative process, where he could adapt many parameters and re-execute the processes to achieve better results.

5.3 HIERARCHICAL KNOWLEDGE BASE GENERATION PROCESS

In this subsection we present our metodology to generate an HKB. It is based on an inductive linguistic rule generation method (LRG-method), that in this contribution will be the WM-method. It also takes as a base a set of input-output data E_{TDS} and a previously defined $DB(t,n)$.

Our HKB generation process has three main steps, that are listed below:

1. **RB(t,n) generation process,** where the rules from the present $DB(t,n)$ are generated.

   An LRG-method is run with the terms defined in the present partitions, that are in $DB(t,n)$*, denoted as* $LRG(DB(t,n), E_{TDS})$.

2. **RB(t+1,2·n-1) generation process,** where the linguistic rules from layer t+1 are generated taking into account $RB(t,n)$, $DB(t,n)$ and $DB(t+1, 2 \cdot n - 1)$.

   (a) *Calculate the error of RB(t,n)*: Compute $MSE\,(E_{TDS}, RB(t,n))$

   (b) *Calculate the error of each individual n-linguistic rule*: Compute $MSE\,(E_i, R_i^n)$.

   (c) *Select the set of n-linguistic rules with bad performance*: Select those bad *n-linguistic rules* which are going to be expanded:
   $IF\ MSE(E_i, R_i^n) \geq \alpha \cdot MSE(E_{TDS}, RB(t,n))\ THEN\ R_i^n \in RB_{bad}(t,n)\ ELSE\ R_i^n \in RB_{good}(t,n)$
   For example, $\alpha = 1.1$ means that an *n-linguistic rule* with an MSE a 10 % higher than the MSE of the entire $RB(t,n)$ should be expanded.

   (d) *Obtain the DB(t+1,2·n−1):* create $DB_{x_j}(t{+}1,2{\cdot}n-1)$ for all input linguistic variables x_j ($j = 1, ..., m$), and $DB_y(t{+}1,2{\cdot}n-1)$ for the output linguistic variable y.

   (e) *Select the (2·n − 1)-linguistic partition terms:* Obtain those terms in DB(t+1,2·n − 1) that are considered in the fuzzy input and output subspaces of the bad rules that are to be expanded: $I(R_i^n), \forall R_i^n \in RB_{bad}(t,n)$.

   (f) *Combine the selected (2·n − 1)-linguistic partition terms to perform (2·n − 1)-linguistic rules:* For each $R_i^n \in RB_{bad}(t,n)$, compute $LRG(I(R_i^n), E_i)$.

3. **HRB summarization process,** where the linguistic rules from the both RBs are joined to obtain the HRB.

   Obtain a set of linguistic rules, joining the group of the new generated (2·n − 1)-linguistic rules and the former good performance n-linguistic rules:

$$HRB = RB_{good}(t,n) \cup RB(t+1, 2 \cdot n - 1)$$

5.4 HIERARCHICAL RULE BASE SELECTION PROCESS

As has been seen in previous sections, the operation mode of the proposed generation method means that in each input subspace, the n-

linguistic rules are created individually from the examples in the input-output data set. This happens without taking into account the cooperation existing among the rules which gives the final model output. That is, no information about the neighbor rules is considered in order to generate them.

In the HRB -where there are coexisting rules with different granularity levels- it may happen that a complete set of *($2{\cdot}n-1$)-linguistic rules,* which replaces an expanded rule, does not produce good results. This means that there will be higher errors. However, a subset of this set of *($2{\cdot}n-1$)-linguistic rules* may work properly, with less rules that have good cooperation between them. Thus, the HRB generated may present redundant or unnecessary rules making the model using this HKB less accurate.

In order to avoid this fact, we will use the linguistic rule selection genetic process described in Section 4.1. with the aim of simplifying the initial linguistic rule set by removing the unnecessary rules from it and generating an HKB with good cooperation.

5.5 USER EVALUATION PROCESS

The application of our metodology could be also considered as an user controlled iterative process. In this sense, the user could adapt the granularity of the initial linguistic partitions and/or the threshold α which determines if an *n-linguistic rule* will be expanded in a set of *($2{\cdot}n-1$)-linguistic rules*, and apply again the process in order to obtain a better model.

This process works in this way: if the error measure of the obtained model (i.e. global error) does not satisfy the user requirements, then the user can adapt the parameter α –item 2.c in the HKB Generation Process– and/or reinitialize the process with a different granularity level for the initial partition of the linguistic variables domain.

5.6 EXPERIMENTAL STUDY

The results obtained by applying the HSLR learning process introduced in this section to the electrical problem are shown in Table 1.4. For a complete experimentation to solve this and other problems, refer to [18]. In view of the data showed in the table, it can be seen that the model generated performs significantly better than the WM-method one, with a short intrepretability lose (28 rules in the KB instead of 24).

Table 1.4 Results obtained with the HSLR methodology

Method	*Granularity*	# Rules	$MSE_{training}$	MSE_{test}
WM-method	7 7 7	24	222622	240018
granularity learning	(WM,5,9)	28	157755	180488

6. APPROACHES TO INDUCE COOPERATION FROM THE INFERENCE SYSTEM

In this section, we propose a new fuzzy reasoning model for a concrete type of linguistic models: FRBCSs, the FRBSs for classification problems. Some specific FRMs included in the general model will be also introduced with the aim of improving the rule cooperation in these kinds of systems.

6.1 IMPROVING THE RULE COOPERATION IN FRBCSS

By using the classical FRBCS reasoning method shown in Section 1.3, the information provided by the other rules that also are compatible (have also been fired) with the example is not considered. In this section, we propose to use FRMs that combine the information given by the different rules fired by a pattern.

To do so, a general reasoning model for FRBCSs [14] is introduced, that in this paper is particularised to an RB composed of rules with a class and its certainty degree in the consequent. This model is described in the following.

In the classification of an example $E^t = (e_1^t, \ldots, e_N^t)$, the RB $R = \{R_1, \ldots, R_L\}$ is divided into M subsets according to the class indicated by its consequent,

$$R = R_{C_1} \cup R_{C_2} \cup \ldots \cup R_{C_M}$$

and the next scheme is followed:

1. **Compatibility degree.** The compatibility degree of the antecedent with the example is computed for all the rules in the RB, applying a t-norm over the membership degree of the values of the example (e_i^t) to the corresponding fuzzy subsets.

$$R^k(E^t) = T(\mu_{A_1^k}(e_1^t), \ldots, \mu_{A_N^k}(e_N^t)), \quad k = 1, \ldots, L$$

2. **Association degree.** The association degree of the example E^t with the M classes is computed according to each rule in the RB.

$$b_i^k = h(R^k(E^t), r^k), \quad k = 1, \ldots, |R_{C_i}| \quad i = 1, \ldots, M$$

3. **Weighting function.** The values obtained are weighted by means of a function g. An expression which promotes the highest values and penalizes the smallest ones seems to be the most adequate choice for this function.

$$B_i^k = g(b_i^k), \quad k = 1, \ \ldots, |R_{C_i}| \quad i = 1, \ldots, M$$

4. **Pattern classification soundness degree for all classes.** To compute this value, an aggregation operator is used which combines, for each class, the positive association degrees computed in the previous step

$$Y_i = f(B_i^k, \ k = 1, \ldots, |R_{C_i}| \ \text{ and } \ B_i^k > 0), \quad i = 1, \ldots, M$$

with f being an aggregation operator that returns a value between the minimum and the maximum.

5. **Classification.** A decision function F is applied to the classification degrees of the example. This function will return the class label corresponding to the maximum value.

$$C_l = F(Y_1, \ldots, Y_M) \ \text{ such that } \ Y_l = \max_{j=1,\ldots,M} Y_j$$

We should note that, in this general model, if we select the function f in the fourth step as the maximum operator, we have the classical FRM:

$$f_0(a_1, \ldots, a_s) = \max_{i=1,\ldots,s} a_i$$

with $a_1, \ldots, a_s$ being the values to aggregate for an example E^t with respect to a class C_j.

According to the general reasoning model, we propose a new kind of inference models. The difference lies on the choice of function $f(\cdot)$ in step 4, due to the fact that we consider FRMs that integrate all fuzzy rules to derive conclusions from a set of fuzzy classification rules and a pattern. This idea is graphically represented in Fig. 1.7.

Some proposals for the function f in FRBCSs belonging to this family are described in Table 1.5. An analysis of these proposals, as well as a review of previous applications of the first function f_1, is to be found in [14].

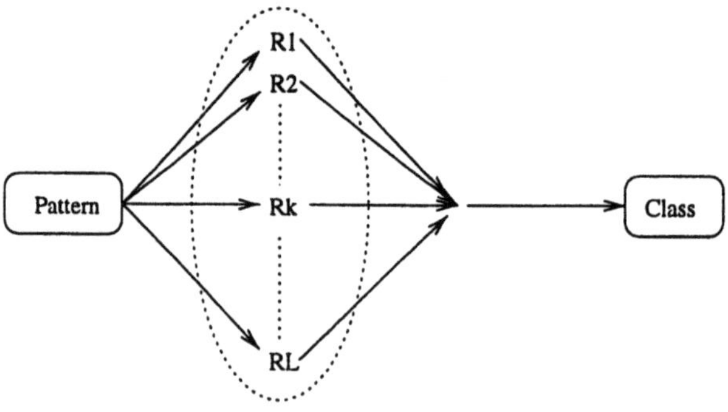

Figure 1.7 FRM integrating all fuzzy rules

Table 1.5 Different proposals for aggregating the association degrees

4. Aggregation functions	
Normalised Sum $f_1(a_1,\ldots,a_s) = \frac{\sum_{i=1}^{s} a_i}{f_{1_{max}}}$ $f_{1_{max}} = \max_{j=1,\ldots,M} \sum_{i=1}^{s} a_i$	**Sowa and-like** $f_4(a_1,\ldots,a_s) = \alpha \cdot a_{min} + (1-\alpha)\frac{1}{s}\sum_{i=1}^{s} a_i$ $\alpha \in [0,1], \quad a_{min} = min\{a_1,\ldots,a_s\}$
Arithmetic Mean $f_2(a_1,\ldots,a_s) = \frac{\sum_{i=1}^{s} a_i}{s}$	**Sowa or-like** $f_5(a_1,\ldots,a_s) = \alpha \cdot a_{max} + (1-\alpha)\frac{1}{s}\sum_{i=1}^{s} a_i$ $\alpha \in [0,1], \quad a_{max} = max\{a_1,\ldots,a_s\}$
Quasiarithmetic Mean $f_3(a_1,\ldots,a_s) = H^{-1}\left[\frac{1}{s}\sum_{i=1}^{s} H(a_i)\right]$ $H(x) = x^p, \quad p \in R$	**Badd** $f_6(a_1,\ldots,a_s) = \frac{\sum_{i=1}^{s} a_i^{\alpha+1}}{\sum_{i=1}^{s} a_i^{\alpha}}, \quad \alpha \in R$
with $a_1,\ldots,a_s$ being the values to aggregate for an example E^t with respect to a class C_j	

6.2 EXPERIMENTAL STUDY

For this brief study, two well known sets of samples, IRIS and PIMA, have been considered. The IRIS data base is a set of 150 examples of iris flowers with three classes and four attributes. PIMA is a set of 768 solved cases of diagnostics of diabetes where eight variables are taken into account and there are two possible classes (having or not having the disease).

Taking into account the characteristics of the example sets, fuzzy partitions constituted by five triangular fuzzy sets have been considered to define the DB in both cases. As regards the RB, two different kinds of fuzzy classification rules have been considered for the experimentation, the ones with only a class and with a class and its certainty degree in the consequent. The RB has been generated by means of the adaption of the WM-method to classification problems shown in the Appendix in both cases.

To calculate an error estimation of an FRBCS, random resampling [32] with five random partitions of the sample bases in training and test sets (70% and 30% respectively) have been considered.

The best results obtained with our different proposals of FRMs are showed in Tables 1.6 and 1.7 for the two types of rules considered. The classification percentages obtained by the classical reasoning method are also shown for comparison purposes. For a complete experimental study including all the FRMs proposed and comparing against different classification techniques, refer to [14].

Table 1.6 Results obtained when using rules with a class in the consequent

Iris			**Pima**		
FRM	Tra	Test	FRM	Tra	Test
Classical (f_0)	90.97	88.25	Classical (f_0)	89.51	64.88
f_1 g_2	98.56	**94.38**	f_1 g_1	83.97	**72.11**
f_5 g_2	90.64	92.27	f_6 g_2	91.32	67.38
f_3 g_2	89.73	91.78	f_5 g_2	85.78	64.56

Table 1.7 Results obtained when using rules with a class and a certainty degree in the consequent

Iris			**Pima**		
FRM	Tra	Test	FRM	Tra	Test
Classical (f_0)	97.31	94.32	Classical (f_0)	85.81	73.23
f_3 g_1	97.31	94.32	f_6 g_2	85.89	**73.53**
f_5 g_1	97.31	94.32	f_6 g_1	85.85	**73.53**
f_6 g_1	97.31	94.32	f_3 g_2	85.78	73.44

In view of the results obtained, the cooperative FRMs have demonstrated a good behaviour with RBs obtained from rule generation processes not considering them. However, it is possible to obtain better results by including them in the FRBCS learning process. To do so, a three-stage genetic fuzzy rule-based classification system considering the rule cooperation induction during the learning stage was introduced in [11].

7. SUMMARY

This contribution has been devoted to the problem of improving the accuracy of linguistic models while maintaining its descriptive power. As shown, linguistic models are human-readable rule-based descriptions of the system modeled but sometimes they are not as accurate as desired due to some problems of the linguistic rule structure.

With this aim, the possibility of improving the way in which the linguistic model performs interpolative reasoning by improving the cooperation between the rules in the KB has been analyzed. Several approaches to do so have been introduced, classified in four different groups according to the FRBS component from which the cooperation is induced: IS, KB, DB and RB. To be precise, the following six approaches have been studied:

- Genetic tuning of the membership functions (DB).
- SA-based Learning of the DB from examples (DB).
- Genetic selection of fuzzy rules (RB).
- The ALM paradigm, based on a double-consequent linguistic rule generation and selection (RB).
- The HALM paradigm, based on a hierarchical linguistic rule generation and selection (KB).
- Cooperative FRMs for classification problems (IS).

The behaviour of the first five has been analyzed in solving a real-world Spanish electrical distribution problem, where all of them have obtained good results, being more accurate than the basic linguistic model generated from the WM-method. On the other hand, the last one has shown good performance with the classical IRIS and PIMA data sets.

As mentioned in the contribution, the different approaches proposed are not mutually exclusive and can be combined to obtain better linguistic models. In fact, the ALM and HALM paradigms make use of other of the approaches introduced, the rule selection genetic process, as one of their components. This fact makes us think that the combination of the different approaches can be a promising research field and our future work will be focused on studying it.

Acknowledgments

This research has been suported by the "Ministerio de Educación y Ciencia" of Spain under project **PB98-1319**.

Appendix: The Wang and Mendel Rule Generation Method

The Wang and Mendel's RB generation method (WM-method) [30] is one of the simplest and most known LM design methods (for more information about the different learning techniques considered for this task and some specific approaches, refer to [4]). In this inductive method, the generation of the RB is put into effect by means of the following steps:

1. *Consider a fuzzy partition of the input variable spaces*: It may be obtained from the expert information (if it is availaible) or by a normalization process. If the latter is the case, perform a fuzzy partition of the input variable spaces dividing each universe of discourse into a number of equal or unequal partitions, select a kind of membership function and assign one fuzzy set to each subspace.

2. *Generate a preliminary linguistic rule set*: This set will be formed by the rule best covering each example (input-ouput data pair) contained in the tarining data set. The structure of these rules is obtained by taking a specific example, i.e., an $n+1$-dimensional real array (n input and 1 output values) and setting each one of the rule variables to the linguistic label associated to the fuzzy set best covering every array component.

3. *Give an importance degree to each rule*: Let $R_l =$ *IF* x_1 *is* A_1 *and* ... *and* x_n *is* A_n *THEN* y *is* B be the linguistic rule generated from the example $e_l = (x_1^l, \ldots, x_n^l, y^l)$. The importance degree associated to it will be obtained as follows:

$$G(R_l) = \mu_{A_1}(x_1^l) \cdot \ldots \cdot \mu_{A_n}(x_n^l) \cdot \mu_B(y^l)$$

4. *Obtain a final RB from the preliminary fuzzy rule set*: The rule with the highest importance degree is chosen for each combination of antecedents.

Appendix: The Wang and Mendel Rule Generation Method for Classification

In [8], an extension of the WM-method was proposed to deal with classification problems. This process starts with a set of input-output data pairs (the training data set) with the following structure:

$$E^1 = (e_1^1, \ldots, e_N^1, o^1),\ E^2 = (e_1^2, \ldots, e_N^2, o^2),\ \ldots,\ E^p = (e_1^p, \ldots, e_N^p, o^p)$$

Table 1.C.1 Variables of the electrical problem

Symbol	Meaning
x_1	Number of inhabitants of the town
x_2	Distance from the center of the town to the three furthest clients
y	Total length of low voltage line installed

where o^h is the class label for the pattern E^h.

The task here is to generate a set of linguistic classification rules from the training data set that describes the relationship between the system variables and determines a mapping D between the feature space S^N and the class set $C = \{C_1, \ldots, C_M\}$.

The method consists of the following steps:

- *Fuzzifying the feature space.* Finding the domain intervals of the attributes and partition each domain into X_i regions ($i = 1, \ldots, N$). A membership function is adopted for each fuzzy region.

- *Generating fuzzy rules from given data pairs.* For each training data $E^h = (e_1^h, \ldots, e_N^h, o^h)$, we have

  - To determine the membership degrees of e_i^h in different input fuzzy subsets.
  - To assign the input $e_1^h, \ldots, e_N^h$ to the region with the maximum membership degree.
  - To produce a fuzzy rule from E^h, with the if-part that represents the selected fuzzy region and the consequent with the class determined by o^h. Repeated fuzzy rules are not considered.

Appendix: Total low voltage line length installed in a rural town

The problem considered is that of finding a model that relates the total length of low voltage line installed in Spanish rural towns [15]. This model will be used to estimate the total length of line being maintained by an electrical company. We were provided with a sample of 495 towns in which the length of line was actually measured and the company used the model to extrapolate this length over more than 10.000 towns with these properties. We will limit ourselves to the estimation of the length of line in a town, given the inputs showed in Table 1.C.1. To develop the different experiments in this contribution, the sample has been randomly divided in two subsets, the training and test ones, with an 80%-20% of the original size respectively. Thus, the training set contains 396 elements, whilst the test one is composed by 99 elements.

References

[1] Aarts, E.H.L., *Simulated Annealing and Boltzman Machines: A Stochastic Approach to Combinatorial Optimization and Neural Computing*, John Wiley & Sons, 1989.

[2] Abe, S., Thawonmas, R., "A Fuzzy Classifier with Ellipsoidal Regions," *IEEE Transactions on Fuzzy Systems*, Volume 5, No. 3, 1997, pp. 358-368.

[3] Alcalá, R., Casillas, J., Cordón, O., Herrera, F., "Approximate Mamdani-type Fuzzy Rule-Based Systems," *Technical Report #DECSAI-990117*, Dept. of Computer Science and A.I., University of Granada, October 1999.

[4] Alcalá, R., Casillas, J., Cordón, O., Herrera, F., Zwir, I., "Techniques for Learning and Tuning Fuzzy Rule-Based Systems for Linguistic Modeling and Their Application," in: C.T. Leondes (ed.), *Knowledge Engineering. Systems, Techniques and Applications*, Academic Press, 1999.

[5] Bardossy, A., Duckstein, L., *Fuzzy Rule-Based Modeling With Application to Geophysical, Biological and Engineering Systems*, CRC Press, 1995.

[6] Bastian, A., "How to Handle the Flexibility of Linguistic Variables with Applications," *International Journal of Uncertainty, Fuzziness and Knowledge-Based Systems*, Volume 2, No. 4, 1994, pp. 463-484.

[7] Carse, B., Fogarty, T.C., Munro, A., "Evolving Fuzzy Rule Based Controllers using Genetic Algorithms," *Fuzzy Sets and Systems*, Volume 80, 1996, pp. 273-293.

[8] Chi, Z., Wu, J., Yan, H., "Handwritten Numeral Recognition Using Self-organizing Maps and Fuzzy Rules," *Pattern Recognition*, Volume 28, No. 1, 1995, pp. 59-66.

[9] Cordón, O., Herrera, F., Peregrín, A., "Applicability of the Fuzzy Operators in the Design of Fuzzy Logic Controllers," *Fuzzy Sets and Systems*, Volume 86, 1997, pp. 15-41.

[10] Cordón, O., Herrera, F., "A Three-stage Evolutionary Process for Learning Descriptive and Approximative Fuzzy Logic Controller

Knowledge Bases from Examples," *International Journal of Approximate Reasoning*, Volume 17, No. 4, 1997, pp. 369-407.

[11] Cordón, O., del Jesus, M.J., Herrera, F., "Genetic Learning of Fuzzy Rule-based Classification Systems Cooperating with Fuzzy Reasoning Methods," *International Journal of Intelligent Systems*, Volume 13, No. 10-11, 1998, pp. 1025-1053.

[12] Cordón, O., Herrera, F., "A Proposal for Improving the Accuracy of Linguistic Modeling," *Technical Report #DECSAI-980113*, Dept. of Computer Science and A.I., University of Granada, May 1998.

[13] Cordón, O., Herrera, F., "ALM: A Methodology to Design Accurate Linguistic Models for Intelligent Data Analysis," *Third Intelligent Data Analysis Conference (IDA'99), L.N.C.S. 1642*, Amsterdam, Holland, 1999, pp. 15-26.

[14] Cordón, O., del Jesus, M.J., Herrera, F., "A Proposal on Reasoning Methods in Fuzzy Rule-based Classification Systems," *International Journal of Approximate Reasoning*, Volume 20, 1999, pp. 21-45.

[15] Cordón, O., Herrera, F., Sánchez, A., "Solving Electrical Distribution Problems Using Hybrid Evolutionary Data Analysis Techniques," *Applied Intelligence*, Volume 10, 1999, pp. 5-24.

[16] Cordón, O., Herrera, F., Villar, P., "Influence of Fuzzy Partition Granularity on Fuzzy Rule-Based System Behaviour", *EUSFLAT-ESTYLF Joint Conference*, Palma de Mallorca, Spain, 1999, pp. 159-162.

[17] Cordón, O., Herrera, F., Villar, P., "Analysis and Guidelines to Obtain a Good Uniform Fuzzy Partition Granularity for Fuzzy Rule-Based Systems using Simulated Annealing," *Technical Report #DECSAI-990116*, Dept. of Computer Science and A.I., University of Granada, September 1999.

[18] Cordón, O., Herrera, F., Zwir, I., "Linguistic Modeling by Hierarchical Systems of Linguistic Rules," *Technical Report #DECSAI-990114*, Dept. of Computer Science and A.I., University of Granada, July, 1999.

[19] Cordón, O., Herrera, F., "Hybridizing Genetic Algorithms with Sharing Scheme and Evolution Strategies for Designing Approximate Fuzzy Rule-Based Systems," *Fuzzy Sets and Systems*, Volume 111, No. 3, 2000.

[20] Goldberg, D.E., *Genetic Algorithms in Search, Optimization, and Machine Learning*, Addison-Wesley, 1989.

[21] González, A., Pérez, R., "Completeness and Consistency Conditions for Learning Fuzzy Rules," *Fuzzy Sets and Systems*, Volume 96, 1998, pp. 37-51.

[22] Herrera, F., Lozano, M., Verdegay, J.L., "Tuning Fuzzy Controllers by Genetic Algorithms," *International Journal of Approximate Reasoning*, Volume 12, 1995, pp. 299-315.

[23] Herrera, F., Lozano, M., Verdegay, J.L., "Fuzzy Connectives Based Crossover Operators to Model Genetic Algorithms Population Diversity," *Fuzzy Sets and Systems*, Volume 92, No. 1, 1997, pp. 21-30.

[24] Ishibuchi, H., Nozaki, K., Tanaka, H., " Distributed Representation of Fuzzy Rules and Its Application to Pattern Classification," *Fuzzy Sets and Systems*, Volume 52, 1992, pp. 21-32.

[25] Ishibuchi, H., Nozaki, K., Yamamoto, N., Tanaka, H., " Selecting Fuzzy If-then Rules for Classification Problems Using Genetic Algorithms," *IEEE Transactions on Fuzzy Systems*, Volume 3, No. 3, 1995, pp. 260-270.

[26] Mandal, D.P., Murthy, C.A., Pal, S.K., "Formulation of a Multivalued Recognition System," *IEEE Transactions on Systems, Man, and Cybernetics*, Volume 22, No. 4, 1992, pp. 607-620.

[27] Michalewicz, Z., *Genetic Algorithms + Data Structures = Evolution Programs*, Third edition, Springer-Verlag, 1996.

[28] Pedrycz, W. (Ed.), *Fuzzy Modelling: Paradigms and Practice*, Kluwer Academic Press, 1996.

[29] Sugeno, M., Yasukawa, T., "A Fuzzy-logic-based Approach to Qualitative Modeling," *IEEE Transactions on Fuzzy Systems*, Volume 1, No. 1, 1993, pp. 7-31.

[30] Wang, L.X., Mendel, J.M., "Generating Fuzzy Rules by Learning from Examples," *IEEE Transactions on Systems, Man, and Cybernetics*, Volume 22, 1992, pp. 1414-1427.

[31] Wang, L.X., *Adaptive Fuzzy Systems and Control*, Prentice-Hall, 1994.

[32] Weiss. S.M., Kulikowski, C.A., *Computer Systems that Learn*, Morgan Kaufmann Publishers, 1991.

[33] Zadeh, L.A., "Outline of a New Approach to the Analysis of Complex Systems and Decision Processes," *IEEE Transactions on Systems, Man, and Cybernetics*, Volume 3, No. 1, 1973, pp. 28-44.

[34] Zadeh, L.A., "The Concept of a Linguistic Variable and its Application to Aproximate Reasoning," *Information Science*, Part I: Volume 8, 1975, pp. 199-249, Part II: Volume 8, 1975, pp. 301-357, Part III: Volume 9, 1975, pp. 43-80.

Chapter 10

LINGUISTIC IF-THEN RULES IN LARGE SCALE APPLICATION OF FUZZY CONTROL

Vilém Novák, Jaromír Kovář
Institute for Research and Applications of Fuzzy Modeling
University of Ostrava
Bráfova 7,
701 03 Ostrava,
Czech Republic

EASY Control Morava
Nábřežní 127
793 51 Břidličná,
Czech Republic

Abstract In this Chapter, we will describe a large scale application of fuzzy control, which has been realized in the Czech Republic. It concerns the enterprise, which is a part of the Metallurgic Plant Břidličná, a.s., an important producer of the high quality aluminium in the Czech Republic. The applications of fuzzy control started in 1995. After good experiences with the first furnace fuzzy control, it was decided to apply it on the other four furnaces one by one, too. At present, the system works in the whole enterprise on all five furnaces. We have used the linguistically oriented fuzzy logic controller, for which it is specific to interpret the IF-THEN rules as linguistically characterized logical implications and the inference is logical deduction based on the formal fuzzy logic in broader sense.

1. INTRODUCTION

In this Chapter, we will describe a large scale application of fuzzy control, which has been realized in the Czech Republic. It concerns the enterprise, which is a part of the Metallurgic Plant Břidličná, a.s., an important producer of the high quality aluminium in the Czech Republic. The enterprise consists of five smelting furnaces TLP9, each of which has the capacity of approximately 10 tons of rough material. The resulting aluminium is used to produce sheets of the highest quality in the continuous process.

The biggest difficulty for control of the production is discontinuity of the process. This means that the sequence "filling the furnace — smelting — emptying" continuously repeats in regular periods of the length approximately 8 hours. It is apparent that during this process, major physical changes occur inside the furnace.

The phase of smelting is extremely slow with great inertia and it is influenced by the latent heat in negative way. After filling the furnace by blocks of rough aluminium, it takes about 3 hours to reach the melting point. Further, it takes still approximately 1 hour until the temperature begins to change because all the heat is absorbed to start to smelt. Only after this, the main phase starts. During this it is necessary to control the temperature (usually in the range 740–800°C). After the required temperature is reached, it is important to keep it. The temperature raises by approximately 0.1°C in 10 sec. In case of the overshoot, the temperature decreases with this rate (without heating) further 5–10 min. The extreme inertia requires to be very careful when adding heat.

Before the application of fuzzy control, several classical control systems have been used, starting from classical PD and PID control systems (including the analogue one). The most advanced has been adaptive control with the continuous identification of the controlled process. Classical PID control turned out to be inconvenient since the temperature had a big overshoot and it was difficult to stabilize it when the conditions changed during the production. Also, the adaptive control, though it promised a lot, finally turned out not to be fully satisfying (see below). Therefore, the final decision was to use the fuzzy control.

The applications of fuzzy control in Břidličná started in 1995. After good experiences with the first furnace fuzzy control, it was decided to apply it on the other four furnaces one by one, too. At present, the system works in the whole enterprise on all five furnaces.

Let us point out that we have used the linguistically oriented fuzzy logic controller, whose main ideas have been described in [7]. We will describe it briefly in Section 4..

2. CHARACTERIZATION OF THE SMELTING FURNACE AS A CONTROLLED SYSTEM

The smelting furnace TLP9 is a highly nonlinear system where the input is the burning power and the output is the temperature of metal. The nonlinearities are caused by physical phenomena proceeding inside the furnace as well as by the construction of the furnace.

The nonlinearities are mainly the following:

(a) Transition of the metal phase from solid to liquid one. The procedure is long and in practice it looks as if breaking the control loop (no influence of burning to the temperature).

(b) Change of the sensors' time constants since in the beginning, they are in the furnace space and later, they are immersed in the liquid metal.

(c) Change of the time-delay due to modification of the distance between the burner flame and the temperature sensor dependingly on the burning power.

(d) Limitation of the range of the power, i.e., it is impossible to have arbitrary small power. This is caused by the way of construction (the ratio of the power control is 1:6).

(e) Temporary change of the sign of the static amplification of the system, which is caused by the exothermic reaction of the refining salt with the smelted metal (in practice, the burners are not in process but the temperature raises).

(f) The temperature of metal is "stratified" — dependingly on the distance from the bottom of the furnace. The highest strata (about 20% of the total height of the melted metal) is significantly warmer than the medium strata (about 65% of the height). Lower stratas change their temperature much slower than the higher ones.

(g) Further rough metal is added to the furnace time to time during the process of melting.

The goal is to design the control algorithm so that it would fulfil the following requirements:

(a) It must be resistant towards all the working states, disregarding whether the operation regime is automatic or by hand (in the case of the occurrence of an error).

(b) It is important to spare the brickwork of the furnace.

(c) The overshoot of the temperature must be at minimum.

(d) The total control time and fuel consumption should be minimized.

3. PROPERTIES OF THE ADAPTIVE CONTROLLER

As mentioned above, several classical control algorithms have been applied in the furnace. The most interesting and promising among them was the adaptive control with continuous process identification. The system has been described in [10].

Before using it in the real system, we have tested it on a small physical model, which was an electrically heated furnace with cca 17 kg of tin. The results of the control were good: small overshoot, satisfactory response on the simulated disturbations as well as changing operation regime.

Unfortunately, the results on the real system were much worse. First of all, it was quite difficult to set up all the parameters. On the other hand, these turned out to be undesirably sensitive in changing conditions. We have also encountered problems with numerical stability.

The main reasons at the adaptive control failure were the following:

(a) The system has great and, moreover, changing transportation delay. Hence, a good identification could be obtained only after more sample periods. However, it was nearly impossible to fulfil this demand in most cases, especially when the smelt material had been rafinated, or the chemical pattern identification had been taken. Of course, if the controller had enough time to recognize the system then the control was satisfactory.

(b) The sample period had to be long (at least 120 sec.). This led to discomfort of the operators since after charging the furnace, "nothing happened" for a long time, though the operators expected an immediate action.

(c) The control algorithm failed in cases when a lot of rafination salt had to be used. The following exothermic reaction caused that the system has been recognized to have negative gain. Though theoretically correct, in practice this caused wrong behaviour of the algorithm and it took too long to recover. In practice, the final control was bad. Some improvement had been obtained after introduction of heuristic rules, which allowed the adaptive controller to work only in convenient moments.

These reasons, and also low robustness of the control system led us to try the fuzzy control. We expected from it especially greater robustness and better control in the mentioned critical phase.

In the next section, we will describe the way of fuzzy control used in the control of smelting furnace. Let us stress that it is different from the classical approaches mostly based on the Mamdani-Assilian formula with the Center of Gravity defuzzification (see, e.g., [2, 5]). Its main characteristics are derivation of the conclusion using fuzzy logical deduction and stress on the linguistic aspect, which requires specific defuzzification method. In Section 5. we provide the reader with some graphs demonstrating behaviour of the fuzzy controller in the reality.

4. LINGUISTIC FUZZY LOGIC CONTROL

It is specific for the fuzzy control system, which we have used that on one side, it deliberately realizes logical deduction based on the interpretation of IF-THEN rules as fuzzy logical implications and on the other side, it puts more stress on the linguistic aspect than the other fuzzy control systems described in the literature (cf., e.g., [2]). In this section, we will briefly characterize its main features.

4.1 FORM OF IF-THEN RULES

The form of the rules used in the linguistic fuzzy logic control is classical, i.e.

$$\mathcal{R}_1 := \mathsf{IF}\ \ x\ \mathsf{is}\ \mathcal{A}_1\ \mathsf{THEN}\ \ y\ \ \mathsf{is}\ \mathcal{B}_1$$

$$\cdots\cdots\cdots\cdots\cdots\cdots$$

$$\mathcal{R}_m := \mathsf{IF}\ \ x\ \mathsf{is}\ \mathcal{A}_m\ \mathsf{THEN}\ \ y\ \ \mathsf{is}\ \mathcal{B}_m.$$

We will call the set $\{\mathcal{R}_1, \ldots, \mathcal{R}_m\}$ of the IF-THEN rules the *linguistic description.*

The rules in the linguistic description are interpreted as linguistically characterized *logical implications.* The symbols $\mathcal{A}_j, \mathcal{B}_j$, $j = 1, \ldots, m$ are called *simple linguistic evaluating syntagms* with sign, which are linguistic expressions of the form

$$[\langle\text{sign}\rangle]\langle\text{linguistic modifier}\rangle\langle\text{atomic evaluating syntagm}\rangle \qquad (1.1)$$

where ⟨sign⟩ is either of the words "positively" or "negatively". Furthermore, ⟨atomic evaluating syntagm⟩ is either of the words "small, medium, big", "zero", or fuzzy number. Finally, ⟨linguistic modifier⟩ is an intensifying adverb such as "very, roughly", etc. Examples of (1.1) may be the expressions, e.g., "positive small, negative very big, quite roughly medium", "approximately z", "about z", "roughly z", etc.

The implemented version of LFLC provides a quite variety of such syntagms (altogether more than 25). The meaning of all the syntagms is represented by fuzzy sets, which are obtained by modification of the basic fuzzy sets having the pre-defined shapes (see Fig. 1.1). The basic motivation for our approach is to make the computer mimic the behaviour of the "human operator" after being explained how to control

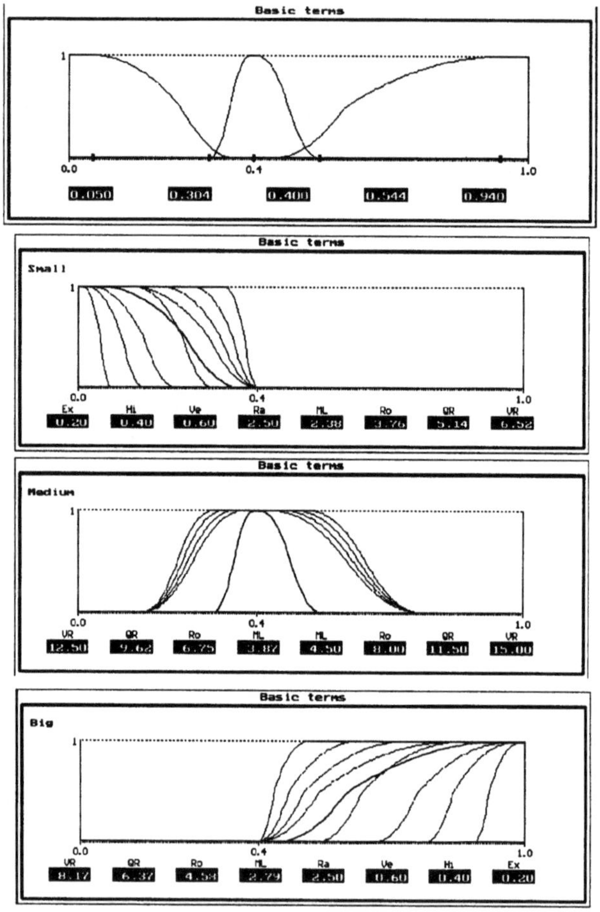

Figure 1.1 Shapes of the fuzzy sets modelling the meaning of simple evaluating expressions

the given system in natural language. Note that we have thus consistently returned to the original idea of fuzzy control, as presented in the early works (cf. [5, 14]).

Unlike the classical approach to fuzzy control where the fuzzy sets are modified independently on their original meaning to obtain as best control course of as possible, in the linguistic approach the fuzzy sets are modified indirectly using the linguistic modifiers. We distinguish modifiers with *narrowing effect*, namely the modifiers

$$\textsf{extremely}, \textsf{significantly}, \textsf{very} \tag{1.2}$$

and those with *widening effect*, namely the modifiers

roughly, more or less, roughly, quite roughly, very roughly. (1.3)

Moreover, we will consider the following *canonical ordering* of the evaluating syntagms:

"negative extremely big"
"negative significantly big"
⋮
"negative very roughly big"
⋮ (1.4)
"negative extremely small"
"negative roughly zero"
"zero"
"positive roughly zero"
"positive extremely small"
⋮ ⋮
"positive extremely big"

These expressions are translated into the formalism of fuzzy logic in broader sense, which will be shortly described in the next subsection (for simplicity, we will not deal with fuzzy numbers).

4.2 INTERPRETATION OF IF-THEN RULES IN THE FORMALISM OF FUZZY LOGIC

The IF-THEN rules are interpreted in the formalism of fuzzy logic in broader sense as follows (for the unexplained details see [9]). We consider the formal language J of the predicate fuzzy logic consisting of all the standard symbols ($\Rightarrow, \wedge, \&, \forall$, etc.) and the following specific ones:

(i) Unary predicate symbols Sm, Me, Bi,

(ii) additional unary connectives

$$\{\lhd_{Ex}, \lhd_{Si}, \lhd_{Ve}, \lhd_{ML}, \lhd_{Ro}, \lhd_{QR}, \lhd_{VR}\}, \tag{1.5}$$

(iii) binary predicate symbol $\leq$,

(iv) sufficiently large sets M_1, M_2 of constants of two sorts.

In the construction of IF-THEN rules applied in fuzzy control, we are interested especially in the atomic formulas $Sm(x), Me(x), Bi(x)$ and simple formulas of the form

$$\lhd_{Ex}(Sm(x)), \lhd_{Si}(Sm(x)), \lhd_{Ve}(Sm(x)), \lhd_{Ro}(Sm(x)), \dots , \tag{1.6}$$

i.e. the latter are atomic ones completed by some of the unary connectives applied to them[1]. The formulas (1.6) are formal representatives of the evaluating syntagms (1.4).

The meaning of the evaluating syntagms is constructed as follows. The ⟨atomic evaluating syntagm⟩ is assigned the corresponding predicate symbol Sm, Me, Bi. Furthermore, each linguistic modifier from (1.2) and (1.3) is assigned the corresponding unary connective from (1.5).

Let $\mathcal{A}$ denote the evaluating syntagm (1.4), $A(x)$ the corresponding formula assigned to it, and let us consider the linguistic expression

$$x \textsf{ is } \mathcal{A} \tag{1.7}$$

where x is a variable standing for some ⟨noun⟩ (we will call (1.7) the evaluating predication). Then the meaning of (1.7) is the fuzzy set of formulas

$$\mathbf{A}_{\langle x\rangle} = \left\{{}^{a_t}/A_x[t] \,\middle|\, t \in M_i\right\} \tag{1.8}$$

where $i = 1$ or $i = 2$ depends on the sort of x (we will call (1.8) the *multiformula*). The symbol ${}^{a_t}/A_x[t]$ in (1.8) is called the *evaluated formula* and it consists of an instance $A_x[t]$ of the formula $A(x)$ (i.e., the variable x has been replaced by the term t in $A(x)$), and its evaluation $a_t \in [0, 1]$.

From the linguistic point of view, the multiformula (1.8) may be understood as the *intension* of the meaning of the linguistic predication (1.7). Using this approach, we will construct the intension of the implication

$$\textsf{IF } x \textsf{ is } \mathcal{A} \textsf{ THEN } y \textsf{ is } \mathcal{B} \tag{1.9}$$

as the multiformula

$$\mathbf{C}_{\langle x,y\rangle} = \mathbf{A}_{\langle x\rangle} \Rightarrow \mathbf{B}_{\langle y\rangle} = \left\{{}^{(a_t \to b_s)}/(A_x[t] \Rightarrow B_y[s]) \,\middle|\, t \in M_{\iota_1}, s \in M_{\iota_2}\right\}. \tag{1.10}$$

When applied in fuzzy control, we furthermore interpret the multiformulas (1.8) or (1.10) in some numerical model whose support is a closed interval of real numbers

$$[p, q] = \{x \mid x \in \mathbb{R}, p \leq x \leq q\} \subset \mathbb{R}. \tag{1.11}$$

Let us now turn our attention to the linguistic modifiers. They have been introduced by L. A. Zadeh in several works (cf. e.g., [12]). According to his proposal, linguistic modifiers are interpreted using a special function, which modifies the membership function representing the meaning of the atomic syntagm. His approach, however, suffers from linguistic non-adequateness and, therefore, other models have been introduced. In [8], the following model was proposed, which had actually been implemented in the LFLC system.

The meaning of all the simple evaluating syntagms (1.4) is constructed on the basis of the following class of membership functions

$$\mathcal{Z}([p,q]) = \{Z, S \mid Z, S : [p,q] \longrightarrow [0,1]\} \tag{1.12}$$

where Z, S are parametric functions given by

$$Z(x, a_Z, c_Z, \sigma_Z) = 1 - \varepsilon_Z(x, a_Z, c_Z, \sigma_Z)x, \tag{1.13}$$

$$S(x, a_S, c_S, \sigma_S) = 1 - \varepsilon_S(x, a_S, c_S, \sigma_S)x \tag{1.14}$$

with the basic parameters $a_Z, c_Z, a_S, c_S \in [p,q]$, and the modifier parameters being some numbers $\sigma_Z, \sigma_S > 0$. Furthermore,

$$\varepsilon_Z(x, a_Z, c_Z, \sigma_Z) : [p,q] \longrightarrow [0,1],$$
$$\varepsilon_S(x, a_S, c_S, \sigma_S) : [p,q] \longrightarrow [0,1]$$

are the following parametric functions:

(a)

$$\varepsilon_Z(x, a_Z, c_Z, \sigma_Z) = \begin{cases} 0 & \text{for } x \le u_Z, \\ \text{is strictly increasing} & \text{for } u_Z < x < v_Z, \\ 1 & \text{for } v_Z \le x \end{cases}$$

where $u_Z = \min\{\sigma_Z c_Z, q\}$ and $v_Z = \min\{\sigma_Z a_Z, q\}$.

(b)

$$\varepsilon_S(x, a_S, c_S, \sigma_S) = \begin{cases} 1 & \text{for } x \le v_S, \\ \text{is strictly decreasing} & \text{for } v_S < x < u_S, \\ 0 & \text{for } u_S \le x \end{cases}$$

where $u_S = \min\{\sigma_S c_S, q\}$ and $v_S = \min\{\sigma_S a_S, p\}$.

Both functions are, moreover, strictly decreasing with respect to the special parameter σ, i.e.

$$\varepsilon(x, a, c, \sigma) > \varepsilon(x, a, c, \sigma')$$

for $\sigma < \sigma'$.

The outcome of such approach is that all the considered membership functions have only two forms taken from the class $\mathcal{Z}([p,q])$ and, moreover, the obtained mathematical model of the meaning of the linguistic syntagms fits the linguistic requirements. To be precise, the functions (1.13), (1.14) are used both to generate the evaluations in the multi-formulas (1.8) as well as to form their interpretation in the numerical model.

In [9], another model has been proposed where the unary connectives are interpreted using special parametric functions and the meaning of the whole evaluating syntagm is obtained from simple linear functions. However, this model has not so far been implemented in LFLC.

4.3 CHARACTERIZATION OF THE INFERENCE METHOD

In this subsection, we will briefly describe the inference method. It should be stressed that it is logical deduction based on the implications (1.10), which lay in the background of the interpretation of IF-THEN rules. The IF-THEN rules forming the linguistic description are understood as special axioms determining some fuzzy theory T. The inference is then realized using the evaluated modus ponens rule

$$r_{MP}: \frac{a/A, b/A \Rightarrow B}{a \otimes b/B}. \tag{1.15}$$

where $a \otimes b \in [0,1]$ is the resulting evaluation of the formula B and $\otimes$ is the operation of Łukasiewicz conjunction (for the details see [9]). It is quite often that one inference step is realized with one rule only. Note, however, that this means that (1.15) is applied to all the evaluated formulas from (1.10).

Instead of formal analysis, we will better demonstrate the behaviour of the used inference method on an example. To do this, let us consider the linguistic description consisting of two rules:

$$\mathcal{R}_1 := \text{IF } x \text{ is small AND } y \text{ is small THEN } z \text{ is big}$$
$$\mathcal{R}_2 := \text{IF } x \text{ is big AND } y \text{ is big THEN } z \text{ is small.}$$

Let x, y, z be interpreted in a model with $[p,q] = [0,1]$. Then one may agree that small values are some values around 0.3 (and smaller) and big ones are some values around 0.7 (and bigger). Furthermore, given the input, e.g., $x_0 = 0.3$ and $y_0 = 0.25$ we expect the result $z \approx 0.7$ due to the rule $\mathcal{R}_1$. Similarly, for $x_0 = 0.75$ and $y_0 = 0.8$ we expect the result

$z \approx 0.25$ due to the rule $\mathcal{R}_2$. The inference method must recognize such situation and in each case, fire the appropriate rule.

The algorithm realizing the inference just described is called *inference with pre-selection of rules* and is implemented in the system LFLC. For the theoretical foundations of the inference method and the interpretation of the used linguistic expressions see the monograph [9].

Another problem, which must be solved in connection with the inference method is *defuzzification*. However, the most often used *center of gravity* defuzzification method fails because the shape of the resulting fuzzy sets after the inference proceeds is similar to those from the Figure 1.1. Therefore, we had to develop a special method called *Defuzzification of Evaluating Expressions* (DOEE) (cf. [7]). This method begins with classification of the resulting fuzzy set into one of the three types, namely "small, medium, big". Then, the defuzzification is realized in such a way that small values are shifted to the left and big values are shifted to the right. Using this method, we are able to find a value (more or less) intuitively fitting the expected one.

4.4 LINGUISTIC CONTEXT

When applying the above described inference method, it is necessary to give the information about the *linguistic context* within which we understood the meaning of the atomic syntagms "small, medium, big". In practice this means to specify the bounds p, q for the interpretation (1.11). This way can be justified as follows.

It is clear that the same word, for example "small", is used in various situations to denote completely different objects. It is obvious that a "small man" is a man about 160 cm while "small mountain" may be about 600 m. Yet, the general shape of the membership function, as demonstrated in many papers, is in both cases more or less the same since people understand the concept of being "small" equally. In our formalism, this general shape is given by the general form of the multi-formula (1.8). When interpreting it, we then specify the support of the model, which is the mentioned interval $(1.11)^2$. Note that in principle, the linguistic context together with the linguistic modifiers provide the necessary modification of the membership functions, otherwise hidden inside the linguistic description.

5. RESULTS OF THE FUZZY CONTROL IN THE REAL PLANT

The fuzzy control of all five furnaces in the Metallurgic Works in Břidličná is based on the above mentioned linguistic approach. The input variables are derived from measuring of the temperature of the metal on two places, which have been chosen to represent the behaviour of the metal temperature in two different stages of its melting — remember that melting begins from the top to down, so that some parts may be already fully melted while the others not yet. Hence, the first temperature sensor is placed in the place which seems to represent the temperature of the whole bulk of metal in the best way. The second sensor is on the place with much higher dynamics, which makes possible to regard the transition of the system when controlling it.

The rules have the following form:

$$\text{IF } E_1 \text{ is } \mathcal{A} \text{ AND } \Delta E_1 \text{ is } \mathcal{B} \text{ AND } E_2 \text{ is } \mathcal{C} \text{ THEN } U \text{ is } \mathcal{D}$$

where $E_1 = w - T_1$ is the error of the first temperature, ΔE_1 is its derivative, $E_2 = w - T_2$ is the error of the second temperature, and U is the control action, which is the amount of the gas brought to the jet.

The form of rules corresponds to the PD-fuzzy controller. Their number is 155. Though fairly high, we decided not to reduce it. First of all, it was not too difficult to prepare all of them and because of their linguistic nature, tuning of the whole linguistic description was quite simple (recall that this means only replacing less suitable terms inside the rules by more suitable ones according to their linguistic meaning). Furthermore, the systems turned out to be able to react correctly also in the emergency situations, such as the work of only one sensor instead of two ones. Of course, the control in such a case is worse, but still acceptable.

Example of the rules from the real process are below:

IF E_1 is +ExBi AND ΔE_1 is ignored AND E_2 is +ExBi THEN U is ExBi
IF E_1 is +Bi AND ΔE_1 is +ExBi AND E_2 is +Bi THEN U is Sm
IF E_1 is +Bi AND ΔE_1 is +Bi AND E_2 is +Bi THEN U is Me
IF E_1 is +Bi AND ΔE_1 is +Me AND E_2 is +Bi THEN U is ExBi

⋮

etc.

From the methodological point of view, it is also useful to have the possibility to define default control action when no specific situation is characterized. This can be accomplished using special syntagms, namely "ignore" and "undefined". For example, let us consider the rules

$$\mathcal{R}_1 := \text{IF } x \text{ is small AND } y \text{ is ignored THEN } z \text{ is big}$$
$$\mathcal{R}_2 := \text{IF } x \text{ is small AND } y \text{ is small THEN } z \text{ is small.}$$

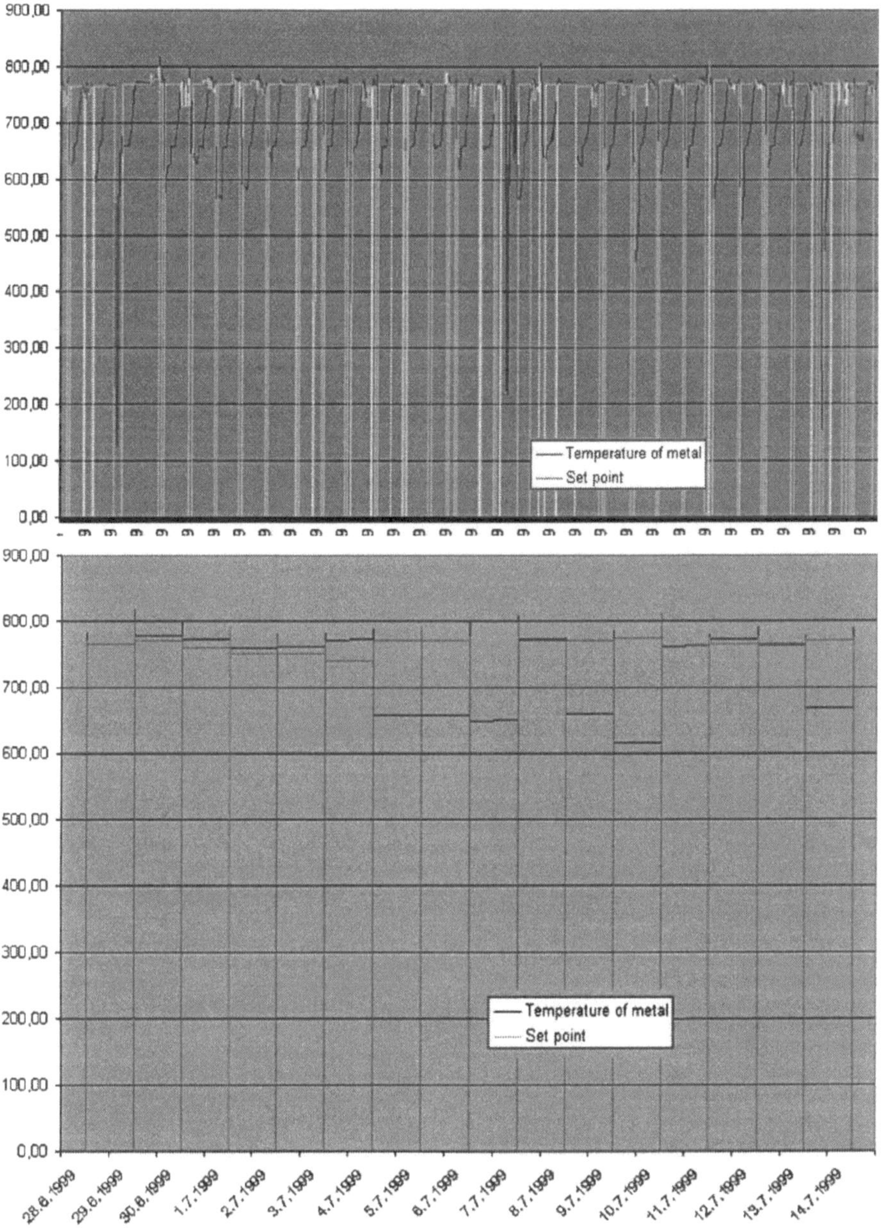

Figure 1.2 Results of fuzzy control on Furnaces 2 (up) and 4 (down) during one month June-July 1999

Then the rule $\mathcal{R}_1$ fires always when the value of x is considered to be "small". Hence, the second rule $\mathcal{R}_2$ is useless since it never fires. On the other hand, if these rules would have the form

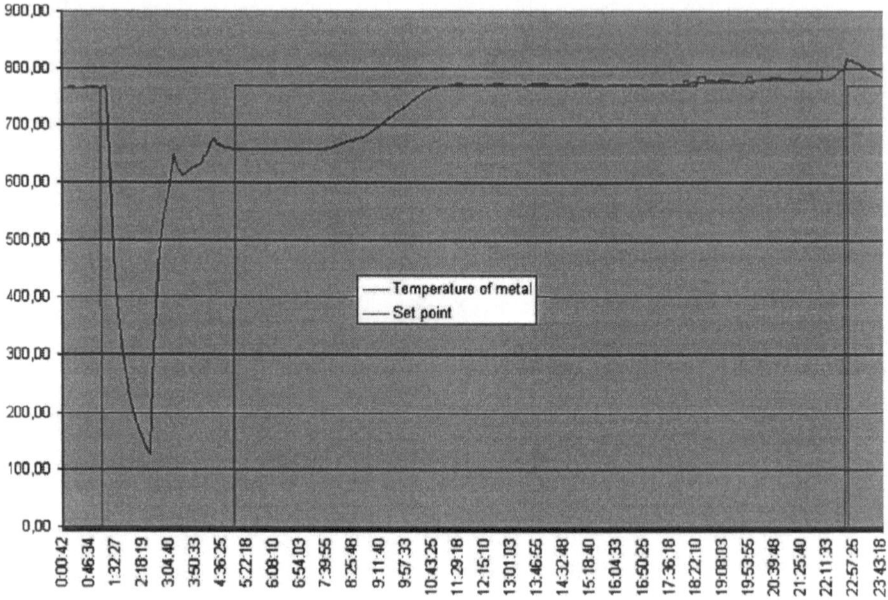

Figure 1.3 Results of fuzzy control on Furnace 2 on June 29, 1999

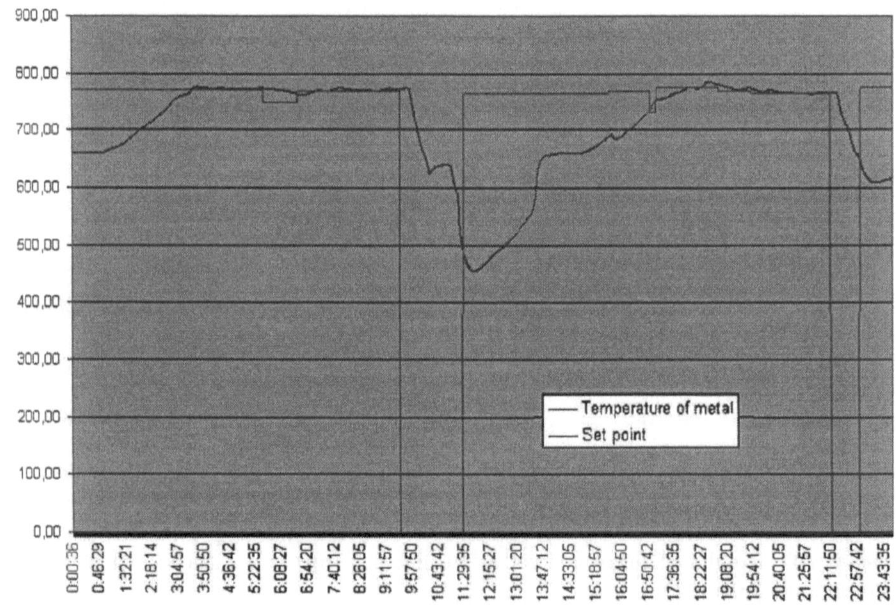

Figure 1.4 Results of fuzzy control on Furnace 2 on July 9, 1999

$$\mathcal{R}_1 := \text{IF } x \text{ is small AND } y \text{ is undefined THEN } z \text{ is big}$$
$$\mathcal{R}_2 := \text{IF } x \text{ is small AND } y \text{ is small THEN } z \text{ is small.}$$

then the rule $\mathcal{R}_2$ fires whenever both x and y are small while the rule $\mathcal{R}_1$ fires if x is small and y is not.

Further, we present some figures demonstrating the behaviour of the fuzzy control on three of the five furnaces. Let us stress that we have chosen the working days quite randomly and the control is practically the same in all times. At the same time, however, the operation situation in the enterprise may vary, sometimes quite significantly. The areas in which the required value is 0 correspond to time periods when the operator has switched out the control system completely. Though improper, it is difficult to persuade the operators not to do it.

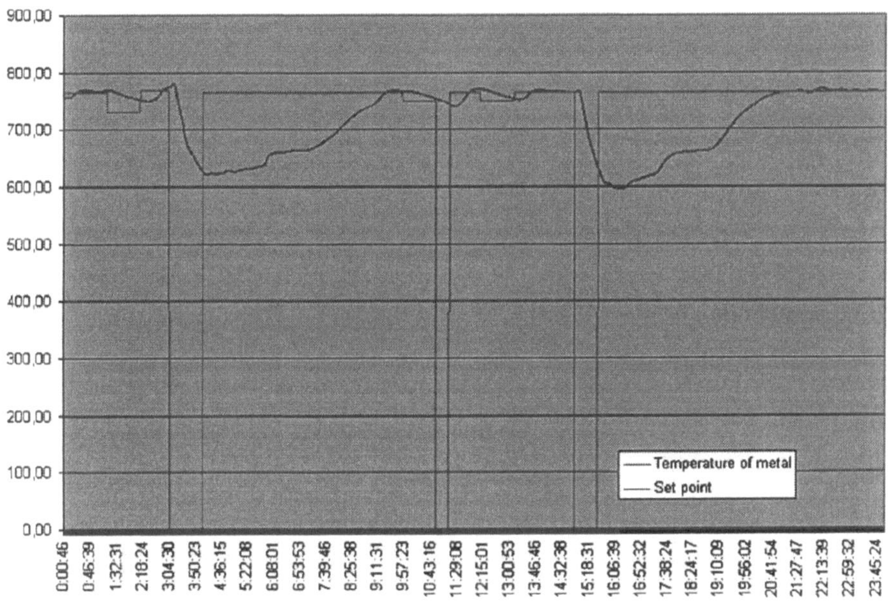

Figure 1.5 Results of fuzzy control on Furnace 4 on June 28, 1999

Figure 1.2 demonstrates the behaviour of Furnaces 2 and 4 during one month. It can be clearly seen that it oscillates in 8 hours periods and the global behaviour is practically stable. The other figures show the behaviour of three different furnaces during randomly chosen periods. Note especially the behaviour of the temperature in the decisive phase when it is necessary to keep it on the required value.

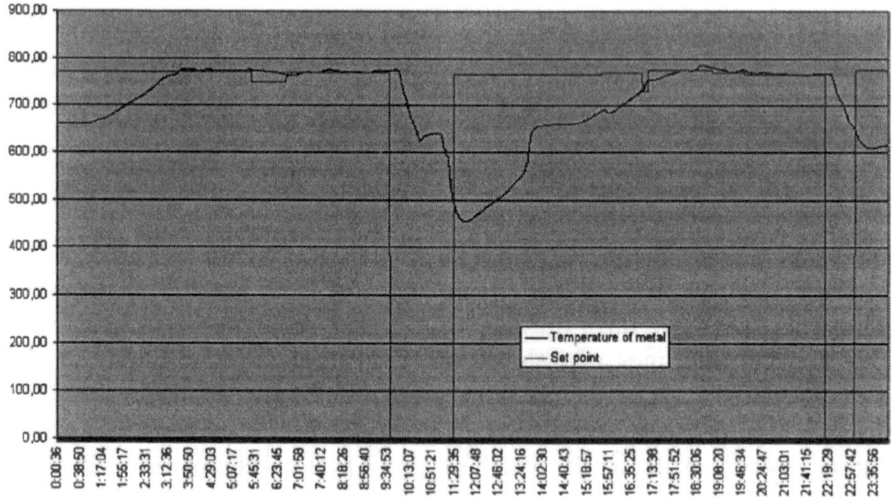

Figure 1.6 Results of fuzzy control on Furnace 4 on July 9, 1999

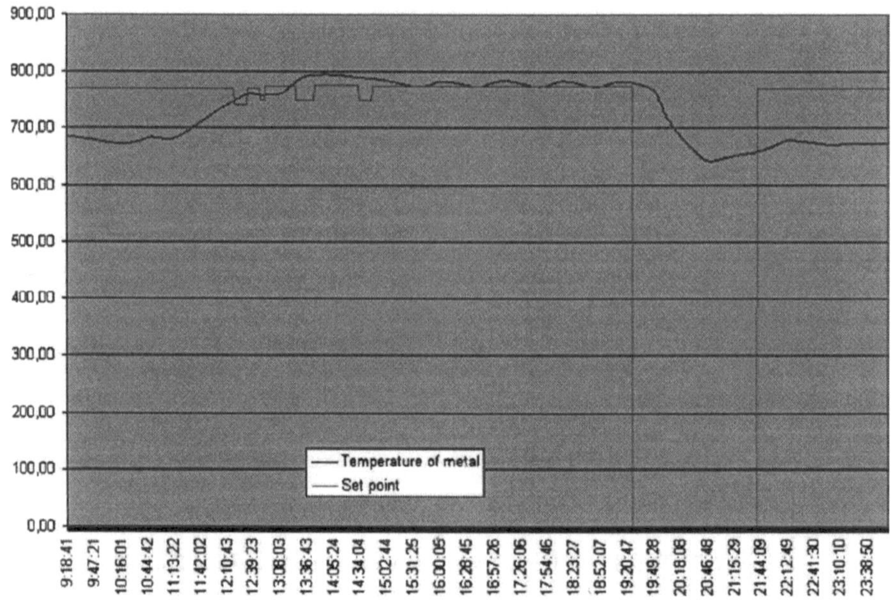

Figure 1.7 Results of fuzzy control on Furnace 5 on June 23, 1998

6. CONCLUSION

This Chapter is a brief presentation of the large scale application of fuzzy control in the Metallurgic plant Břidličná. The fuzzy control is used on five aluminium smelting furnaces, which produce aluminium for further processing in the other plants of the company.

We have briefly described problems with control of the furnaces and our experience with classical controllers used before, as well as the reasons which led us to use fuzzy control. We described principles of the construction of the linguistic description, using which the fuzzy control is realized. We also presented figures demonstrating behaviour of three of five furnaces chosen at random, and in the periods chosen by chance, too.

At the end, let us mention that besides control of the temperature of the metal, we have solved also another control problem of the furnace — to control its empty space after the brickwork is replaced. The structure of rules is similar to those presented above with the exception that the rules correspond to the PI-fuzzy ones. The results of this control are also very satisfactory.

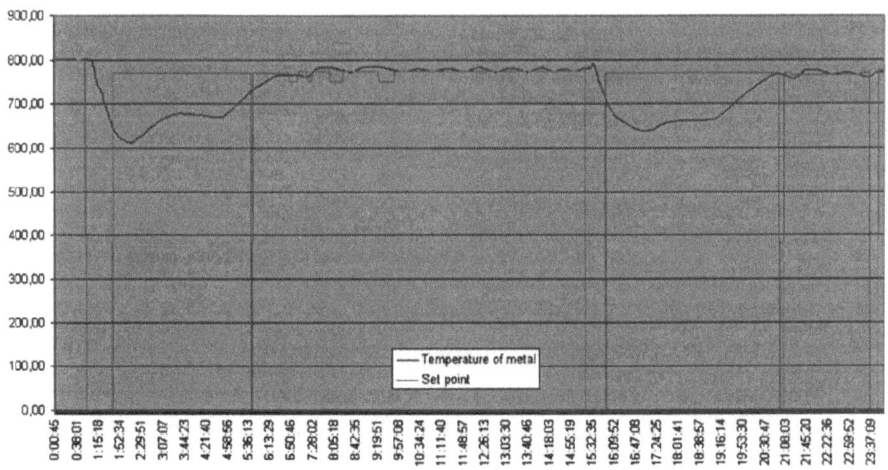

Figure 1.8 Results of fuzzy control on Furnace 5 on July 2, 1999

References

[1] Bělohlávek, R., and V. Novák. Learning Rule Base of the Linguistic Expert Systems. Int. J. of Soft Computing (submitted).

[2] Driankov, D., Hellendoorn, H., and M. Reinfrank. (1993). *An Introduction to Fuzzy Control.* Berlin: Springer.

[3] Klir, G.J., and Yuan, B. (1995). *Fuzzy Sets and Fuzzy Logic: Theory and Applications.* New York: Prentice-Hall.

[4] Kruse, R., Gebhardt, J., and Klawonn, F. (1994). *Foundations of Fuzzy Systems*, Chichester: Wiley.

[5] Mamdani, E.H. (1974). "Applications of fuzzy algorithms for simple dynamic plant," Proc. IEE **121**, 1585–1588.

[6] Novák, V. (1989). *Fuzzy Sets and Their Applications.* Bristol: Adam Hilger.

[7] Novák, V. (1995). "Linguistically Oriented Fuzzy Logic Controller and Its Design," Int. J. of Approximate Reasoning 1995, **12**, 263–277.

[8] Novák, V., and I. Perfilieva (1999). "Evaluating Linguistic Expressions and Functional Fuzzy Theories in Fuzzy Logic," In: L. A. Zadeh and J. Kacpryk (eds.): *Computing with Words in Information/Intelligent Systems 1.* Springer-Verlag, Heidelberg, 383–406.

[9] Novák, V., Perfilieva I., and J. Močkoř (1999). *Mathematical Principles of Fuzzy Logic.* Kluwer, Boston/Dordrecht 1999.

[10] Prokop, R., Turisová, Z., and Mészaros, A. (1992). "Unified Self-Tuners for Industrial Applications," Proc. of IFAC Symposium on LCA'92, Vienna, Austria.

[11] Zadeh, L. A. (1965). "Fuzzy Sets," Inf. Control. **8**, 338–353.

[12] Zadeh, L.A. (1972). "A fuzzy-set-theoretic interpretation of linguistic hedges," J. Cyber. **2**, 2–34.

[13] Zadeh, L.A. (1973). "Outline of a New Approach to the Analysis of Complex Systems and Decision Processes," IEEE Trans. Syst. Man and Cybern, **1**, 28–44.

[14] Zadeh, L.A. (1972). "A rationale for fuzzy control". J. of Dynamical Systems, Measurement, and Control **94**, 3–4.

Chapter 11

FUZZY RULES EXTRACTION-BASED LINGUISTIC AND NUMERICAL HETEROGENEOUS DATA FUSION FOR INTELLIGENT ROBOTIC CONTROL

Changjiu Zhou and Da Ruan*
Department of Electronics & Communication Engineering
Singapore Polytechnic, 500 Dover Road
Singapore 139651

**Belgian Nuclear Research Centre (SCK•CEN)*
Boeretang 200, B-2400 Mol
Belgium
Key words: Fuzzy if-then rules extraction, information integration, data fusion, intelligent control, robot control, biped robot.

Abstract: Autonomous robots, which achieve tasks without human operators, are required in many fields. They have to be intelligent to determine their own motions in unknown environments based on both sensory data (numerical) and expert knowledge (linguistic), that is, the linguistic-numerical heterogeneous data (LNHD). This chapter looks at the recent development in the area of data fusion approach in conjunction with the concept of fuzzy if-then rule for intelligent robotic control. To make the most use of the LNHD available for robot control, we classify the LNHD into four categories: direct data (DD), direct rules (DR), indirect data (ID), and indirect rules (IR). To fuse the LNHD with inverse mapping properties, based on inverse learning, we convert the ID, IR, DD, and DR into a uniform domain. Hence, some data fusion operators are used to fuse the LNHD with direct and indirect forms. Then we propose a fuzzy if-then rules extraction-based LNHD fusion method. It can construct a uniform fuzzy rule base through the fusion of the fuzzy rule bases obtained from both experts and numerical data extraction. To illustrate the proposed method, we demonstrate how this method can be used for gait synthesis of a biped robot. The simulation results show that the gait can be significantly improved by the fuzzy rules extraction-based LNHD fusion through the integration of linguistic rules obtained from human intuitive walking knowledge, and numerical data received from biped locomotion experiments. Current problems and future directions in this area are also addressed in this chapter.

1. INTRODUCTION

In recent years there has been a rapid growing interest in the need for information integration to solve real-world problems in control and planning for robotic systems [10, 16]. Sensor-based operation of autonomous robots in unstructured and/or outdoor environments has been proved to be an extremely challenging problem, mainly because of the many uncertainties that are always present in the real world. In robot control, these uncertainties are primarily due to sensor imprecision and unpredictability of the environment, that is, lack of full knowledge of the environment characteristics and dynamics. Humans, on the other hand, seem to cope very well with uncertain and unpredictable environments, often relying on approximate or qualitative (fuzzy) data and reasoning to make decisions and to accomplish their objectives. To meet such challenges, it is important to develop robot control systems that can integrate various types of sensory information and human knowledge in order to carry out tasks with or without need for human intervention. However, most of information obtained for robot control are hybrid, that is, their components are not homogeneous but a blend of numerical and linguistic, direct and indirect, and etc. We call the data of above types or vectors whose components are any of this, Linguistic-Numerical Heterogeneous Data (LNHD). As an example, the heterogeneous sensor fusing of sonar and laser measurements in a mobile robot can be described as a following fuzzy IF-THEN rule:

IF	the front space is wide *enough*	(laser)
THEN	*center* the robot	(sonars)

To center the robot, that is, to steer away from close obstacles sensed by the sonars on either side of the robot, we can use the following fuzzy IF-THEN rules:

IF left-sonar is *close* AND right-sonar is *not close* THEN steer *right*
IF left-sonar is *not close* AND right-sonar is *close* THEN steer *left*
IF left-sonar is *close* AND right-sonar is *close* THEN steer *center*
IF left-sonar is *not close* AND right-sonar is *not close* THEN steer *center*

It is unlikely that one technique or one architecture can provide a uniformly superior solution for robot control. Neural networks (NN) and conventional control theory are useful if sufficient numerical data is available or measurable, while fuzzy logic (FL) and artificial intelligence (AI) are appropriate if sufficient linguistic knowledge is available. Therefore, there is a need to use hybrid intelligent techniques [20, 35] to integrate the LNHD for robot control.

There exists voluminous literature on the subject of fuzzy IF-THEN rules, hybrid intelligent techniques and multisensor fusion to make use of either sensory information (numerical data) or expert knowledge (linguistic rules) for robotic planning and control [2, 10, 11, 16, 19, 24, 26, 27, 37]. However, very few papers are found that report the linguistic and numerical heterogeneous data (LNHD) fusion for intelligent robotic control.

Several successful applications have been reported on the sensor fusion approach in conjunction with fuzzy behavior fusion and fuzzy IF-THEN rule concept for intelligent robotic systems. Beom and Cho [2] proposed a sensor-based navigation method that utilizes fuzzy logic and reinforcement learning for navigation of a mobile robot in the uncertain environments. The fuzzy rules for each behavior are learned through reinforcement learning; therefore, the fuzzy rule base can be easily constructed for more complex environments. Goodrige *et al.* [11] implemented a multilayered fuzzy behavior fusion based reactive control system on an autonomous mobile robot, MARGE, with great success. The MARGE won first price in the event III of the 1993 Robot Competition sponsored by the America Association for Artificial Intelligence. Matia and Jimenez [19] studied the fusion of heterogeneous sensors in a mobile robot to reduce uncertainty while it navigates, avoids obstacles, integrates new observations to refresh environment maps, and coordinates vision systems to improve location estimation. Pin [24] proposed a fuzzy behaviorist approach (FBA) which provides formalism for the development of fuzzy IF-THEN rule systems for control of autonomous robots. The FBA forms the fuzzy rule bases to embody "human-like" behaviors in sensor-based decision-making systems. Shi *et al.* [26] discussed the issues of behavior integration and action fusion for excavation control of a robotic front-end-loader-type machine. They also proposed a fusion procedure to combine bucket actions at a particular state using some fuzzy logic reasoning and defuzzification methods. Yung and Ye [33] presented a fuzzy navigator that performs well in complex and unknown environments using a fuzzy IF-THEN rule base which is learned from a simple corridor-like environment.

There are also a few literature of multisensor fusion with fuzzy logic for other applications. Fillippidis [9] applied the Dempster-Shafer, fuzzy reasoning, and neural networks to data fusion for automatic target recognition using both sensory data from measurements and *a priori* information obtained from flight timetables. Hong and Wang [13] discussed centralized integration of multisensor noisy and fuzzy data (IF-THEN rules) using the Kalman filtering and fuzzy arithmetic. Stover *et al.* [27] developed a general-purpose fuzzy logic architecture that provides the capability for fusion of multisensor data to achieve high-level inferences about objects and situations. This fuzzy logic architecture has been applied to implement an autonomous vehicle that is capable of self-direction, obstacle avoidance, and mission completion. There have been several reviews of sensor fusion and its application. Kam *et al.* [16] reviewed some techniques for sensor fusion in mobile robot navigation.

Most of the results on data fusion for robot planning and control were obtained under some assumptions, such as the complete knowledge of the working environment is usually assumed. However, one of the greatest challenges facing the research community is to extend the domains of application of intelligent robotic control to the general class of unstructured environments, that is, the environments that are generally dynamic, not fully known *a priori*, and typically unpredictable. Humans, on the other hand, seem to cope very well with uncertain and unpredictable environments. Therefore, the real challenge in this area is to integrate both sensory information and human knowledge, that is, the LNHD, for intelligent robotic systems.

In reviewing the existing literature, very few papers are found those report the fusion of LNHD with different forms, such as direct and indirect information. Pedrycz *et al.* [12, 23] proposed two heterogeneous fuzzy data (HFD) models: parametric and non-parametric models, and their applications to pattern recognition. The basic idea of their models is to transform all the data types (real numbers, real intervals, and real fuzzy sets) into a uniform set of generalized coordinates of the HFD model. However, this model should be modified for control applications.

To make the most use of the LNHD available for robot control, we classify the LNHD into four categories: direct data (DD), direct rules (DR), indirect data (ID), and indirect rules (IR). The direct data/rules can be used to describe the controller behavior, while the indirect data/rules can only describe the robots' behavior. In our previous work [35, 36, 37], we studied three kinds of methods to integrate the ID, IR, DD, and DR. In the following sections, in order to fuse the LNHD with inverse mapping properties, we convert the ID, IR, DD, and DR into a uniform domain based on inverse learning [32]. Then we propose some simple and intuitive methods to integrate the LNHD by means of data fusion operators [4, 5, 7]. In the last part of this chapter, we show how the proposed fuzzy rules extraction-based LNHD fusion method can be used for the gait synthesis of a biped robot.

2. LINGUISTIC-NUMERICAL HETEROGENEOUS DATA (LNHD) IN INTELLIGENT ROBOTIC CONTROL

2.1 The Information Available for Intelligent Robotic Systems

There has been a growing interest in the development of intelligent robotic systems. The intelligence of a robotic system can be characterized by three functional abilities. First, the robotic system should be controlled directly as a task level; that is, it should take task-level commands directly, without any planning type decomposition to joint level commands. Second, the control systems of robots should be designed for a large class of tasks rather than a specific task. Finally, the robotic system should be able to handle some unexpected or uncertain events. Intelligent robotic control depends on a large extent of the capability of the robotic system to acquire, to process, and to utilize both *sensory information* and *human knowledge* to plan and execute actions in the presence of various changing or uncertain events in the robot-working environment.

Figure 1 shows a functional architecture for an intelligent robotic controller with an interface to an autonomous robot involving sensing and actuation, and an interface to humans and other systems [25]. The organizer provides for the supervision of lower level functions and for managing the interface to the humans. The coordinator provides for tuning, scheduling, supervision, and redesign of the control algorithms; crisis management; planning and learning capabilities for the coordination of the execution-level tasks; and higher-level symbolic decision making. The controller has low-level numeric signal processing and control algorithms. Based on the principle of increasing precision with decreasing intelligence (IPDI) [25], the highest level is assigned with the highest machine intelligence and smallest complexity (size of database) and the lowest level with the lowest machine intelligence and largest complexity. As an example, for the biped robot [30, 37], at the organization level, the system reasons symbolically for strategic plans or schedules of the biped motion using the knowledge base. This level should also include the common sense knowledge for the biped motion. At the organizational level, a human operator mainly gives the knowledge in a top-down manner. There are different kinds of information available for the coordination level (gait synthesizer), such as the sampled data from the biped experiments (numerical); the intuitive walking knowledge (linguistic); and the walking knowledge which has been obtained from the biomechanics studies of human walking (linguistic + numerical). At the control level, the information is mainly given by measuring instruments in a bottom-up manner (numerical). Moreover, the servo control knowledge (linguistic) is also useful for the joint controller design and tuning. However, each of above sets of information alone is usually incomplete.

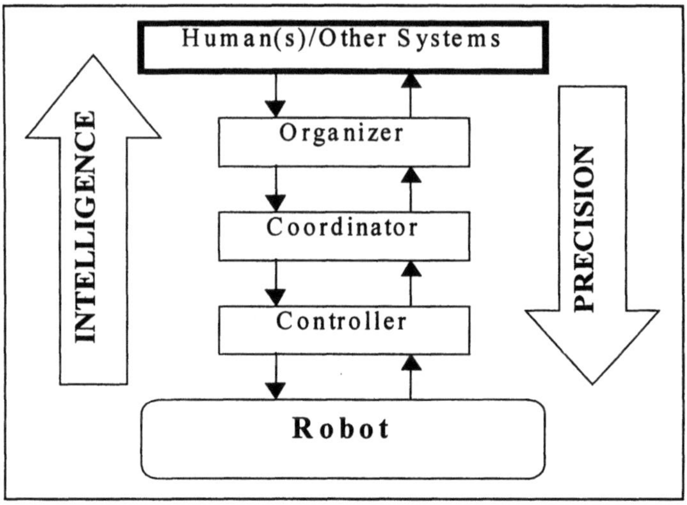

Figure 1. Hierarchical intelligent robotic controller.

2.2 Linguistic-Numerical Heterogeneous Data (LNHD)

From the above discussion we found that the information available for the intelligent robotic systems may be contained in two types of data: numerical data received from sensor measurements, and linguistic data obtained from human operators and domain experts. We call the data of above types or vectors whose components are any of this, Linguistic-Numerical Heterogeneous Data (LNHD).

In the real-world situations, the numerical data may be noisy, inconsistent, and incomplete, and the linguistic information is usually imprecise and vague. Therefore, a fuzzy set is a naturally good choice to fuse the LNHD. As an example, let's consider the speed measurement of a mobile robot (see Figure 2). The speed can be expressed as a real number if the sensors can precisely measure it. If the speed is evaluated non-numerically by a human observer, it may be interpreted as the linguistic term, such as the speed is "fast." It can also be represented as a single interval. Uncalibrated instruments could lead to this situation.

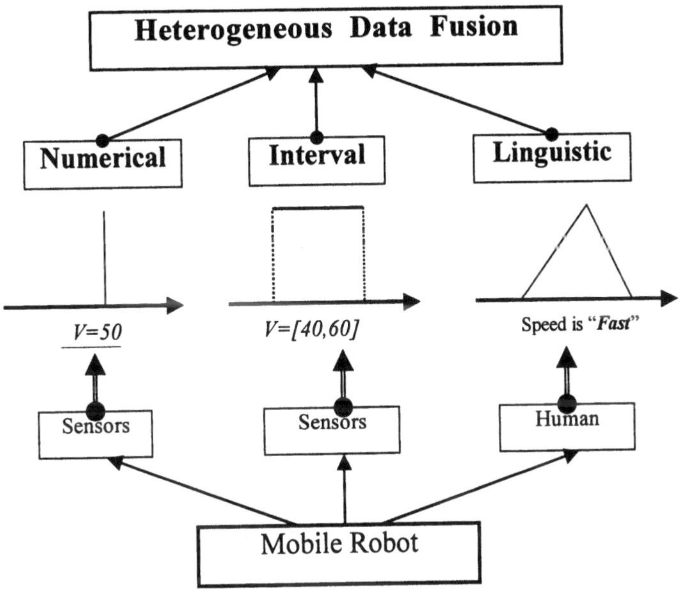

Figure 2. Heterogeneous data fusion for a mobile robot.

In [12, 23], Pedrycz *et al.* proposed two heterogeneous fuzzy data (HFD) models: parametric and nonparametric models, and their applications to the pattern recognition. Let $\Re$ be the real numbers, $I(\Re)=I$ be all real closed intervals such as $[a,b]$, and $F(\Re)=F$ be the real fuzzy subsets of $\Re$. An element of F is a membership function $m:\Re \mapsto [0,1]$ whose values $\{m(x): x \in \Re\}$ are the grades of membership of each x in the fuzzy set m. Every real number and interval have crisp membership functions. Thus the HFD (real numbers, real intervals, and fuzzy sets) are all the elements of F. In this way, the most general form of the HFD becomes a collection of n $p-$tuples $M=\{m_1, m_2, \cdots, m_n\} \subset F^p$, where p is the number of features observed in the chosen representation of the HFD. The above method assumes all the sensors and experts have the same importance, that is, the same degree of truth. However, for some applications, each sensor and expert may not have the same contribution to the final fusion decision. On the other hand, for robot control systems, there are some special problems to be considered.

From a high-level conceptual point of view, any robot control system consists of at least a planner and a controller (see Figure 3). For controller design, there are two types of expert knowledge. One is the *robot knowledge* that describes the behavior of the robots; another is *control knowledge* that can be described as fuzzy IF-THEN rules that state in which situations what control actions should be taken. The control knowledge can be *directly* used to design the controller, while the robot knowledge must be used in an *indirect* way. The control knowledge may also be classified into two categories: *conscious* knowledge and *subconscious* knowledge. By conscious knowledge we mean the knowledge that can be explicitly expressed in words, therefore, we can simply ask the human experts to express it in terms of fuzzy IF-THEN rules. By subconscious knowledge we refer to the situations where the human experts know what to do but cannot express exactly in words how to do it, what we can do is to ask the human experts to demonstrate, and collect a set of input-output numerical data pairs.

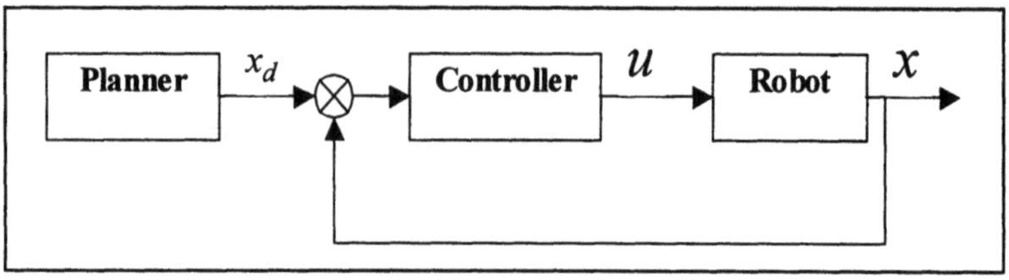

Figure 3. Robot planning and control system.

Based on the above consideration, in order to make the most use of the LNHD available for robot control, we classify the LNHD into four categories: direct data (DD), direct rules (DR), indirect data (ID), and indirect rules (IR).

2.3 Representation of the LNHD

In this chapter, we consider a single-output fuzzy rule based system in the n-dimensional input space $[0,1]^n$. Let us assume that human experts give the following fuzzy IF-THEN rules:

$$\begin{aligned} R^{(l)}: & IF \;\; x_1 \;\; is \;\; F_1^l \;\; and \cdots and \;\; x_n \;\; is \;\; F_n^l \\ & THEN \;\; y \;\; is \;\; G^l \end{aligned} \tag{1}$$

where F_i^l $(i = 1,2,\cdots,n)$ and G^l are fuzzy sets, $l = 1,2,\cdots,M$, and M is the number of rules in the fuzzy rule base. $X = (x_1,\cdots,x_n)^T \in U \subset \Re^n$ and $y \in V \subset \Re$ are input and output linguistic variables, respectively.

We also assume that the following input-output pairs are given as numerical data from measuring instruments:

$$(x_1^p, x_2^p, \cdots, x_n^p; y^p) \tag{2}$$

where $p = 1,2,\cdots,P$, and P is the number of input-output data pairs. $X_p = (x_1^p, x_2^p, \cdots, x_n^p)$ is the input vector of the pth input-output pair and y^p is the corresponding output.

As an example (see Figure 3), we assume that a robot has only one state $x(k)$ and one input $u(k)$, the control objective is to design a controller such that the robot state $x(k)$ can follow a desired trajectory $x_d(k)$ as closely as possible. If we assume that the order of the model is given, and all the state variables are measurable, and the inverse dynamics of the model exist, then we have

$$\begin{aligned} x(k+n) &= F(x(k),u(k)) \\ u(k) &= G(x(k),x(k+n)) \end{aligned} \tag{3}$$

For the above robot control system, the LNHD can be described as following.

- Direct Rules (DR): $IF\ \ x(k)\ \ AND\ \ x(k+n)\ \ THEN\ \ u(k)$
- Indirect Rules (IR): $IF\ \ x(k)\ \ AND\ \ u(k)\ \ THEN\ \ x(k+n)$
- Direct Data (DD): $(x(k), x(k+n); u(k))$
- Indirect Data (ID): $(x(k), u(k); x(k+n))$

3. LINGUISTIC AND NUMERICAL HETEROGENEOUS DATA (LNHD) FUSION

3.1 Information Integration

In recent years, multisensor data fusion technology has been rapidly evolved. Numerous prototype systems have been developed [4, 5, 7, 16]. Robot designers can choose from a large number of sensor types and sensing modules. These are sometimes complementary, sometimes redundant, and there also exist architectures where sensors are used in both fashions. Kam *et al.* [16] classified the sensor fusion techniques into two categories. The low-level fusion is used for direct integrating sensory data, resulting in parameter and state estimates. The high-level fusion is used for indirect integration of sensory data in hierarchical architectures, through command arbitration and integration of control signals suggested by different modules.

When pieces of information issued from several sources have to be integrated, each of them represented as a degree of belief in a given event, these degrees are combined in the form $F(x_1, x_2, \cdots, x_n)$, where x_i denotes the representation of information issued from sensor i. A large variety of information combination operators F for data fusion have been proposed.

Let us consider a function F acting only two pieces of information x and y. Where x and y denote two real variables representing the degree of belief to be combined, they take values into the interval $I \subset [0,1]$. Under closure property assumption, $F(x,y)$ also has values in the interval I. Based on the definitions for fuzzy operators, the fusion operator F can be classified into three categories:

(1) F is conjunctive if $F(x,y) \le \min(x,y)$.

(2) F is disjunctive if $F(x,y) \ge \max(x,y)$.

(3) F behave like a compromise if $\min(x,y) \le F(x,y) \le \max(x,y)$.

In the framework of fuzzy sets and possibility theory, triangular norms (T-norms), triangular conorms (T-conorms), and mean operators are typical examples of conjunctive, disjunctive, and compromise fusion operators respectively. Therefore, we may choose some fuzzy operators for information integration. The following are some most common used fusion operators.

- T-norm operators: $\min(x,y)$, xy, $\max(0, x+y-1)$.
- T-conorm operators: $\max(x,y)$, $x+y-xy$, $\min(1, x+y)$.
- Mean operators: the median operators $m_\alpha(x,y) = med_\alpha(x,y,\alpha)^2$, the harmonic mean $2xy/(x+y)$, the geometrical mean $\sqrt{xy}$, the arithmetical mean $(x+y)/2$, the quadratic mean $\sqrt{(x^2+y^2)/2}$, the ordered weighted operators.

The choice of the fusion operators depends on the aim of the fusion. Bloch [4] proposed a classification of fusion operators. The Context Independent Constant Behavior operators (CICB) have the same behavior whatever the values of the information to be combined. It can be proved that the T-norms, T-conorms, and mean operators discussed above are all the CICB operators [4]. The Context Independent Variable Behavior operators (CIVB) are more or less severe, depending on the degrees of belief. The Context Dependent operators (CD) are computed not only from the degrees of belief but also depend on a global knowledge or measure on the sources to be fused, such as the degree of conflict, source reliability [8], special constraints [29]. Most of the above fusion operators assume that all the information sources including sensors and expert have the same importance or degree of belief. However, for heterogeneous data fusion, each sensor or expert does not have the same contribution for the final fusion decision [7].

3.2 The LNHD Fusion

There are various approaches to integrate the linguistic rules and numerical data by means of the hybrid intelligent techniques. In [35] we studied different kinds of methods to integrate the ID, IR, DD and DR for controller design and learning. As an example, the comparison of the above LNHD fusion methods for the integration of the direct rules (DR) and indirect data (ID) is shown in Figure 4.

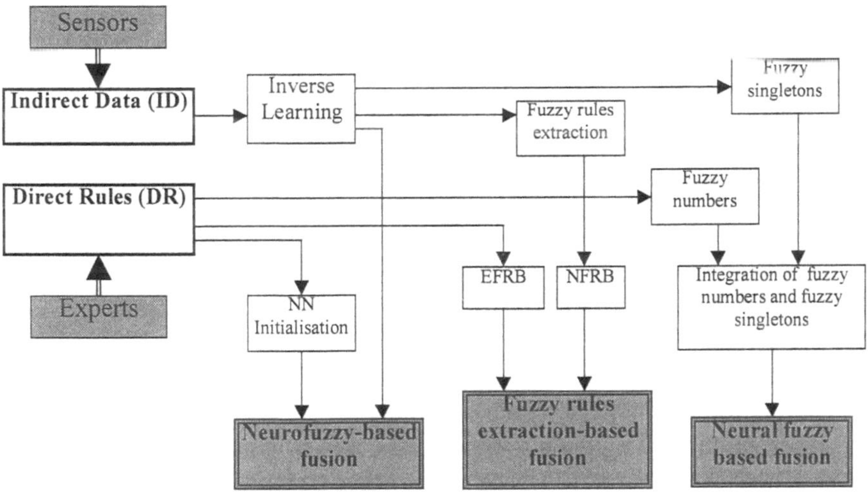

Figure 4. The comparison of the LNHD fusion methods.

The straightest way is to construct a neural-network-based fuzzy system using the fuzzy rules obtained from experts, and then fine-tune its parameters by the numerical data. This is the most common method used for neuro-fuzzy systems [17]. Another approach is to develop a neural fuzzy system with fuzzy supervised learning to process the integration of numerical and fuzzy information [14, 18]. The inputs, outputs, and weights of a neural network, which is the connectionist realization of a fuzzy inference system, can be fuzzy numbers of any shape. The training data can be hybrid of fuzzy numbers and numerical numbers through the use of fuzzy singletons. In order to handle linguistic information in NN, the fuzzy data are viewed as convex and normal fuzzy sets represented by α-level sets. Since the α-level sets of fuzzy numbers are intervals, the operations of real numbers in the traditional NN are extended to closed intervals. A notable deficiency in this approach is that the computations become complex (e.g. multiplication of interval) and time-consuming. In this chapter, we will address the fuzzy rules extraction-based fusion method. In this approach, fuzzy IF-THEN rules are first generated from input-output numerical data pairs [1, 6, 15, 22, 28]. We then construct the numerical data based fuzzy rule base (NFRB) with the degree of belief (DOB) using direct and heuristic matching method. We acquire experts-based fuzzy rules (EFRB) with the

DOB from experts. The final uniform fuzzy rule base (UFRB) can be constructed through the fusion of NFRB with EFRB. After normalizing NFRB and EFRB to ensure the completeness and consistency of the fuzzy rule bases, we propose some simple and intuitive methods to integrate the LNHD through fusing the DOB.

In this chapter, we only discuss the fuzzy IF-THEN rules extraction based LFHD fusion method.

4. FUZZY IF-THEN RULES EXTRACTION FROM NUMERICAL DATA

In many applications of fuzzy rule-based systems, fuzzy if-then rules are obtained from human experts. Recently, several different methods for extracting fuzzy IF-THEN rules from numerical data have been developed. Most of these methods involve iterative learning procedures or complicated rule generation mechanisms such as gradient decent learning method [15], NN based method [15], genetic algorithms (GAs)-based method, fuzzy NN method, heuristic method [22], and many other methods.

For the applications with high-dimensional input space, the above methods usually need a long rules requisition (RA) time. The major restriction of the NN based method is that the number of rules must be predefined before training. In [15, 31], fuzzy rules are defined for multiple subspaces obtained by partitioning the input space. These methods therefore need a long RA time for the applications with a high-dimensional input space. In [6], fuzzy rules are extracted based on clusters. It can group the input-output pairs into clusters and use one rule for one cluster, that is, the number of rules equals to the number of clusters, and thereby the number of fuzzy rules is not fixed before learning. This method can effectively handle applications with high-dimensional input space; however, its RA still grows according to the order of the number of data used for extracting fuzzy rules. In [1, 28], fuzzy rules are defined according to the hyperboxes of input data. A hyperbox is represented by the minimum and maximum values of each input variable for the corresponding data. Due to a compact representation of the hyperboxes, these methods have an extremely low RA time.

Construction of fuzzy logic rules from numerical data for control application consists of two phases: structure learning and parameter identification. Some of the most common used structure learning methods are listed below:

- ♦ Cluster-oriented approaches [6]
    - Find clusters in the data space using fuzzy cluster analysis methods, such as fuzzy C-means.

   - Each cluster represents a fuzzy rule.
- Hyperbox-oriented approaches [1, 28]
   - Search hyperboxes in the data space.
   - Create fuzzy rules by projecting the hyperboxes.
- Structure-oriented approaches [31]
   - Use predefined fuzzy sets to structure the data space, and then the data space is partitioned by a fuzzy grid.
   - Create fuzzy rules from the grid cells.

For the parameter identification of the fuzzy IF-THEN rules, the most popular method is gradient descent algorithm. However, it is only applicable if the fuzzy operators selected are differentiable. If the differentiation is impossible, some special heuristic procedures that do not use gradient information can be employed for the parameter identification. The learning algorithms are usually based on the idea of backpropagation.

In this section, we extend a structure-orientated approach (Wang and Mendel [31]) to extract fuzzy IF-THEN rules from numerical data. It is an efficient rule generation method with no iterative. This is a one-pass build-up procedure that avoids time consuming, the approach is very straightforward; however, the input space must be divided in advance. It divides the input space according to the partition of the training data. As an example, the fuzzy rules can be extracted as following.

Step 1. Partition the training data. To avoid under-fitting or over-fitting of the training data, we partition the training data using a heuristic method. The partition process will be terminated before over-fitting.

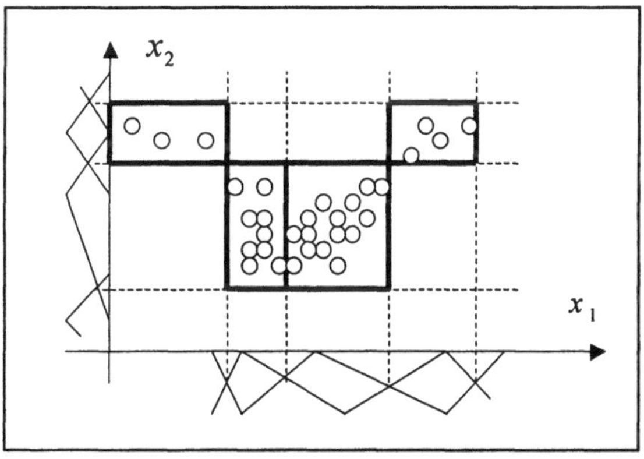

Figure 5. An example of dividing input space into fuzzy regions and the corresponding membership functions according to the partition of the training data.

Step 2. Divide the input and output spaces into fuzzy regions according to the data partition results. The domain intervals of $x_1, x_2, \cdots, x_n$ and y are $[x_1^-, x_1^+]$, $[x_2^-, x_2^+]$,..., $[x_n^-, x_n^+]$ and $[y^-, y^+]$ respectively. Figure 5 shows an example how to divide fuzzy regions and assign fuzzy membership function according to the data partition.

Step 3. Generate fuzzy rules from numerical data pairs.

Step 4. Assign a truth degree to each rule. The following product and maximum strategies are used to assign a truth degree to the rule: " IF x_1 is A_1 and x_2 is A_2 and ... and x_n is A_n THEN y is B."

• Product strategy

$$w^N = \mu_{A_1}(x_1)\mu_{A_2}(x_2)\cdots\mu_{A_n}(x_n)\mu_B(y) \tag{5}$$

• Maximum strategy

$$w^N = Max(\mu_{A_1}(x_1), \mu_{A_2}(x_2), \cdots, \mu_{A_n}(x_n), \mu_B(y)) \tag{6}$$

Using the above fuzzy rules extraction algorithm, we can construct a numerical data based fuzzy rule base (NFRB) with the DOB. If the numerical database has the format as $(x_1^p, x_2^p; y^p)$, $p = 1,2,\cdots,P$, here P is the number of the data pairs, then we can construct the NFRB with the following fuzzy rules:

$$R_i^N : IF \;\; x_1 \;\; is \;\; A_i \;\; AND \;\; x_2 \;\; is \;\; B_i \;\; THEN \;\; y \;\; is \;\; C_i^N \quad (w_i^N)$$

where w_i^N is the DOB of the ith rule in the NFRB.

5. FUZZY RULES EXTRACTION-BASED LNHD FUSION

5.1 Experts Control Knowledge Acquisition

In the above section, we have build an NFRB from the numerical data obtained from the sensors. The next task is to construct an expert knowledge-based fuzzy rule base (EFRB). Expert knowledge acquisition is one of the greatest challenges facing the AI community. For robot control knowledge acquisition, the most common problem is that the human experts know how to control the robot but cannot express in words how to do it. Thus, transferring human empirical knowledge to a controller may turn out to be a difficult task. We refer this as *subconscious* knowledge. In this case, we may extract fuzzy rules from the action of an experienced manual operator. Modeling characteristics of the human operator (HO) as an element in control systems has been the subject of much research efforts since the late 1940's [34]. Early studies have focused on identifying linear transfer functions. More recently, neural networks [21] and fuzzy logic [34] have been applied in this field.

Conventional models such as transfer functions and even neural networks, although useful to reproduce human control actions, often fail to give insight in actual strategy adopted by the operator. Fuzzy IF-THEN rules can help visualizes, in a more natural way, the logical relations between a set of cause-effect data.

We consider two situations for the expert knowledge acquisition. If the control expert can explicitly express the knowledge (conscious), we can acquire the knowledge in terms of fuzzy IF-THEN rules. For the subconscious knowledge, we can extract the rules from experimental human operator data.

Using above knowledge acquisition methods, we can construct an expert knowledge-based fuzzy rule base (EFRB) with the DOB. For two inputs one output controller, the fuzzy rule has the following forms:

$$R_i^E : IF \;\; x_1 \;\; is \;\; A_i \;\; AND \;\; x_2 \;\; is \;\; B_i \;\; THEN \;\; y \;\; is \;\; C_i^E \qquad (w_i^E)$$

where w_i^E is the DOB of the i th rule in the EFRB.

5.2 Fusion of NFRB and EFRB

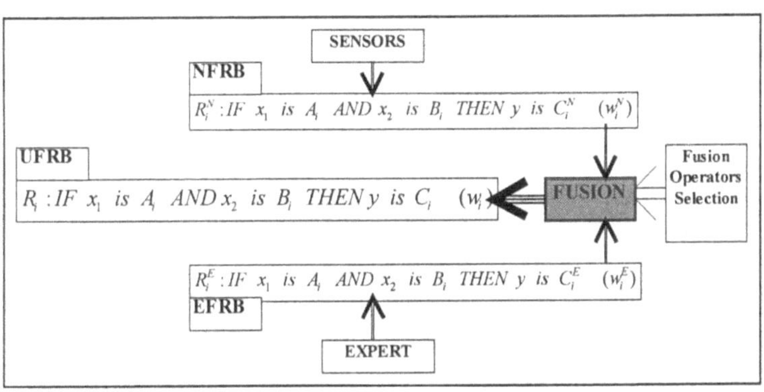

Figure 6. Fusion of NFRB and EFRB.

The proposed fusion scheme can be outlined as shown in Figure 6. It can be summarized as follows.

Step 1. Define the final uniform fuzzy rule base UFRB structure in advance. The input and output space are divided into fuzzy regions, and then we assign each region a fuzzy membership function according to expert knowledge. This predefined fuzzy system structure will be used for both fuzzy rules extraction from numerical data and fuzzy rules acquisition from experts. In this way, we design a fuzzy rule base formed as a look-up table for next two steps.

Step 2. Extract fuzzy rules from numerical data and construct the NFRB.

Step 3. Acquire fuzzy rules from experts and construct the EFRB.

Step 4. Construct the UFRB through fusing the NFRB with the EFRB. Because the EFRB, NFRB and UFRB have the same structure, we only consider the rules with same IF part for combination. By means of the fusion operators [4], a simple and intuitive fusion method is given below.

Let us assume the fuzzy IF-THEN rules in the UFRB has the following forms:

$$R_i : IF \;\; x_1 \;\; is \;\; A_i \;\; AND \;\; x_2 \;\; is \;\; B_i \;\; THEN \;\; y \;\; is \;\; C_i \qquad (w_i)$$

where w_i is the DOB of the i th rule in the UFRB.

The fusion of EFRB and NFRB can be described as follows.

Step 1. Convert the EFRB and NFRB into the complete fuzzy rule base.

If there are no rules in some of the fuzzy grids, we will take the fuzzy rules with same IF parts from another fuzzy rule base, and then a zero DOB will be given. As an example, if there is no rule in "$IF \;\; A_i \;\; and \;\; B_i$" grid in the NRRB, then we can "borrow" a rule from the EFRB. That is, let $w_i^N = 0$, and $C_i^N = C_i^E$.

Step 2. Choose rules with maximum DOB if conflict.

If there are conflicting rules in the EFRB and the NFRB, that is, rules with the same IF parts but different THEN parts ($C_i^E \neq C_i^N$), then we can choose the rule that has maximum DOB.

$$C_i = \begin{cases} C_i^E & if \quad w_i^E > w_i^N \\ C_i^E & otherwise \end{cases}, \quad w_i = \max(w_i^E, w_i^N) \tag{6}$$

Step 3. Fusing the DOB through fusion operators if consistent.

If the fuzzy rules are consistent, that is, the THEN parts are same for the same IF condition, set $C_i = C_i^E$ or C_i^N. The DOB of UFRB can be calculated by different ways using fusion operators [4].

$$C_i = C_i^E \;\; or \;\; C_i^N, \qquad w_i = FUSION(w_i^E, w_i^N) \tag{7}$$

where $FUSION$ is a data fusion function. For example, we can use a T conorm operator to choose the maximum DOB, that is, $w_i = \max(w_i^E, w_i^N)$. If a global measure is needed, we can choose a mean operator, for example, a weighted averaging operator, $w_i = \alpha w_i^E + (1-\alpha) w_i^N$. Where α is a weighted factor. If experts knowledge plays an important role in the integrated information, we can choose a bigger α, otherwise, choose a smaller one. As an example, if only numerical information from measuring instruments is used, set $\alpha = 0$, otherwise, if we only use linguistic rules obtained from experts, let $\alpha = 1$. We can set $\alpha = 0.5$ for processing the LNHD if the expert's knowledge is as important as the measuring numerical data.

5.3 Inverse Learning-Based Fusion

As we discussed in Section 2, the LNHD robot control has different forms, such as, linguistic and numerical, direct and indirect, and so on. For the LNHD with inverse mapping properties, one of the straightest ways to integrate is to convert the DD, DR, ID, IR into a uniform domain by means of inverse learning [32]. The basic idea of inverse control is to derive the plant with a signal from a controller whose transfer function is the inverse of that of the plant itself. Since the robot model is generally unknown, it is necessary to adapt or adjust the parameters of the controller in order to create a true robot inverse. The development of inverse learning for the designing of controller involves two phases. In the learning phase, the inverse dynamics of the robot is modeled (see Figure 7). The inverse model is then used to generate control actions in the application phase.

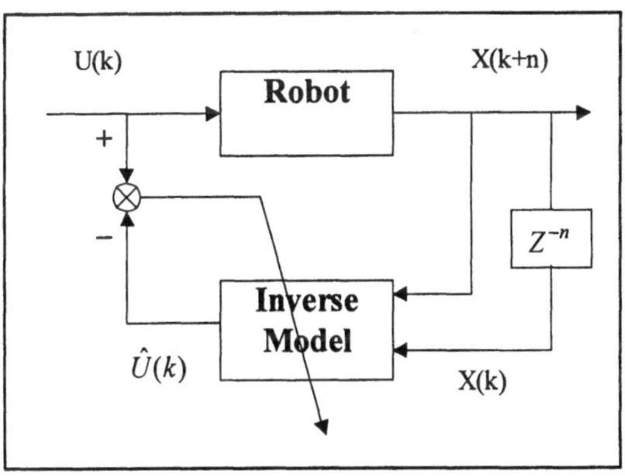

Figure 7. Inverse learning.

If a robot and its controller are described as Eq. (3). By assuming all the state variables are measurable and the numerical data available is in the form of Indirect Data (ID): $[X(k)^T, U(k)^T; X(k+n)^T]$. In order to fuse the LNHD in the form of ID and DR, or DD and IR, for the robot controller design, we must extract *direct rules* from *indirect* data, or extract *indirect* rules from *direct* data. If the inverse dynamics of the plant do exist, then the problem now becomes how to find the inverse mapping G in Eq. (3). Although the inverse mapping G doesn't always have an analytically close form (even may not exist), we may use the training data pairs $[X^T(k), X^T(k+n); U^T(k)]$ to approximate the inverse mapping G. Then we have an estimated $\hat{U}$.

$$\hat{u}(k) = \hat{G}(x(k), x_d(k+n)) \tag{8}$$

Compare Eq. (8) with Eq. (3), if $\hat{G} = G$, then after n steps, the control sequence can bring the state $x(k)$ to the desired state $x_d(k+n)$. The rationale behind inverse learning seems straightforward. However, there are some problems for this inverse learning. Some of the following problems can be solved by adaptive inverse control theory [32].

- The existence of inverse dynamics for a robot is not generally valid.
- If the model has internal delay, the inverse controller may have difficulty in overcoming it. The controller would need to have a predictor.
- If the model were in a nonminimum phase, then the inverse controller would want to have poles in the right half of the s-plane or outside the unit circle in the z-plane. Such an inverse would be unstable.
- Minimize $\|u(k) - \hat{u}(k)\|^2$ does not guarantee minimize $\|x_d(k) - x(k)\|^2$.

6. LNHD FUSION FOR BIPED GAIT SYNTHESIS

Biped robot control is a challenging problem due to its high order nonlinear dynamics and uncertainty. In this section, we demonstrate how to use the LNHD fusion to improve the gait of a biped robot. The dynamic equation of the biped robot in workspace can be derived from the Lagrangian approach as [30].

$$D(\theta)\ddot{\theta} + C(\theta,\dot{\theta})\dot{\theta} + G(\theta) = \tau \tag{9}$$

where $\tau \in \Re^n$ is the vector of generalized torque, $\theta \in \Re^n$ is the vector of the joint angular position, τ_i is corresponding to the variable θ_i, $D(\theta)$ is the positive symmetric $n \times n$ inertia matrix, $C(\theta,\dot{\theta})$ is the $n \times n$ matrix which includes terms from the centrifugal and Coriolis toques, and $G(\theta)$ is the n-dimensional vector which includes the gravitational torque. The zero moment point (ZMP) [30] is used as a criterion to evaluate the dynamic walking of the biped robot.

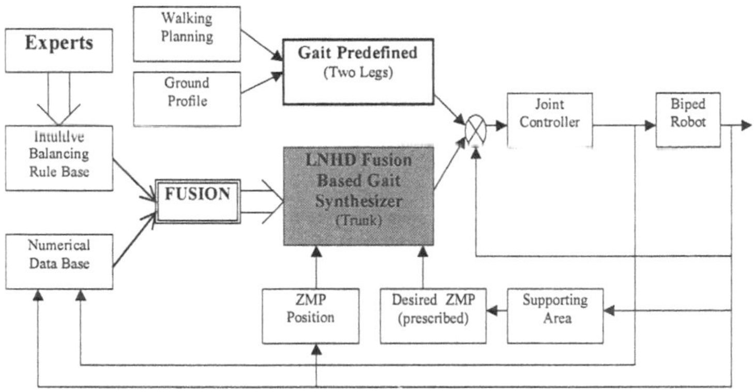

Figure 8. The LNHD fusion-based gait synthesizer.

The proposed LNHD fusion based gait synthesizer is shown in Figure 8. Based on intuitive balancing knowledge, 25 gait synthesis fuzzy rules for sagittal plane and another 25 rules for the frontal plane are obtained [37, 38]. An NFRB is obtained from the successful biped control experiments.

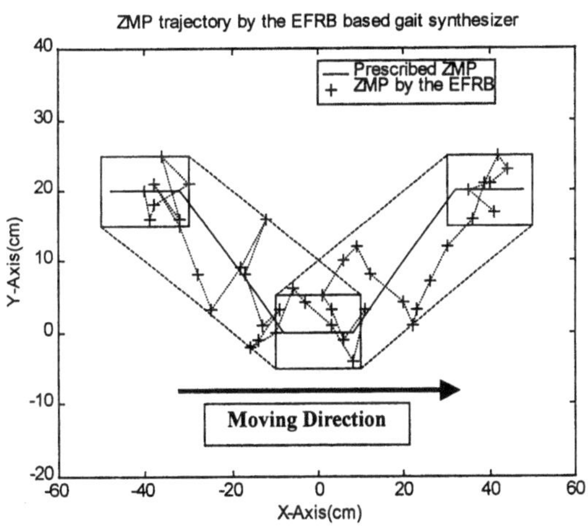

Figure 9. Result by the EFRB based gait synthesizer.

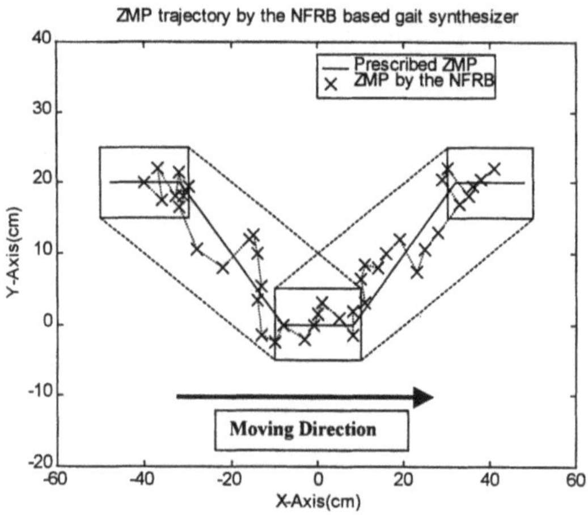

Figure 10. Result by the NFRB based gait synthesizer.

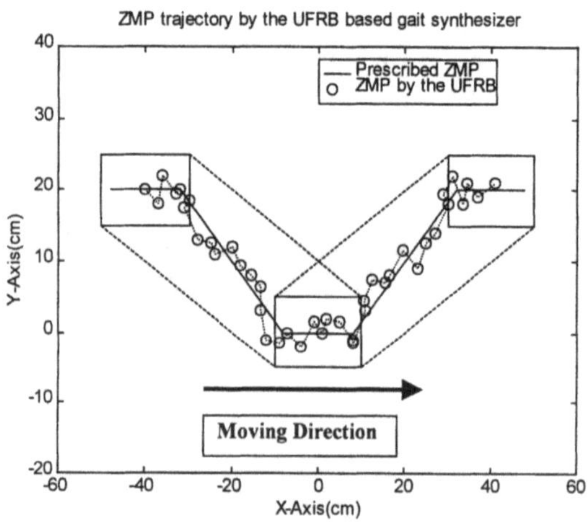

Figure 11. Result by the UFRB based gait synthesizer.

Figure 9 shows the gait synthesizing result by the EFRB based gait synthesizer. After some manually tuning of the parameters of the fuzzy gait synthesizer, we found that the actual ZMP can roughly track the desired trajectory, however, the ZMP tracking error is very big. The result by the NFRB based gait synthesizer is shown in Figure 10. From Figure 11, we can see that the ZMP is much closer to the desired trajectory by using the UFRB which is the integration of the EFRB obtained from human balancing knowledge, and the NFRB which is extracted from the biped control experiments.

7. DISCUSSION AND CONCLUSION

In this chapter we review the recent development in the area of data fusion approach in conjunction with fuzzy if-then rule concept for intelligent robotic control. Then a fuzzy if-then rules extraction-based linguistic and numerical heterogeneous data fusion method is discussed. We also demonstrate how to use the LNHD fusion method for the gait synthesis of a biped robot. The simulation results show that the biped gait can be significantly improved by the fusion of the linguistic rules obtained from human intuitive walking knowledge, and the numerical data received from biped control experiments. Following are some further points:

1) The hybrid information for robot control usually has different forms, symbolic and numerical, precise and fuzzy, direct and indirect, and etc., however, the proposed method can only process a few simple forms of hybrid information. Therefore, there is a need to develop a more general LNHD fusion method for intelligent robotic control.
2) For extracting the fuzzy if-then rules from the numerical data as well as from experts, some constraints must be considered, such as the fuzzy sets must be normal, convex, overlap, and must not exchange their relative positions.
3) The proposed method may allow the combination of fuzzy rules extracted from different experts and numerical databases. Research should be conducted on whether and when the LNHD fusion yields better results than simply averaging the different kinds of information.
4) The existence of the inverse of the plant or controller is not valid in general. Some results from adaptive inverse control [32] should be useful for the general fusion of the LNHD with inverse mapping properties.
5) The LNHD fusion method discussed in this chapter needs sufficient "precise" data for information integration. However, such detailed and precise data are usually difficult or even impossible to obtain for robot control. Therefore, reinforcement learning [3, 38] may be helpful to improve the LNHD fusion for intelligent robotic control.

ACKNOWLEDGMENTS

We would like to thank Miss S.C. Oh and Mr S.H. Ker at the SRG Lab of Singapore Polytechnic for their support in biped hardware development.

REFERENCES

[1] Abe, S. and Lan, M.S., "Fuzzy rules extraction directly from numerical data for function approximation," *IEEE Trans. on Systems, Man, and Cyberne*tics 25 (1995), 119-129.

[2] Beom, H.K. and Cho, H.S., "A sensor-based navigation for a mobile robot using fuzzy logic and reinforcement learning, " *IEEE Trans. Systems Man Cybernetics* 25 (1995), 464-477.

[3] Berenji, H.R. and Khedkar, K., "Learning and tuning fuzzy logic controllers through reinforcements, " *IEEE Trans. Neural Networks* 3 (1992), 724-740.

[4] Bloch, I., "Information combination operators for data fusion: a comparative review with classification," *IEEE Trans. on Systems, Man, and Cybernetics* 26 (1996) 52-67.

[5] Bouchon-Meunier, B. (Ed.), *Aggregation and Fusion of Imperfect Information*, Physica-Verlag, 1998.

[6] Chiu, S.L., "Fuzzy model identification based on cluster estimation*," J. Intelligent Fuzzy Systems* 2 (1994), 267-278.

[7] Cron, G. and Dubuisson, B., "A weighted fuzzy aggregation method," In*: Proc. IEEE Conference Fuzzy Systems*, 1998, pp. 675-680.

[8] Dubois, D. and Prade, H., "Combination of fuzzy information in the framework of possibility theory," In: *Data Fusion and Machine Intelligence*, Academic Press (1992), 481-505.

[9] Filippidis, A., "Data fusion using sensor data and a priori information," *Control Eng. Practice* 4 (1996), 43-53.

[10] Ghosh, B.K., Xi, N., and Tarn, T.J. (Eds.), *Control in Robotics and Automation: Sensor-Based Integration*, Academic Press, 1999.

[11] Goodridge, S.G., Kay, M.G., and Luo, R.C., "Multilayered fuzzy behavior fusion for real-time reactive control of systems with multiple sensors," *IEEE Trans. Industrial Electronics* 43 (1996), 387-394.

[12] Hathaway, H.J., Bezdek, J.C., and Pedrycz, W., "A parametric model for fusing heterogeneous fuzzy data," *IEEE Trans. on Fuzzy Systems* 4 (1996), 270-281.

[13] Hong, L. and Wang, G.-J., "Centralised integration of multisensor noisy and fuzzy data," *IEE Proc. –Control Theory Appl.* 142 (1995), 459-465.

[14] Ishibuchi, H., Fujioka, H., and Tanaka, H., "Neural networks that learn from fuzzy if-then rules," *IEEE Trans. Fuzzy Syst*em 1 (1993), 85-97.

[15] Jang, J.-S.R., "ANFIS: Adaptive-network-based fuzzy inference systems," *IEEE Trans. Systems Man Cybernetics* 23 (1993), 665-685.

[16] Kam, M., Zhu, X., and Kalata, P., "Sensor fusion for mobile robot navigation," *Proc. IEEE* (85), pp. 108-119.

[17] Lin, C.T. and Lee, C.S.G., *Neural Fuzzy Systems: A Neuro-Fuzzy Synergism to Intelligent Systems*, Prentice-Hall, Englewood Cliffs, NJ, 1996.

[18] Lin, C.T. and Lu, Y.C., "A neural fuzzy system with fuzzy supervised learning," *IEEE Trans. Systems, Man, and Cyberne*tics 26 (1996), 744-763.

[19] Matia, F. and Jimenez, A., "Multisensor fusion: an autonomous mobile robot," *J. Intelligent and Robotic Systems* 22 (1998), 129-141.

[20] Medsker, L.R., *Hybrid Intelligent Systems*, Kluwer Academic Publisher, 1995.

[21] Nechyba, M.C. and Xu, Y., "Stochastic similarity for validating human control strategy models," *IEEE Trans. Robot Automation* 14 (1998), 437-451.

[22] Nozaki, K., Ishibuchi, H., and Tanaka, H., "A simple but powerful heuristic method for generating fuzzy rules from numerical data," *Fuzzy Sets and Systems* 86 (1997), 251-270.

[23] Pedrycz, W., Bezdek, J.C., and Hathaway, R.J., "Two nonparametric models for fusing heterogeneous fuzzy data," *IEEE Trans. on Fuzzy Systems* 6 (1998), 411-425.
[24] Pin, F.G. and Watanabe, Y., "Navigation of mobile robots using a fuzzy behaviorist approach and custom-designed fuzzy inferencing boards," *Robotica* 12 (1994), 491-503.
[25] Saridis, G.N., "Intelligent robotic control," *IEEE Trans. Automatic Control* 28 (1983), 547-557.
[26] Shi, X., Lever, P.J.A., and Wang, F.-Y., "Fuzzy behavior integration and action fusion for robotic excavation," *IEEE Trans. Industrial Electronics* 43 (1996), 395-402.
[27] Stover, J.A. and Gibson, R.E., "A fuzzy-logic architecture for autonomous multisensor data fusion," *IEEE Trans. Industrial Electronics* 43 (1996), 403-1996.
[28] Thawonmas, R. and Abe, S., "Function approximation based on fuzzy rules extraction from partitioned numerical data," *IEEE Trans. Systems, Man, and Cybernetics* B29 (1999), 525-534.
[29] Torra, V., "Combining fuzzy sets: the geometric consensus function family," *Fuzzy Sets and Systems* 74 (1995), 335-342.
[30] Vokobratovic, M., Borovac, B., Surla, D., and Stokic, D., *Biped Locomotion: Dynamics, Stability, Control and Application*, Springer-Verlag, 1990.
[31] Wang, L.X. and Mendel, J.M., "Generating fuzzy rules by learning from examples," *IEEE Trans. Systems, Man, and Cybern*etics 22 (1992), 1414-1427.
[32] Widraw, B. and Walach, E., *Adaptive Inverse Control.* Prentice-Hall, Englewood Cliffs, NJ, 1996.
[33] Yung, N.H. and Ye, C., "An intelligent mobile vehicle navigator based on fuzzy logic and reinforcement learning," *IEEE Trans. Systems Man Cybernetics* B29(1999), 314-321.
[34] Zapata, G.O.A., Galvao, R.K.H., and Yoneyama, T. "Extraction fuzzy control rules from experimental human operator data," *IEEE Trans. Systems, Man, and Cybernetics* B29 (1999), 398-406.
[35] Zhou, C., Jagannathan, K., and Meng Q., "Integration linguistic and numerical information for hybrid intelligent control," in: *Proc. IEEE Int. Conf. on Systems, Man and Cybernetics,* 1998, 800-805.
[36] Zhou, C. and Meng, Q., "Fuzzy rules extraction based integration of linguistic and numerical information for hybrid intelligent systems,"*Lecture Notes in Artificial Intelligence* 1531 (1998), Springer, 282-293.
[37] Zhou, C. and Ruan, D., "Integration of linguistic and numerical information for biped control," *Robotics and Autonomous Systems* 28 (1999), 53-70.
[38] Zhou, C., "Neuro-fuzzy gait synthesis with reinforcement learning for a biped walking robot," *Soft Computing* (2000), to appear.

Chapter 12

FUZZY IF-THEN RULES FOR PATTERN CLASSIFICATION

Hisao Ishibuchi, Tomoharu Nakashima, and Manabu Nii
Department of Industrial Engineering
Osaka Prefecture University
{hisaoi,nakashi,manabu}@ie.osakafu-u.ac.jp

Abstract This chapter illustrates how fuzzy if-then rules can be used for pattern classification problems. First we describe a heuristic method for automatically generating fuzzy if-then rules for pattern classification problems from training patterns. The heuristic method uses a simple fuzzy grid for partitioning a pattern space into fuzzy subspaces. A fuzzy if-then rule is generated in each fuzzy subspace. Using the heuristic rule generation method, we examine some basic aspects of fuzzy rule-based classification systems such as the shape of membership functions, the definition of the compatibility grade, and the choice of a fuzzy reasoning method. Next we describe a fuzzy rule selection method for designing compact fuzzy rule-based systems with high classification ability. A small number of fuzzy if-then rules are selected from a large number of candidate rules by a genetic algorithm. Finally we describe two genetics-based machine learning algorithms for designing fuzzy rule-based systems for high-dimensional pattern classification problems. In those methods, fuzzy rule-based systems are evolved by genetic operations such as selection, crossover, and mutation. Simulation results on some well-known data sets are shown for illustrating our approaches to the design of fuzzy rule-based systems.

Keywords: Pattern classification, rule generation, genetic algorithms, machine learning.

1. INTRODUCTION

Fuzzy systems based on fuzzy if-then rules have been successfully applied to various problems in the field of fuzzy control. Since fuzzy rule-based systems are universal approximators of nonlinear functions, they can handle highly nonlinear plants. They also have high comprehensibility because human users can easily understand the meaning of each fuzzy if-then rule through its linguistic interpretation. Recently fuzzy rule-based systems are also applied to pattern classification problems. The merits of the use of fuzzy rule-based systems for pattern classification problems are also the nonlinearity and the comprehensibility as in their applications to control problems. Corresponding to these two merits, there are two research directions in the field of fuzzy rule-based pattern classification. One direction is to try to design compact fuzzy rule-based systems with high comprehensibility. In this direction, the main criterion in the design of fuzzy rule-based systems is their comprehensibility. That is, extraction of comprehensible linguistic knowledge in the form of fuzzy if-then rules is stressed. Linguistic values (e.g., $small$ and $large$) given by human users are usually used for generating fuzzy if-then rules with no modification of their membership functions. Several approaches have been proposed for generating compact fuzzy rule-based systems with high comprehensibility from numerical data (for example, see [7, 10, 14]). In those approaches, the number of fuzzy if-then rules is minimized for constructing compact fuzzy rule-based systems. The other research direction is to try to maximize the classification ability of fuzzy rule-based systems. In this direction, usually their comprehensibility is not stressed. Fuzzy if-then rules often have no linguistic interpretation. Several approaches have been proposed for generating fuzzy if-then rules and tuning membership functions for pattern classification problems (for example, see [1, 2, 3, 5, 16, 17, 18, 19, 21]). In those approaches, fuzzy rule generation and learning are performed to minimize error rates on training patterns.

Pattern classification problems usually involve much more input variables than control problems. While fuzzy control problems can be usually handled by fuzzy if-then rules in a tabular form, we cannot use such a tabular form representation for high-dimensional pattern classification problems because the number of fuzzy if-then rules exponentially increases with the number of input variables. Thus we need some tricks to handle high-dimensional pattern classification problems by fuzzy if-then rules. One approach is to use multi-dimensional antecedent fuzzy sets that are directly defined on pattern spaces [1, 2, 3, 21]. Clustering techniques are used for determining the multi-dimensional antecedent fuzzy set of each fuzzy if-then rule. In this approach, the number of fuzzy if-then rules does not depend on the dimensionality of pattern classification problems. High-dimensional pattern classification problems can be handled by a small number of fuzzy if-then rules. Another approach is to use hierarchical fuzzy rule-based systems where a number of low-dimensional fuzzy rule tables are hierarchically connected [6]. Outputs from fuzzy rule tables in lower layers are used as inputs to those in higher layers. Since each fuzzy rule table has only a small number of input variables, the exponential increase can be avoided. These two approaches are promising in the design of compact fuzzy rule-based systems with high classification ability. The common drawback of these two approaches is the difficulty of the linguistic interpretation of each fuzzy if-then rule. Another hierarchical approach is to use tree structures of fuzzy if-then rules (i.e., fuzzy decision trees [15]). The most straightforward approach to the design of comprehensible fuzzy rule tables for high-dimensional pattern classification problems is to select only a few input variables. While fuzzy rule-based systems with only two input variables surprisingly work well on some real-world pattern classification problems [10], this is not always the case. The input selection works well on some problems, and does not work well on other problems.

In this chapter, we show how compact fuzzy rule-based classification systems with high comprehensibility can be designed from numerical data. That is, we stress the linguistic interpretation of each fuzzy if-then rule. First we describe a heuristic rule generation method for designing fuzzy rule-based systems for low-dimensional pattern classification problems. Fuzzy if-then rules are generated by simple fuzzy grids and represented in a tabular form. Next we describe a fuzzy rule selection method for designing compact fuzzy rule-based systems with high classification ability. A small number of fuzzy if-then rules are selected from a large number of candidate rules by a genetic algorithm. Finally we describe two genetics-based machine learning (GBML) algorithms for designing fuzzy rule-based systems for high-dimensional pattern classification problems. In those GBML methods, fuzzy rule-based systems are evolved by genetic operations such as selection, crossover, and mutation.

2. SIMPLE FUZZY RULE-BASED SYSTEMS

2.1 PATTERN CLASSIFICATION PROBLEMS

Our problem in this chapter is a c-class pattern classification problem with n continuous attributes. We assume that m labeled patterns (i.e., training patterns) are given from the c classes as $\mathbf{x}_p = (x_{p1}, \cdots, x_{pn})$, $p = 1, 2, \cdots, m$. For simplicity of explanation, we also assume that all attribute values x_{pi}'s have already been normalized into real numbers in the unit interval $[0, 1]$. This means that the pattern space of our problem is the n-dimensional unit hypercube $[0, 1]^n$.

Many approaches such as statistical techniques, neural networks, and machine learning algorithms can be applied to our pattern classification problem. Fuzzy rule-based systems in this chapter are viewed as an alternative to those methods. That is, they are used as a general-purpose pattern classification method. While fuzzy rule-based systems have an inherent advantage in the handling of uncertainty over other classification methods, we do not discuss such ability because it can be utilized only in special situations. We discuss their advantages as a general-purpose pattern classification method.

For generating fuzzy if-then rules that can be easily understood by human users, we assume that a set of linguistic values (e.g., *small*, *medium*, and *large*) and their membership functions have already been given for describing each of the n attributes. Human users are supposed to specify the membership function of each linguistic value according to their knowledge and intuition. The given linguistic values are used as antecedent fuzzy sets for linguistically describing antecedent conditions of each fuzzy if-then rule. When human users cannot specify the exact shape of the membership function for each linguistic value, it may be roughly specified as a triangular or trapezoidal membership function. This chapter shows that we can obtain good classification results by simple fuzzy grids with homogeneous fuzzy partitions by triangular membership functions. In Figure 1.1, we show an example of such a fuzzy grid together with given training patterns.

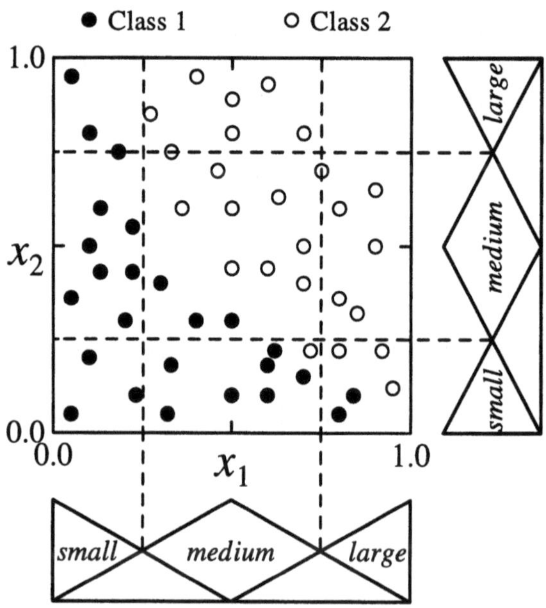

Figure 1.1 A simple fuzzy grid and training patterns.

2.2 FUZZY IF-THEN RULES

For our n-dimensional pattern classification problem, we use fuzzy if-then rules of the following form:

$$\text{Rule } R_j : \text{ If } x_1 \text{ is } A_{j1} \text{ and} \ldots \text{and } x_n \text{ is } A_{jn}$$
$$\text{then Class } C_j \text{ with } CF_j, \quad j = 1, 2, \cdots, N, \quad (1.1)$$

where $\mathbf{x} = (x_1, \cdots, x_n)$ is an n-dimensional pattern vector, A_{ji} is a linguistic value such as *small* and *large* ($i = 1, 2, \cdots, n$), C_j is a consequent class (i.e., one of the given c classes), CF_j is a certainty grade ($0 \leq CF_j \leq 1$), and N is the number of fuzzy if-then rules. The main characteristic feature of our fuzzy if-then rules in (1.1) is the use of the certainty grade CF_j. We will compare our fuzzy if-then rules in (1.1) with the following fuzzy if-then rules, which have no certainty grade, for examining the effect of the certainty grade.

$$\text{Rule } R_j : \text{ If } x_1 \text{ is } A_{j1} \text{ and} \ldots \text{and } x_n \text{ is } A_{jn} \text{ then Class } C_j, \quad j = 1, 2, \cdots, N. \quad (1.2)$$

Our pattern classification problem can be also handled by the following fuzzy if-then rules, which are usually used for control problems and function approximation problems.

$$\text{Rule } R_j : \text{ If } x_1 \text{ is } A_{j1} \text{ and} \ldots \text{and } x_n \text{ is } A_{jn} \text{ then } y_1 \text{ is } f_{j1}(\mathbf{x}) \text{ and} \ldots \text{and } y_c \text{ is } f_{jc}(\mathbf{x}), \quad j = 1, 2, \cdots, N, \quad (1.3)$$

where y_k denotes the membership grade of the pattern vector $\mathbf{x} = (x_1, \cdots, x_n)$ to Class k ($k = 1, 2, \cdots, c$), and $f_{jk}(\mathbf{x})$ is a linear function of $\mathbf{x}$, a real number, or a linguistic value. These fuzzy if-then rules are used for approximately realizing an unknown nonlinear mapping from $\mathbf{x} = (x_1, \cdots, x_n)$ to $\mathbf{y} = (y_1, \cdots, y_c)$. As in the learning of neural networks, we can define a target vector $\mathbf{t}_p = (t_{p1}, \cdots, t_{pc})$ for each training pattern $\mathbf{x}_p$ as

$$t_{pk} = \begin{cases} 1 & \text{if } \mathbf{x}_p \text{ belongs to Class } k, \\ 0 & \text{otherwise,} \end{cases} \quad k = 1, 2, \cdots, c. \quad (1.4)$$

The given m training patterns are used as the m input-output pairs $(\mathbf{x}_p, \mathbf{t}_p)$, $p = 1, 2, \cdots, m$. In this manner, our pattern classification problem can be handled by the fuzzy if-then rules in (1.3) as a function approximation problem with the training data $(\mathbf{x}_p, \mathbf{t}_p), p = 1, 2, \ldots, m$. In this case, we can use various fuzzy modeling techniques for generating and tuning the fuzzy if-then rules in (1.3).

Each of the fuzzy if-then rules in (1.1) and (1.2) directly indicates the consequent class in the corresponding fuzzy subspace defined by the Cartesian product $A_{j1} \times \cdots \times A_{jn}$. On the other hand, each fuzzy if-then rule in (1.3) indirectly shows the class in the corresponding fuzzy subspace by approximately realizing a nonlinear mapping from the n-dimensional pattern space to a c-dimensional space. Moreover, the fuzzy if-then rules in (1.3) have c linear functions in the consequent part while those in (1.1) and (1.2) have a single class label. From these discussions, we can see that the fuzzy if-then rules in (1.1) and (1.2) may be more easily understood by human users than those in (1.3). Thus in this chapter we use the fuzzy if-then rules in (1.1) and (1.2).

2.3 HEURISTIC FUZZY RULE GENERATION

Let us consider the pattern classification problem in Figure 1.1. From this figure, we can intuitively see that the consequent class is Class 1 if x_1 and x_2 are *small*. Thus we have the following fuzzy if-then rule:

$$\text{If } x_1 \text{ is } small \text{ and } x_2 \text{ is } small \text{ then Class 1 with } CF_j. \tag{1.5}$$

Since all patterns compatible with the antecedent part of this fuzzy if-then rule are from Class 1, we can also intuitively see that its certainty grade CF_j is very high (i.e., its maximum value 1.0). Such an intuitive idea was formulated as the following heuristic rule generation procedure in [13]:

[Heuristic Rule Generation Procedure]

Step 1: Calculate the compatibility grade $\mu_j(\mathbf{x}_p)$ of each training pattern $\mathbf{x}_p = (x_{p1}, \cdots, x_{pn})$ with the fuzzy if-then rule R_j by the product operation as

$$\mu_j(\mathbf{x}_p) = \mu_{j1}(x_{p1}) \times \cdots \times \mu_{jn}(x_{pn}), \tag{1.6}$$

where $\mu_{ji}(\cdot)$ is the membership function of the antecedent fuzzy set A_{ji}.

Step 2: For each class, calculate the sum of the compatibility grades of the training patterns with the fuzzy if-then rule R_j:

$$\beta_{\text{Class } h}(R_j) = \sum_{\mathbf{x}_p \in \text{Class } h} \mu_j(\mathbf{x}_p), \quad h = 1, 2, \cdots, c. \tag{1.7}$$

Step 3: Find Class C_j that has the maximum value of $\beta_{\text{Class } h}(R_j)$:

$$\beta_{\text{Class } C_j}(R_j) = \max\Big(\beta_{\text{Class } 1}(R_j), \cdots, \beta_{\text{Class } c}(R_j)\Big). \tag{1.8}$$

If the consequent class (i.e., Class C_j) of the fuzzy if-then rule R_j cannot be uniquely determined by (1.8), we do not generate the fuzzy if-then rule R_j. For example, if $\beta_{\text{Class } h}(R_j) = 0$ for all classes, we cannot determine Class C_j.

Step 4: Specify the certainty grade CF_j as follows:

$$CF_j = \Big(\beta_{\text{Class } C_j}(R_j) - \overline{\beta}\Big) \Big/ \sum_{h=1}^{c} \beta_{\text{Class } h}(R_j), \tag{1.9}$$

where

$$\overline{\beta} = \sum_{\substack{h=1 \\ h \neq C_j}}^{c} \beta_{\text{Class } h}(R_j) \Big/ (c-1). \tag{1.10}$$

Using this heuristic procedure, we can easily determine the consequent class C_j and the certainty grade CF_j of each fuzzy if-then rule R_j. In Figure 1.2, we show fuzzy if-then rules that are generated from Figure 1.1 by the heuristic procedure. Each real number in parentheses denotes the certainty grade of the corresponding fuzzy if-then rule. From the comparison between Figure 1.1 and Figure 1.2, we can see that the generated fuzzy if-then rules coincide with our intuition.

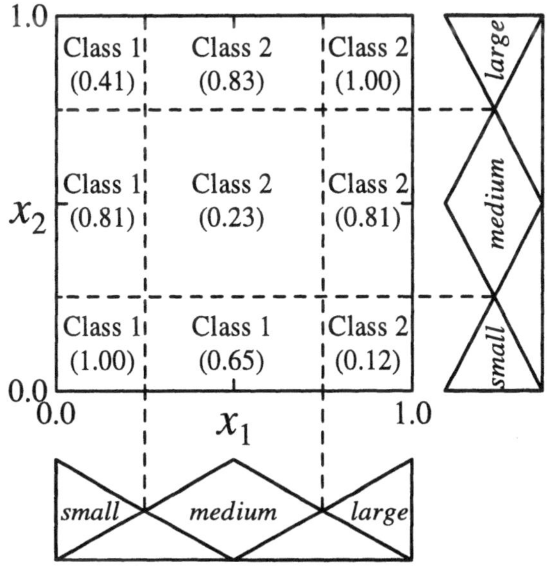

Figure 1.2 Generated fuzzy if-then rules.

2.4 FUZZY REASONING METHODS

As we can see from Figure 1.2, each fuzzy if-then rule locally indicates the most possible class in the corresponding fuzzy subspace. Thus the basic idea of fuzzy reasoning is that a new pattern should be classified by fuzzy if-then rules whose fuzzy subspaces cover that pattern. When a new pattern is covered by a single fuzzy if-then rule, the classification result is the same as the consequent class of that fuzzy if-then rule. Since fuzzy subspaces of neighboring fuzzy if-then rules overlap with each other, multiple fuzzy if-then rules usually cover a new pattern with different compatibility grades. We use a single winner method for classifying a new pattern [13]. The winner rule for a new pattern $\mathbf{x}_p$ has the maximum product of the compatibility grade $\mu_j(\mathbf{x}_p)$ and the certainty grade CF_j. Let us denote the set of fuzzy if-then rules in a fuzzy rule-based classification system by S. In this case, the winner rule R_{j*} for the new pattern

$\mathbf{x}_p$ is determined as follows:

$$\mu_{j*}(\mathbf{x}_p) \cdot CF_{j*} = \max\{\mu_j(\mathbf{x}_p) \cdot CF_j \mid R_j \in S\}. \qquad (1.11)$$

When we use the fuzzy if-then rules in (1.2) with no certainty grades, the winner rule R_{j*} for the new pattern $\mathbf{x}_p$ is simply determined by their compatibility grades as

$$\mu_{j*}(\mathbf{x}_p) = \max\{\mu_j(\mathbf{x}_p) \mid R_j \in S\}. \qquad (1.12)$$

In our fuzzy reasoning method in (1.11) and (1.12), the single winner rule R_{j*} is responsible for the classification result of the new pattern $\mathbf{x}_p$. This means that each fuzzy if-then rule has its own decision area in which new patterns are classified by that fuzzy if-then rule. In Figure 1.3 (a), we illustrate such a decision area of each fuzzy if-then rule in the rule table in Figure 1.2. From Figure 1.2 and Figure 1.3 (a), we can see that the size of the decision area depends on the certainty grade of the corresponding fuzzy if-then rule. In general, the larger the certainty grade is, the larger the decision area is. The classification boundary constructed by the decision areas in Figure 1.3 (a) is shown in Figure 1.3 (b) where two patterns are misclassified. The number of misclassified patterns can be decreased by using finer fuzzy partitions. For example, if eight linguistic values are given for each axis of the two-dimensional pattern space as in Figure 1.4, all the given training patterns are correctly classified.

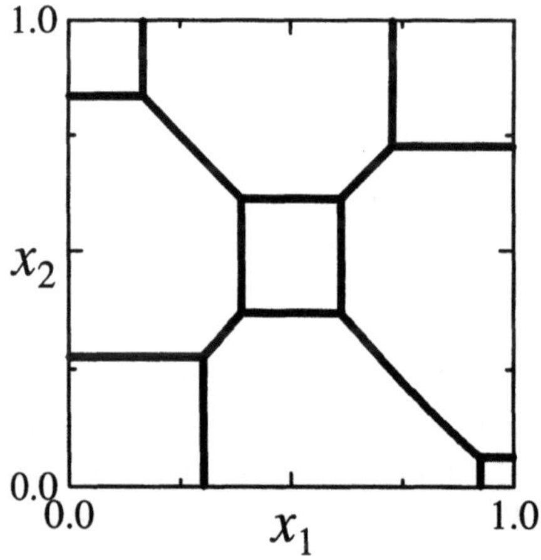

(a) Decision area of each rule.

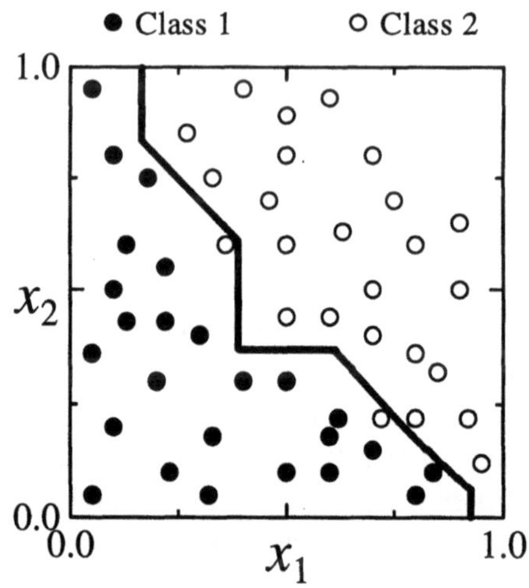

(b) Classification boundary.

Figure 1.3 Classification results by the fuzzy rule table in Figure 1.2.

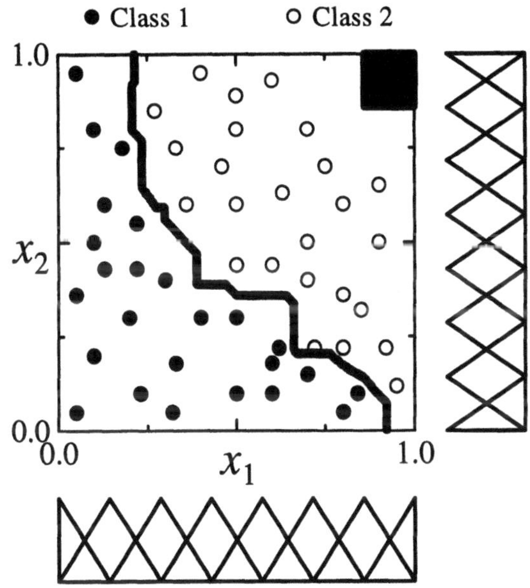

Figure 1.4 Classification boundary obtained by a finer fuzzy partition.

2.5 COMPARISON OF VARIOUS SPECIFICATIONS

We have already illustrated the heuristic rule generation method and the single winner fuzzy reasoning method. In this section, we examine various specifications in those methods.

2.5.1 Comparison between Crisp and Fuzzy Partitions. If we divide each axis of the pattern space into non-fuzzy intervals as shown in Figure 1.5, the pattern space is divided into multiple cells (or patches). Each cell can be viewed as a non-fuzzy if-then rule. In this case, classification boundaries are always parallel to the axes of the pattern space. In Figure 1.5, no rule can be generated by the local information in the shaded cell because that cell includes no training pattern. On the other hand, the corresponding fuzzy if-then rule was generated in Figure 1.2 because some training patterns are compatible with its antecedent part "If x_1 is $large$ and x_2 is $large$."

For examining the difference between crisp and fuzzy partitions further, let us consider how many rules can be generated by a single training pattern. In the case of a crisp partition, only a single rule can be generated from a single training pattern. In Figure 1.6 (a), a crisp if-then rule in the shaded area is generated by the given training pattern. On the other hand, multiple fuzzy if-then rules can be generated by a single training pattern. In Figure 1.6 (b), four fuzzy if-then rules

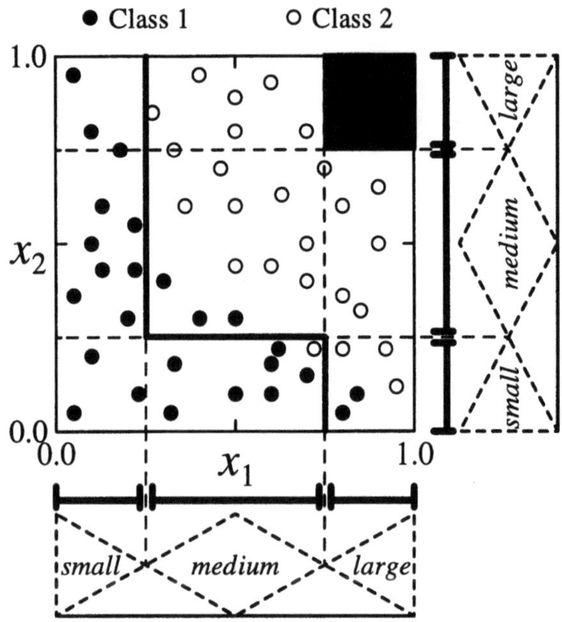

Figure 1.5 Pattern classification by a crisp partition.

in the shaded fuzzy subspaces are generated. The generated four fuzzy if-then rules cover the square region indicated by dotted lines in Figure 1.6 (b). From Figure 1.6, we can see that the effect of the single training pattern expands into the large region in the pattern space in the case of the fuzzy partition while it is limited within the single rule in the case of the crisp partition. Such expansion of the effect of training patterns may lead to high generalization ability of fuzzy rule-based classification systems.

Using the well-known iris data, we compared the crisp partition in Figure 1.6 (a) with the fuzzy partition in Figure 1.6 (b). The iris data set includes 150 samples with four continuous attributes from three classes. In our computer simulations, first all the attribute values were normalized into real numbers in the unit interval $[0, 1]$. That is, the iris data set was handled as a three-class pattern classification problem in the four-dimensional unit hypercube $[0, 1]^4$. As in Figure 1.6, each axis was divided into five intervals in the case of the crisp partition and five triangular fuzzy sets in the case of the fuzzy partition. The performance of each partition on training data was evaluated by designing a classification system using all the 150 samples and testing the designed classification system by the same 150 samples. On the other hand, the performance of each partition on test data (i.e., generalization ability) was evaluated by the leaving-one-out (LV1) procedure [20, 22]. In the LV1 procedure, the design

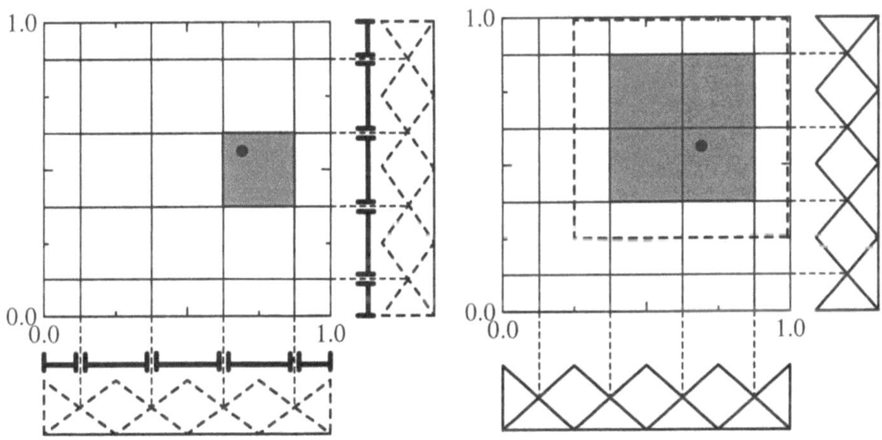

(a) Crisp partition. (b) Fuzzy partition.

Figure 1.6 Generated rules by a single training pattern.

of a classification system by 149 samples and its test by the remaining sample were iterated 150 times until all the 150 samples were used as test data. We also examined coarser partitions (e.g., Figure 1.1 with three triangular fuzzy sets on each axis and Figure 1.5 with three intervals on each axis). Simulation results are summarized in Table 1.1 and Table 1.2 where K denotes the number of intervals or triangular fuzzy sets on each axis. From Table 1.1, we can see that the performance of the crisp partitions on training data is almost the same as that of the fuzzy partitions. On the contrary, these two types of partitions have different generalization ability. As shown in Table 1.2, the performance of the crisp partitions on test data is inferior to that of the fuzzy partitions. These results show that the fuzzy partitions have higher generalization ability than the crisp partitions.

Table 1.1 Comparison between the crisp and fuzzy partitions by classification rates on training data.

	$K = 3$	$K = 4$	$K = 5$	Average
Crisp partitions	94.7%	86.7%	96.7%	92.7%
Fuzzy partitions	94.0%	92.7%	96.0%	94.2%

Table 1.2 Comparison between the crisp and fuzzy partitions by classification rates on test data.

	$K = 3$	$K = 4$	$K = 5$	Average
Crisp partitions	94.0%	81.3%	86.0%	87.1%
Fuzzy partitions	93.3%	89.3%	95.3%	92.6%

2.5.2 Comparison between Product and Minimum Operations. Traditionally the following minimum operation has been used for calculating the compatibility grade between the antecedent part of each fuzzy if-then rule and the input vector $\mathbf{x} = (x_1, \cdots, x_n)$:

$$\mu_j(\mathbf{x}) = \min\Big(\mu_{j1}(x_1), \cdots, \mu_{jn}(x_n)\Big). \tag{1.13}$$

Recently the product operation is often used especially in the field of fuzzy modeling by neurofuzzy techniques.

In Figure 1.7, we show the contour lines of the compatibility grade of the antecedent part "If x_1 is *medium* and x_2 is *medium*" for the minimum and product operations. As shown in Figure 1.7, the compatibility grade by the minimum operation is a pyramid shape while it is similar to a circular cone shape in the case of the product operation. Different fuzzy if-then rules are generated from the same training patterns by different operations. As a result, different classification boundaries are obtained by the generated fuzzy if-then rules. The choice of an appropriate operation is problem-dependent. It also depends on parameter specifications such as the choice of a fuzzy partition.

In the same manner as in the previous subsection, we compared the minimum operation with the product operation by computer simulations on the iris data. Simulation results are summarized in Table 1.3 and Table 1.4. From these tables, we can see that slightly better results were obtained by the minimum operation on the average for the iris data.

2.5.3 Comparison between Single Winner and Voting Methods. In our fuzzy reasoning method in (1.11) and (1.12), a single winner rule is selected for classifying a new pattern. This means that the other fuzzy if-then rules have no contribution to the classification of that pattern. Another fuzzy reasoning method is based on the voting by all fuzzy if-then rules [11]. In the voting method, each fuzzy if-then rule votes for its consequent class with its own weight defined by the product of the compatibility grade and the certainty grade. For example, the weight of the fuzzy if-then rule R_j is $\mu_j(\mathbf{x}) \cdot CF_j$ when a new

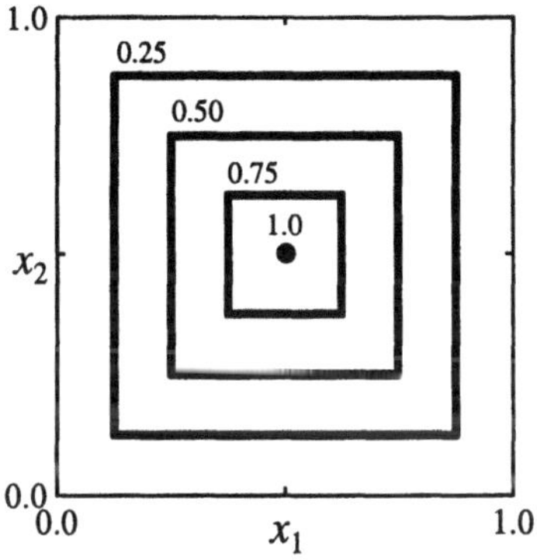

(a) Minimum operation.

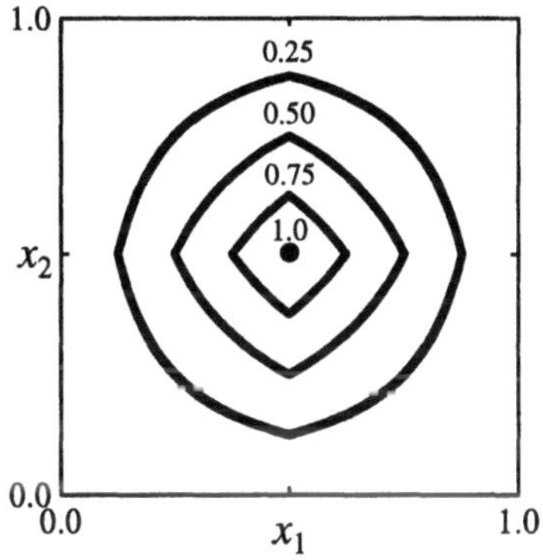

(b) Product operation.

Figure 1.7 Contour lines of the compatibility grade.

Table 1.3 Comparison between the minimum and product operations by classification rates on training data.

	$K = 3$	$K = 4$	$K = 5$	Average
Minimum	96.0%	92.0%	96.0%	94.7%
Product	94.0%	92.7%	96.0%	94.2%

Table 1.4 Comparison between the minimum and product operations by classification rates on test data.

	$K = 3$	$K = 4$	$K = 5$	Average
Minimun	95.3%	91.3%	95.3%	94.0%
Product	93.3%	89.3%	95.3%	92.6%

pattern $\mathbf{x}$ is to be classified. This weight is given to the consequent class C_j. The classification result of the new pattern $\mathbf{x}$ is the class with the maximum total weight.

In Figure 1.8, we illustrate the difference between these two fuzzy reasoning methods. The solid line in Figure 1.8 is the classification boundary by the single winner method, which is the same as Figure 1.3 (b). The dotted line is the classification boundary by the voting method. In Figure 1.8, we use the fuzzy rule table shown in Figure 1.2 for the two fuzzy reasoning methods. As we have already shown in Figure 1.3, each fuzzy if-then rule has its own decision area in the case of the single winner method. The classification of new patterns in each decision area is performed by the corresponding fuzzy if-then rule. On the other hand, in the case of the voting method, multiple fuzzy if-then rules are responsible for the classification of a new pattern. In the voting method, it is possible for many weaker fuzzy if-then rules with the same consequent class to classify a new pattern even if no single rule dominates the winner rule with a different consequent class.

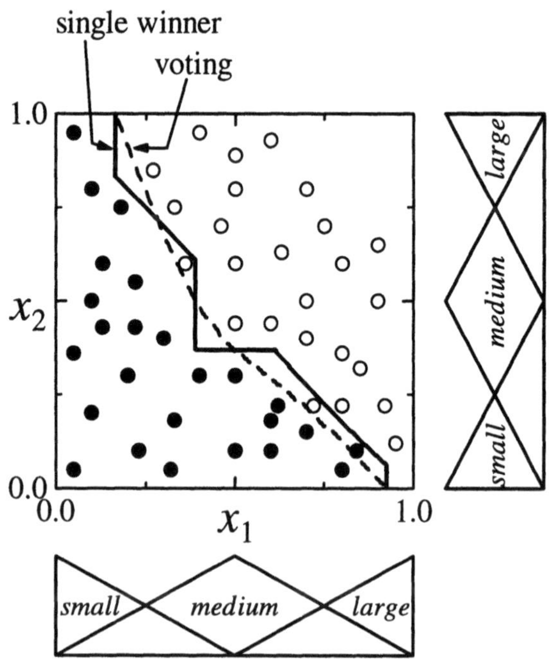

Figure 1.8 Classification boundaries by the single winner and voting methods.

As in the previous subsections, we compared the single winner method with the voting method by computer simulations on the iris data. Simulation results are summarized in Table 1.5 and Table 1.6. From these tables, we can see that better results were obtained by the voting method.

While the voting method seems to outperform the single winner method, this is not always the case. The choice of an appropriate fuzzy reasoning method is problem-dependent. The single winner method has several advantages over the voting method because of its simplicity. For example, the learning of fuzzy if-then rules based on a reward-punishment scheme is straightforward in the

Table 1.5 Comparison between the single winner and voting methods by classification rates on training data.

	$K=3$	$K=4$	$K=5$	Average
Single winner	94.0%	92.7%	96.0%	94.2%
Voting	96.0%	96.7%	96.7%	96.5%

Table 1.6 Comparison between the single winner and voting methods by classification rates on test data.

	$K=3$	$K=4$	$K=5$	Average
Single winner	93.3%	89.3%	95.3%	92.6%
Voting	96.0%	96.0%	96.0%	96.0%

case of the single winner method because a single winner rule is responsible for the classification result of each pattern [19].

In this chapter, we only discussed the voting by fuzzy if-then rules. The idea of voting can be used for integrating the classification results by multiple fuzzy rule-based classification systems [11].

2.5.4 Effect of Certainty Grades. One characteristic feature of our fuzzy if-then rules in (1.1) is the use of the certainty grade CF_j. If we do not use the certainty grade, classification boundaries by the single winner method are usually parallel to the axes of the pattern space as in the case of crisp fuzzy partitions. For example, the fuzzy if-then rules in Figure 1.2 generate the same classification boundary as the crisp partition in Figure 1.5 if we do not take account of the certainty grade of each fuzzy if-then rule. Fuzzy if-then rules with no certainty grades, however, are not the same as crisp if-then rules because of the expansion of the effect of training patterns described in Figure 1.6.

As in the previous subsections, we examined the effect of the certainty grade by computer simulations on the iris data. Simulation results are summarized in Table 1.7 and Table 1.8. From these tables, we can see that better results were obtained by fuzzy if-then rules with certainty grades on the average. While the effect of certainty grades is not so clear in Table 1.7 and Table 1.8, classification rates can be drastically improved by adjusting the certainty grade of each fuzzy if-then rule. For example, an average classification rate 98.0% on test data was reported for $K = 4$ as a result of adjusted certainty grades using a simple reward-punishment learning scheme in [19].

Table 1.7 Comparison between fuzzy rules with and without certainty grades by classification rates on training data.

	$K = 3$	$K = 4$	$K = 5$	Average
With certainty	94.0%	92.7%	96.0%	94.2%
Without certainty	92.0%	82.7%	94.7%	89.8%

Table 1.8 Comparison between fuzzy rules with and without certainty grades by classification rates on test data.

	$K = 3$	$K = 4$	$K = 5$	Average
With certainty	93.3%	89.3%	95.3%	92.6%
Without certainty	92.0%	78.7%	94.7%	88.5%

3. DESIGN OF COMPACT FUZZY SYSTEMS

In the previous section of this chapter, we used K^n fuzzy if-then rules for our n-dimensional pattern classification problem where K is the number of linguistic values on each axis of the n-dimensional pattern space. Except for the case of $n = 2$ (e.g., Figure 1.2), those fuzzy if-then rules cannot be easily presented to human users in an understandable manner (i.e., in a tabular form). For example, in the case of the iris data with four continuous attributes, it is not an easy task for human users to manually examine all the $3^4 = 81$ fuzzy if-then rules generated by three linguistic values on each axis. In this section, we show two methods for finding a small number of fuzzy if-then rules that can be easily understood by human users.

3.1 INPUT SELECTION

For the iris data, it is well known that the third and fourth attributes are important for the classification task. In Table 1.9 and Table 1.10, we examined the performance of fuzzy rule-based systems with the original four attributes and the selected two attributes. From these tables, we can see that the classification ability did not deteriorate when we removed the first and second attributes from the iris data. In some cases, input selection can improve the generalization ability of fuzzy rule-based systems because it may avoid their overfitting to training patterns.

Table 1.9 Performance of fuzzy rule-based systems with the original four attributes and the selected two attributes on training data.

	$K = 3$	$K = 4$	$K = 5$	Average
Four attributes	94.0%	92.7%	96.0%	94.2%
Two attributes	94.0%	94.7%	95.3%	94.7%

Table 1.10 Performance of fuzzy rule-based systems with the original four attributes and the selected two attributes on test data.

	$K = 3$	$K = 4$	$K = 5$	Average
Four attributes	93.3%	89.3%	95.3%	92.6%
Two attributes	94.0%	90.7%	95.3%	93.3%

Input selection can be performed by a heuristic sequential forward selection method for fuzzy rule-based systems [10]. Let us denote the set of all the n attributes by $A_{\mathrm{ALL}} = \{a_1, \cdots, a_n\}$. We also denote the set of the selected attributes by A where $A \subset A_{\mathrm{ALL}}$. In the heuristic forward selection method, first the best combination of two attributes is selected by calculating the classification rates on training data for all combinations of two attributes. That is, the best combination of two attributes is found by the exhaustive search. Now A includes the selected two attributes. Then for each remaining attribute $a_i \in A_{\mathrm{ALL}} - A$, the performance of the attribute set $A + \{a_i\}$ is examined. The selected attribute set A is updated by the best attribute set including three attributes (i.e., two in A and another attribute). In this manner, the remaining attributes that are not included in A are examined and added to A.

Some simulation results in [10] demonstrated high classification performance of simple fuzzy rule-based systems with two attributes on some data sets. Other simulation results in [10] showed that more than two attributes were required in other data sets. That is, the usefulness of the input selection is problem-dependent.

3.2 FUZZY RULE SELECTION

In many cases, fuzzy rule-based systems generated by fuzzy grids with all attributes include redundant fuzzy if-then rules. The size of those fuzzy rule-based systems can be decreased by removing redundant fuzzy if-then rules with no deterioration in their classification performance. Furthermore some fuzzy if-then rules can be combined into a single rule. For example, in Figure 1.2, the left three fuzzy if-then rules with the same condition on x_1 (i.e., x_1 is $small$) but different conditions on x_2 can be merged into the following single fuzzy if-then rule because they have the same consequent class:

$$\text{If } x_1 \text{ is } small \text{ then Class 1 with } CF_j. \tag{1.14}$$

The certainty grade of this rule is to be recalculated by the heuristic rule generation method from training patterns compatible with the antecedent condition "x_1 is $small$." This fuzzy if-then rule has no antecedent condition on x_2. Such a fuzzy if-then rule can be generated by considering "$don't\ care$" as an additional linguistic value. The membership function of "$don't\ care$" is always unity for all attribute values. Since the pattern space of our pattern classification problem is the n-dimensional unit hypercube $[0, 1]^n$, "$don't\ care$" can be viewed as the entire domain interval $[0, 1]$ of each axis of the pattern space. When we use "$don't\ care$" in addition to the given K linguistic values for each axis of our n-dimensional pattern classification problem, the total number of possible fuzzy if-then rules is $(K+1)^n$. In Figure 1.9, we show fuzzy subspaces corresponding to those fuzzy if-then rules in the case of $K = 5$ and $n = 2$. As we can see from Figure 1.9, many fuzzy if-then rules overlap with each other. This means that many fuzzy if-then rules can be removed with no deterioration in classification performance.

The concept of fuzzy rule selection was proposed for designing a compact fuzzy rule-based classification system with high comprehensibility by selecting a small number of fuzzy if-then rules from a large number of candidate rules [7, 14]. The basic idea of the fuzzy rule selection is to simultaneously perform the minimization of the size of a fuzzy rule-based system and the maximization of its classification performance. While only the number of fuzzy if-then rules was usually considered as the size of a fuzzy rule-based system, we also take account of the number of antecedent conditions in each fuzzy if-then rule (i.e., the length of each fuzzy if-then rule). The length of a fuzzy if-then rule is defined by the number of its antecedent conditions excluding "$don't\ care$" conditions. For example, the length of the following fuzzy if-then rule is two:

$$\text{If } x_1 \text{ is } small \text{ and } x_4 \text{ is } large \text{ then Class 1 with } CF_j. \tag{1.15}$$

Short fuzzy if-then rules with only a few antecedent conditions are general rules, which can be easily understood by human users. Such a short fuzzy if-then

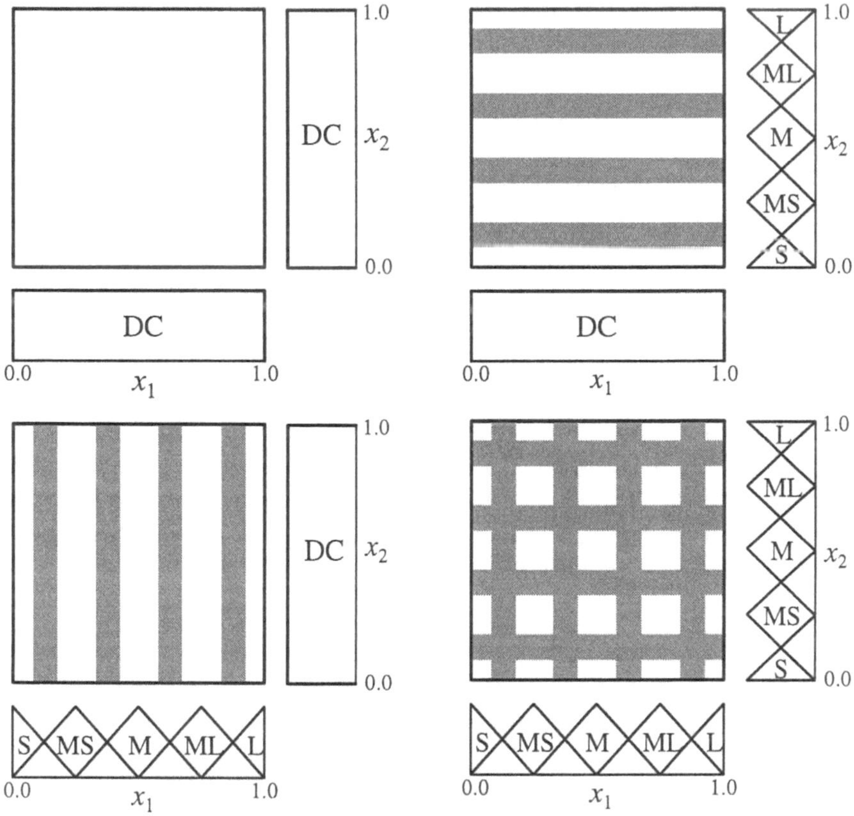

Figure 1.9 Fuzzy subspaces corresponding to 36 fuzzy if-then rules.

rule can cover a large area in the pattern space. On the other hand, long fuzzy if-then rules with many antecedent conditions are specific rules that can cover small areas in the pattern space. It is not easy for human users to intuitively understand the meaning of a long fuzzy if-then rule with many conditions.

Our goal in this section is to design a compact fuzzy rule-based system with high comprehensibility that has also high classification ability. An objective function for designing such a fuzzy rule-based system can be formulated as follows:

$$\begin{aligned} \text{Maximize } f(S) = {} & w_1 \cdot (\text{Classification ability of } S) \\ & - w_2 \cdot (\text{Total number of fuzzy if-then rules in } S) \\ & - w_3 \cdot (\text{Total number of antecedent conditions in } S), \quad (1.16) \end{aligned}$$

where S is a fuzzy rule-based system and w_i is a positive weight ($i = 1, 2, 3$). The sum of these three weights is not necessary to be equal to 1. We measure the classification ability of S by the number of correctly classified training patterns by S.

Let us denote the entire set of fuzzy if-then rules by S_{ALL}, which is generated from $(K+1)^n$ fuzzy subspaces where K is the number of given linguistic values for each of the n attributes. It should be noted that all the $(K+1)^n$ fuzzy if-then rules are not always generated. For example, some fuzzy if-then rules have no compatible training patterns. The rule set S_{ALL} includes all the generated fuzzy if-then rules. The rule generation is performed by the heuristic rule generation method in the previous section. Our task is to find a subset S of S_{ALL} to maximize the objective function $f(S)$ in (1.16).

It is impossible to examine all subsets of S_{ALL} because the total number of those subsets is terribly large. We use a genetic algorithm to efficiently search for a good rule set S, which is not always the optimal solution of our rule selection problem in (1.16) but at least its near optimal solution. In our genetic algorithm, each rule set S (i.e., a subset of S_{ALL}) is denoted by the following binary string of the length N where N is the number of fuzzy if-then rules in S_{ALL}:

$$S = s_1 s_2 \cdots s_N, \tag{1.17}$$

where $s_j = 1$ and $s_j = 0$ mean that the j-th fuzzy if-then rule in S_{ALL} is included in and excluded from the rule set S, respectively.

The fitness value of the binary string S is defined by the objective function in (1.16). We can apply standard genetic algorithms to our rule selection problem because its search space consists of a number of binary strings. We use the random generation of an initial population, the roulette wheel selection with a linear scaling, the uniform crossover, the bit-change mutation, and the elite strategy with a single elite solution.

We modify only the specification of the mutation probability in the above standard implementation of genetic algorithms. Our genetic algorithm biases the mutation probability for efficiently decreasing the number of fuzzy if-then rules included in each string. A larger probability is assigned to the mutation from $s_j = 1$ to $s_j = 0$ than the mutation from $s_j = 0$ to $s_j = 1$. That is, the mutation probability was biased towards the decrease in the number of 1's in each string. The biased mutation works very well especially when the entire rule set S_{ALL} includes many fuzzy if-then rules.

We applied our rule selection method to the iris data using five linguistic values on each axis of the four-dimensional pattern space as in Figure 1.9. All the 150 samples in the iris data set were used as training data. Since we use "*don't care*" as an additional linguistic value for generating candidate fuzzy if-then rules, the total number of possible fuzzy if-then rules is $6^4 = 1296$. Among those fuzzy if-then rules, 587 rules were actually generated for the iris data. The generated fuzzy if-then rules were used as candidate rules in the fuzzy rule selection. That is, the entire rule set S_{ALL} consisted of the generated 587 fuzzy if-then rules. We applied our genetic algorithm with the following parameter specifications to the fuzzy rule selection problem:

Population size: 50 strings,
String length: 587,
Stopping condition: 1000 generations,
Crossover probability: 0.8,
Mutation probabilities: 0.1 for the mutation from $s_j = 1$ to $s_j = 0$,
0.01 for the mutation from $s_j = 0$ to $s_j = 1$,
Weights in the objective function: $w_1 = 1$ or 10, $w_2 = 5$, $w_3 = 1$.

We examined two specifications of weight values (i.e., $\mathbf{w} = (w_1, w_2, w_3) = (1, 5, 1), (10, 5, 1)$). One specification with the small weight value $w_1 = 1$ for the classification performance was for minimizing the size of fuzzy rule-based systems. The other specification with the large weight value $w_1 = 10$ for the classification performance was for maximizing the classification performance of fuzzy rule-based systems. Our computer simulation was iterated 20 times for each specification. Average results are summarized in Table 1.11 where classification rates were measured on all the 150 samples used as training data. From Table 1.11, we can see that our genetic algorithm could find compact fuzzy rule-based systems with high classification performance. Moreover, we can also see that the tradeoff between the size of fuzzy rule-based systems and their classification performance was handled by the choice of the weight value.

When we try to apply the rule selection method to high-dimensional pattern classification problems, it is impractical to examine all the $(K + 1)^n$ combinations of antecedent linguistic values for generating candidate fuzzy if-then rules. Here we describe how the rule selection method can be applied to high-dimensional pattern classification problems. As an example of a high-dimensional pattern classification problem, we use wine data. The wine data set is a 13-dimensional pattern classification problem with 178 patterns from three

Table 1.11 Average simulation results of the fuzzy rule selection for iris data.

Weight value (w_1)	Classification rate (%)	Number of selected rules	Average length of selected rules
1	94.7	3.0	1.00
10	97.7	4.6	1.36

classes. This data set, which is available from the UC Irvine machine learning database, was used in [4] for examining their non-fuzzy genetics-based machine learning algorithm. If we use five linguistic values and "$don't\ care$" for each of the 13 attributes of the wine data set, the total number of possible combinations of antecedent linguistic values is $(5+1)^{13} \approx 1.3 \times 10^{10}$ (about 13 billion). First we generate a tractable number of candidate fuzzy if-then rules by examining only a tiny proportion of those combinations. That is, we need a prescreening procedure for generating candidate fuzzy if-then rules.

One simple prescreening procedure is to generate only general fuzzy if-then rules with only a few antecedent conditions as candidate rules. For example, if we restrict the rule length within two or less, the number of combinations of antecedent linguistic values is calculated as $1+5 \times {}_{13}C_1 + 5^2 \times {}_{13}C_2 = 2016$ for the wine data set with five linguistic values. By examining those combinations, we generated 1834 candidate fuzzy if-then rules. Thus the string length in our genetic algorithm is 1834. In the same manner as in the application to the iris data, we applied the rule selection method to the wine data. Average simulation results were summarized in Table 1.12. From this table, we can see that compact fuzzy rule-based systems with high classification performance were found by the rule selection method. We also examined coarser fuzzy partitions. Better results were obtained using three linguistic values (see Figure 1.1). Average simulation results are summarized in Table 1.13. Since the wine data set is a simple pattern classification problem with no overlaps between different classes in the pattern space, a small number of general fuzzy if-then rules generated by the coarse fuzzy partition can classify almost all the given samples. Of course, this is not always the case. Some pattern classification problems need specific fuzzy if-then rules as well as general fuzzy if-then rules. In such cases, it is very difficult to prescreen a tractable number of promising fuzzy if-then rules as candidate rules. In the next section, we describe fuzzy genetics-based machine learning (GBML) methods that search for fuzzy rule-based systems with no prescreening procedure. That is, while the application of the rule selection method to high-dimensional pattern classification problems requires the reduction of the search space, the fuzzy GBML methods search for fuzzy rule-based systems in the entire search space defined by the combinations of antecedent linguistic values.

4. FUZZY GENETICS-BASED MACHINE LEARNING

As we have already described, the design of a compact fuzzy rule-based system can be viewed as finding a small number of useful fuzzy if-then rules from the $(K+1)^n$ combinations of antecedent linguistic values. In this section, we describe fuzzy genetics-based machine learning (GBML) algorithms for this task. As in non-fuzzy GBML algorithms, fuzzy GBML algorithms can be

Table 1.12 Average simulation results of the fuzzy rule selection for wine data with five linguistic values.

Weight value (w_1)	Classification rate (%)	Number of selected rules	Average length of selected rules
1	93.6	13.9	1.75
10	98.6	19.2	1.80

Table 1.13 Average simulation results of the fuzzy rule selection for wine data with three linguistic values.

Weight value (w_1)	Classification rate (%)	Number of selected rules	Average length of selected rules
1	93.8	4.6	1.83
10	98.9	8.3	1.75

classified into two categories: One is the Michigan approach and the other is the Pittsburgh approach. In this section, first we describe the Michigan-style fuzzy GBML algorithm. Then we describe the Pittsburgh-style algorithm.

4.1 MICHIGAN APPROACH

In the Michigan approach, a single rule is handled as an individual called a classifier. Thus Michigan-style GBML methods are often referred to as classifier systems. In our Michigan-style fuzzy GBML algorithm [8, 12], a single fuzzy if-then rule is coded as a string, and a population consists of a prespecified number of fuzzy if-then rules. Thus the search for a compact fuzzy rule-based system with high classification ability corresponds to the evolution of a population of fuzzy if-then rules. Since the consequent class and the certainty grade of each fuzzy if-then rule can be easily specified from the given training patterns by the heuristic rule generation procedure, each rule is coded by its antecedent linguistic values. For example, if we use five linguistic values and "*don't care*," we use the following six symbols (i.e., 1, 2, 3, 4, 5, #) for coding each fuzzy if-then rule:

$$\begin{array}{llll} small & \to 1, & medium\ small & \to 2, \\ medium & \to 3, & medium\ large & \to 4, \\ large & \to 5, & don't\ care & \to \#. \end{array}$$

For example, the following fuzzy if-then rule for a six-dimensional pattern classification problem is coded as "#1#2#5":

$$\text{If } x_2 \text{ is } small \text{ and } x_4 \text{ is } medium\ small \text{ and } x_6 \text{ is } large \\ \text{then Class } C_j \text{ with } CF_j, \quad (1.18)$$

where "*don't care*" conditions on x_1, x_3, and x_5 are omitted.

First our fuzzy classifier system randomly generates a prespecified number of fuzzy if-then rules (say, N_{rule} fuzzy if-then rules) as an initial population. For our n-dimensional pattern classification problem, N_{rule} strings of the length n are generated by randomly selecting each of the six symbols with the probability 1/6.

Next the fitness value of each fuzzy if-then rule in the current population is evaluated. Let S be the set of the fuzzy if-then rules in the current population. The evaluation of each fuzzy if-then rule is performed by classifying all the given training patterns by the rule set S using the single winner method in (1.11). The fitness value $fitness(R_j)$ of each fuzzy if-then rule R_j in S is defined by the number of correctly classified training patterns by R_j.

Then new fuzzy if-then rules are generated from the existing fuzzy if-then rules in the current population by genetic operations. As parent strings, two fuzzy if-then rules are selected from the current population by the roulette wheel selection with a linear scaling. From the selected two strings, two new strings are generated by the uniform crossover with a prespecified crossover probability. Then each symbol of the generated strings by the crossover operation is randomly replaced with a different symbol using a prespecified mutation probability. The selection, crossover, and mutation are iterated until a prespecified number of new strings (say, N_{replace} strings) are generated.

Finally the worst N_{replace} strings with the smallest fitness values in the current population are removed, and the newly generated N_{replace} strings are added to the remaining strings to form a new population.

The above procedures are applied to the new population again. The generation update is iterated until a prespecified stopping condition is satisfied. In computer simulations of this chapter, we use the total number of generation updates as the stopping condition.

4.2 PITTSBURGH APPROACH

In Pittsburgh-style GBML algorithms, a rule set is handled as an individual. In the case of our Pittsburgh-style fuzzy GBML algorithm, a set of fuzzy if-then rules is handled as an individual and represented by a string. As in the previous subsection, each fuzzy if-then rule is coded as a string of the length n. Thus a rule set with N_{rule} fuzzy if-then rules is coded as a concatenated string of the length $n \times N_{\text{rule}}$ where each substring of the length n represents a fuzzy if-then rule.

First a prespecified number of rule sets (say, N_{pop} rule sets) with N_{rule} fuzzy if-then rules are randomly generated in the same manner as in the Michigan-style fuzzy GBML algorithm. The generated N_{pop} rule sets comprise an initial population where N_{pop} is the population size. Next each rule set is evaluated by classifying the given training patterns. The fitness value of each rule set S_i in the current population is defined by the number of correctly classified training patterns by S_i.

Next new rule sets are generated from the existing rule sets in the current population by genetic operations. As parent strings, two rule sets are selected from the current population using the roulette wheel selection with a linear scaling. From the selected two strings, two new strings are generated by a uniform crossover with a prespecified crossover probability. The uniform crossover in the Pittsburgh-style algorithm exchanges substrings between the two parent strings. Then each symbol of the generated strings by the crossover operation is randomly replaced with a different symbol using a prespecified mutation probability. The selection, crossover, and mutation are iterated until $(N_{\text{pop}} - 1)$ rule sets are generated. Finally the best rule set in the current population is added to the newly generated rule sets as an elite rule set to form a new population including N_{pop} rule sets.

The generation update is iterated until a prespecified stopping condition is satisfied. As in the Michigan-style fuzzy GBML algorithm, we use the total number of generation updates as the stopping condition.

4.3 COMPARISON OF TWO APPROACHES

For comparing the two fuzzy GBML algorithms with each other, we performed computer simulations on some commonly used data sets in the literature. We applied the two fuzzy GBML algorithms to the following data sets.

Iris data : 150 samples with four attributes from three classes.
Appendicitis data : 106 samples with seven attributes from two classes.
Cancer data : 286 samples with nine attributes from two classes.
Glass data : 214 samples with nine attributes from six classes.
Wine data : 178 samples with 13 attributes from three classes.

Credit data : 690 samples with 14 attributes from two classes.

The iris data, appendicitis data, and cancer data were used for comparing various classification methods in Weiss & Kulikowski [22]. The iris data, glass data, wine data, and credit data are available from the UC Irvine machine learning database.

We applied our two fuzzy GBML algorithms to these six data sets using the following parameter values:

The number of fuzzy rules: $N_{\text{rule}} = 20$,
The number of rule sets: $N_{\text{pop}} = 50$ (Pittsburgh),
Crossover probability: 0.9,
Mutation probability: 0.1,
Stopping condition: 500 generations,
The number of replaced fuzzy if-then rules: $N_{\text{replace}} = 4$ (Michigan).

From these parameter values, we can see that our task is to design a fuzzy rule-based system with 20 fuzzy if-then rules by each algorithm for each data set. We can also see that only 500 rule sets and 25000 rule sets are examined by the Michigan-style and the Pittsburgh-style fuzzy GBML algorithm, respectively. The number of examined rule sets in each algorithm is very small if compared with the total number of possible fuzzy if-then rules (i.e., $(5+1)^n$ rules).

We applied the Michigan-style algorithm 20 times and the Pittsburgh-style algorithm 10 times to each data set. Average classification rates by each algorithm, which were the classification performance on training data, are summarized in Table 1.14. From Table 1.14, we can see that higher classification rates were obtained by the Michigan-style algorithm for high-dimensional problems (e.g., the wine data) than the Pittsburgh-style algorithm. On the contrary, for low-dimensional problems (e.g., the iris data), better results were obtained by the Pittsburgh-style algorithm. These observations suggest that the Michigan-style algorithm has high search ability to find good fuzzy if-then rules in large search spaces. The Michigan-style algorithm, however, cannot directly optimize fuzzy rule-based systems because it only measures the performance of each fuzzy if-then rule. That is, the optimization of fuzzy rule-based systems is indirectly performed by finding good fuzzy if-then rules. The lack of the direct optimization ability leads to inferior results by the Michigan-style algorithm on low-dimensional problems. On the other hand, the Pittsburgh-style algorithm seems to have high search ability to find good combinations of fuzzy if-then rules because it measures the performance of each rule set. The Pittsburgh style algorithm, however, does not have high search ability to find good fuzzy if-then rules in large search spaces. These observations suggest that we can implement a better algorithm by hybridizing the two algorithms [9]. In the hybrid algorithm, the Michigan-style algorithm is used for generating good fuzzy if-then rules, and the Pittsburgh-style algorithm is used for finding good combinations of generated fuzzy if-then rules.

Table 1.14 Average classification rates on training data.

Data sets	Number of attributes	Number of samples	Classification rates (%)	
			Michigan	Pittsburgh
Iris	4	150	96.1	97.4
Appendicitis	7	106	84.1	88.5
Cancer	9	286	76.4	76.6
Glass	9	214	65.5	56.9
Wine	13	178	97.1	69.9
Credit	14	690	86.5	85.6

5. CONCLUDING REMARKS

In this chapter, we described several methods for designing compact fuzzy rule-based systems with high comprehensibility from numerical data. If we do not require high comprehensibility for human users, we can use many fuzzy if-then rules, and adjust the membership function of each antecedent fuzzy set. In this case, we can also use hierarchical fuzzy rule-based systems and multi-dimensional antecedent fuzzy sets because the linguistic interpretation of each fuzzy if-then rule is not required. On the other hand, when our main aim is to extract comprehensible linguistic knowledge from numerical data in the form of fuzzy if-then rules, we may not modify the meaning of each antecedent linguistic value. That is, it is not desired to improve the classification ability of fuzzy rule-based systems by adjusting the membership function of each antecedent linguistic value. In this case, our task is to generate a small number of comprehensible fuzzy if-then rules by combining the given linguistic values. From the viewpoint of comprehensibility, the number of fuzzy if-then rules and the length of each rule are minimized.

While fuzzy if-then rules in a tabular form have high comprehensibility for two-dimensional pattern classification problems, some trick is required for designing comprehensible fuzzy rule-based systems for high-dimensional problems. Input selection may be a simple but useful approach to the design of such a comprehensible fuzzy rule-based system. When we cannot find a small number of attributes with high classification ability by the input selection, we need another trick. Fuzzy rule selection may work well on pattern classification problems with four or five attributes. When pattern classification problems include many attributes, we need a prescreening procedure for generating a tractable number of candidate fuzzy if-then rules in the fuzzy rule selection. While the use of only short fuzzy if-then rules worked well on wine data, such a prescreening procedure does not always work well. Some pattern classification problems may need long fuzzy if-then rules with many antecedent conditions as well as short rules. As general-purpose algorithms for generating compact fuzzy rule-based systems for high-dimensional pattern classification problems, we described two fuzzy GBML algorithms. Each algorithm has its own advantages and disadvantages. Hybridization of these algorithms may lead to a better fuzzy GBML algorithm [9].

References

[1] Abe, S. and Lan, M.-S. (1995). A method for fuzzy rules extraction directly from numerical data and its application to pattern classification. *IEEE Trans. on Fuzzy Systems*, 3(1):18–28.

[2] Abe, S., Lan, M.-S., and Thawonmas, R. (1996). Tuning of a fuzzy classifier derived from data. *International J. of Approximate Reasoning*, 14(1):1–24.

[3] Abe, S. and Thawonmas, R. (1997). A fuzzy classifier with ellipsoidal regions. *IEEE Trans. on Fuzzy Systems*, 5(3):358–368.

[4] Corcoran, A. L. and Sen, S. (1994). Using real-valued genetic algorithms to evolve rule sets for classification. In *Proc. of 1st IEEE International Conf. on Evolutionary Computation, Orlando, FL*, pages 120–124.

[5] Gordon, O., del Jesus, M. J., and Herrera, F. (1998). Genetic learning of fuzzy rule-based classification systems cooperating with fuzzy reasoning methods. *International Journal of Intelligent Systems*, 13:1025–1053.

[6] Holve, R. (1997). Rule generation for hierarchical fuzzy systems. In *Proc. of 16th annual Meeting of the North American Fuzzy Information Processing Society, Syracuse, NY*, pages 444–449.

[7] Ishibuchi, H., Murata, T., and Turksen, I. B. (1997). Single-objective and two-objective genetic algorithms for selecting linguistic rules for pattern classification problems. *Fuzzy Sets and Systems*, 89(2):135–149.

[8] Ishibuchi, H. and Nakashima, T. (1999). Improving the performance of fuzzy classifier systems for pattern classification problems with continuous attributes. *IEEE Trans. on Industrial Electronics*, 46(6):1057–1068.

[9] Ishibuchi, H., Nakashima, T., and Kuroda, T. (1999). A hybrid fuzzy genetics-based machine learning algorithm: Hybridization of Michigan ap-

proach and Pittsburgh approach. In *Proc. of IEEE International Conference on Systems, Man and Cybernetics, Tokyo, Japan*, pages 296–301.

[10] Ishibuchi, H., Nakashima, T., and Morisawa, T. (1997). Simple fuzzy rule-based classification systems perform well on commonly used real-world data sets. In *Proc. of 16th annual Meeting of the North American Fuzzy Information Processing Society, Syracuse, NY*, pages 251–256.

[11] Ishibuchi, H., Nakashima, T., and Morisawa, T. (1999). Voting in fuzzy rule-based systems for pattern classification problems. *Fuzzy Sets and Systems*, 103(2):223–238.

[12] Ishibuchi, H., Nakashima, T., and Murata, T. (1999). Performance evaluation of fuzzy classifier systems for multi-dimensional pattern classification problems. *IEEE Trans. on Systems, Man, and Cybernetics - Part B: Cybernetics*, 29(5):601–618.

[13] Ishibuchi, H., Nozaki, K., and Tanaka, H. (1992). Distributed representation of fuzzy rules and its application to pattern classification. *Fuzzy Sets and Systems*, 52(1):21–32.

[14] Ishibuchi, H., Nozaki, K., Yamamoto, N., and Tanaka, H. (1995). Selecting fuzzy if-then rules for classification problems using genetic algorithms. *IEEE Trans. on Fuzzy Systems*, 3(3):260–270.

[15] Janikow, C. Z. (1996). A genetic algorithm method for optimizing the fuzzy components of a fuzzy decision tree. In Pal, S. K. and Wang, P. P., editors, *Genetic Algorithms for Pattern Recognition*, CRC Press, pages 253–281.

[16] Mitra, S. and Kuncheva, L. I. (1995). Improving classification performance using fuzzy MLP and two-level selective partitioning of the feature space. *Fuzzy Sets and Systems*, 70:1–13.

[17] Murata, T., Ishibuchi, H., Nakashima, T., and Gen, M. (1998). Fuzzy partition and input selection by genetic algorithms for designing fuzzy rule-based classification systems. *Springer-Verlag Lecture Notes in Computer Science Vol. 1447 : Evolutionary Programming IV*, pages 407–416.

[18] Nauck, D. and Kruse, R. (1999). A neuro-fuzzy method to learn fuzzy classification rules from data. *Fuzzy Sets and Systems*, 89:277–288.

[19] Nozaki, K., Ishibuchi, H., and Tanaka, H. (1996). Adaptive fuzzy rule-based classification systems. *IEEE Trans. on Fuzzy Systems*, 4(3):238–250.

[20] Quinlan, J. R. (1993). *C4.5: Programs for Machine Learning.* Morgan Kaufmann Publishers.

[21] Uebele, V., Abe, S., and Lan, M. S. (1995). A neural-network-based fuzzy classifier. *IEEE Trans. on Systems, Man, and Cybernetics*, 25:353–361.

[22] Weiss, S. M. and Kulikowski, C. A. (1991). *Computer Systems That Learn.* Morgan Kaufmann Publishers.

Chapter 13

EXPERIMENTS ON FUZZY LOGIC SYSTEMS BASED INDIRECT ADAPTIVE CONTROL OF A FLEXIBLE LINK MANIPULATOR

James X. Lee and George Vukovich

Directorate of Spacecraft Engineering, Canadian Space Agency

St. Hubert, QC J3Y 8Y9, Canada

{james.lee, george.vukovich}@space.gc.ca

Abstract Experiments are demonstrated where a newly developed fuzzy logic systems based, indirect adaptive control algorithm is applied to trajectory control of a flexible link manipulator. The manipulator is characterized by a very flexible link, complex nonlinear joint friction, and an unknown internal speed regulating loop, and thus serves as a challenging testbed. Experimental results are presented for two different desired trajectories. All the results are quite satisfactory, indicating that the fuzzy logic systems based adaptive control algorithm is effective in dealing with complex, nonlinear systems, and can be useful in practical applications.

1. INTRODUCTION

Many impressive successes have been produced by fuzzy logic approaches for control of complex and nonlinear industrial systems. It has also been widely recognized that this approach provides a systematic and efficient mathematical framework to incorporate linguistic descriptions of human expert knowledge in the form of IF-THEN statements, or IF-THEN rules, into design of automatic controllers. To date fuzzy logic controllers have been designed mainly by directly utilizing linguistic information, and an overwhelming body of literature dealt with a form of nonlinear fuzzy proportional-derivative-integral control or its various subsets for set point regulation problems [9, 10, 14, 19]. However, this kind of fuzzy logic control approach generally lacks formal synthesis techniques which guarantee the basic requirements of global stability and acceptable performance, and the controller parameters often must be manually adjusted by trial and error. There is thus a need to develop systematic design and analysis approaches such that the resulting fuzzy logic control systems not only incorporate linguistic information from human experts, but also automatically adjust their parameters to maintain system stability and consistent performance in the presence of uncertainties and variations.

One of the successful efforts to address this need is the introduction of adaptive schemes into fuzzy logic control systems with the help of a Lyapunov synthesis approach. This results in so-called adaptive fuzzy logic control systems which are further separated into direct [12, 15, 18] and indirect [12, 13, 16] schemes. In the former, fuzzy logic systems are used as controllers whose parameters are directly adjusted to reduce some norm of the output errors between the plants and the reference models. Linguistic IF-THEN rules describing control actions from given system states can be directly incorporated into controllers of this kind. In the latter, fuzzy logic systems are used to model the plants whose parameters are estimated, and the controllers are chosen assuming that the estimated parameters represent the true values of the plant parameters. The fuzzy IF-THEN rules describing the behavior of the plant can then be incorporated into the indirect adaptive fuzzy controllers.

However, the fuzzy logic systems in all the cited literature are static in nature. As the physical systems of interest in control engineering are generally dynamic, a so-called Dynamic Fuzzy Logic System (DFLS) was introduced in [6], which includes dynamical elements into fuzzy logic systems to take advantage of the intrinsic dynamics. Based on this dynamic fuzzy logic system, a stable indirect adaptive control algorithm was developed in [7] and was further generalized in [8]. However, it is noted that both FLS and DFLS based adaptive fuzzy logic control schemes mentioned above are only applied in computer simulation examples and experimental verification and practical application of these control algorithms on real physical systems would be extremely useful and informative.

In this chapter, we apply the DFLS-based indirect adaptive controller in [8] to control the trajectory of a real flexible link manipulator. This system poses significant challenges in mathematical modeling and control system design. First, the direct drive motor exhibits complex nonlinear friction which significantly affects system dynamics [1]. Thus far, friction other than viscous damping, such as stiction and coulomb friction and which are intrinsically nonlinear, has been generally ignored, to the frequent detriment of the associated dynamic models [1]. Second, the beam of this manipulator is very flexible, and therefore, the restriction of "small deflections" which has been imposed on much published work in this area to avoid model nonlinearities may be untenable. Third, this motor has an internal control loop which regulates driving torque on the rotor according to rotor speed and command voltage, whose characteristics are unclear. In addition, this system represents an important practical system and responds to the increasing need for high performance robotic manipulators of light weight which move faster and require smaller actuators with lower energy consumption [1].

In the following section, the DFLS and a corresponding indirect adaptive control scheme are briefly introduced. In Section 3, the control algorithm of Section 2 is applied to control the trajectory of a flexible link mechanical manipulator and the experimental results are presented. The concluding remarks are presented in Section 4.

2. DFLS AND INDIRECT ADAPTIVE CONTROL

2.1 DESCRIPTION OF DFLS

Mathematical expression of a *dynamic fuzzy logic system*, abbreviated as DFLS, is [6]

$$\dot{\hat{y}} = -\alpha\hat{y} + \psi(\mathbf{z}) , \tag{1.1}$$

where $\hat{y}$ is a scalar variable, α is a positive constant, the vector $\mathbf{z} = \{z_1, \ldots, z_P\}^T$ is the input of this fuzzy logic system, $z_p \in \mathcal{Z}_p \subset R$, $p = 1, \ldots, P$, and $\mathcal{Z}_p$ are the corresponding *universes of discourse* [4]. The second term on the right hand side of above equation, $\psi(\mathbf{z})$, represents the output of a static fuzzy logic system, or fuzzy logic system (FLS) in short, whose fuzzy rule base is defined by a collection of IF-THEN rules of the form,

$$F^i : IF\ [(z_1\ is\ Z_1^i)\ AND\ \ldots\ AND\ (z_P\ is\ Z_P^i)]\ THEN\ [\psi\ is\ \Psi^i], \tag{1.2}$$

where for i=1,...,I and $p = 1, \ldots, P$, F^i represents the ith rule and I represents the size of the rule base, $Z_p^i \in \{Z_{pj_p} \mid Z_{pj_p} \in \mathcal{Z}_p,\ j_p = 1, \ldots, J_p\}$ stands for a primary fuzzy set in $\mathcal{Z}_p$ and appears in the rule F^i, and Z_{pj_p} is the j_pth *primary fuzzy set* [4] defined in $\mathcal{Z}_p$. In the THEN part of (1.2), $\psi \in \Omega \subset R$, and $\Psi^i \in \{\Psi_k \mid \Psi_k \in \Omega,\ k = 1, \ldots, K\}$, where Ω is the universe of discourse of the FLS output and Ψ_k represents the kth primary fuzzy set in Ω.

Considering an FLS with fuzzy rule base (1.2), the product inference engine, singleton fuzzifier, center average defuzzifier, and Gaussian membership functions for primary fuzzy sets of the input linguistic variables, its mathematical expression is given as [17, Lemma 9.1],

$$\psi(\mathbf{z}) \triangleq \frac{\sum_{i=1}^{I} \overline{\psi}_i[\prod_{p=1}^{P} \mu_{Z_p^i}(z_p)]}{\sum_{i=1}^{I}[\prod_{p=1}^{P} \mu_{Z_p^i}(z_p)]} , \tag{1.3}$$

where

$$\mu_{Z_p^i}(z_p) \triangleq \exp[-\frac{1}{2}(\frac{z_p - \overline{z}_p^i}{\sigma_p^i})^2] , \tag{1.4}$$

$\overline{\psi}_i$ represents the center of the fuzzy set Ψ^i, $\overline{z}_p^i$ and σ_p^i are the parameters characterizing respectively the position and shape of the membership function of the fuzzy set Z_p^i.

$$\theta_i(\mathbf{z}) \triangleq \frac{\prod_{p=1}^{P} \exp[-\frac{1}{2}(\frac{z_p-\overline{z}_p^i}{\sigma_p^i})^2]}{\sum_{i=1}^{I}\prod_{p=1}^{P} \exp[-\frac{1}{2}(\frac{z_p-\overline{z}_p^i}{\sigma_p^i})^2]} \tag{1.5}$$

is called a fuzzy basis function [16], and Eq.(1.3) can be written as

$$\psi(\mathbf{z}) = \sum_{i=1}^{I} \overline{\psi}_i \cdot \theta_i(\mathbf{z}) = \mathbf{\Theta}^T(\mathbf{z}) \cdot \overline{\mathbf{\Psi}} , \tag{1.6}$$

where $\mathbf{\Theta}$ and $\overline{\mathbf{\Psi}}$ are vectors defined as

$$\left\{ \begin{array}{rcl} \mathbf{\Theta}(\mathbf{z}) & \triangleq & \{\theta_1(\mathbf{z}), ..., \theta_I(\mathbf{z})\}^T , \\ \overline{\mathbf{\Psi}} & \triangleq & \{\overline{\psi}_1, ..., \overline{\psi}_I\}^T . \end{array} \right. \tag{1.7}$$

Using Eq.(1.6) in Eq.(1.1) results in

$$\dot{\hat{y}} = -\alpha\hat{y} + \mathbf{\Theta}^T(\mathbf{z}) \cdot \overline{\mathbf{\Psi}} , \tag{1.8}$$

which possesses the universal approximation property [6] and has been used as on-line identifiers for unknown nonlinear systems [6, 7]. The parameters α and $\mathbf{\Theta}$ in Eq.(1.8) are determined in an off-line design process, leaving only $\overline{\mathbf{\Psi}}$ as free parameter vector which is adaptively tuned on-line.

2.2 CONTROL OBJECTIVE

Consider a class of nonlinear dynamic systems which can be modeled as:

$$\left\{ \begin{array}{rcl} x^{(N)} & = & f(\mathbf{x}) + g(\mathbf{x})u , \\ y & = & x , \end{array} \right. \tag{1.9}$$

where $u \in R$ and $y \in R$ are the input and output of the system, $\mathbf{x} \triangleq \{x, \dot{x}, \ldots, x^{(N-1)}\}^T \in R^N$ is the state vector which is assumed to be available for measurement, and $x^{(N)}$ represents the Nth order derivative of x with respect to time. $f : R^N \to R$ and $g : R^N \to R$ are unknown continuous nonlinear functions defined in certain controllability region $\mathcal{X} \subset R^N$. In state space form,

$$\begin{cases} \dot{x}_1 & = & x_2 , \\ \vdots & & \\ \dot{x}_{N-1} & = & x_N , \\ \dot{x}_N & = & f(\mathbf{x}) + g(\mathbf{x})u , \end{cases} \tag{1.10}$$

where

$$\begin{cases} x_1 & \triangleq & x, \\ x_2 & \triangleq & \dot{x}, \\ \vdots & & \\ x_N & \triangleq & x^{(N-1)}, \end{cases} \tag{1.11}$$

and $\mathbf{x} = \{x_1, \ldots, x_N\}^T$.

For the system to be controllable, the gain function g is assumed to satisfy the following conditions: $\forall\ \mathbf{x} \in \mathcal{X}$ and $0 \leq t < \infty$, it has $0 < \varepsilon_g \leq \mid g(\mathbf{x}) \mid \leq M_g$, where M_g is a large positive constant and ε_g is a small positive constant. In addition, the sign of $g(\mathbf{x})$ is known.

Let q be a desired trajectory of system output y, and denote

$$\mathbf{q} \triangleq \{q_1, q_2, \ldots, q_N\}^T \triangleq \{q, \dot{q}, \ldots, q^{(N-1)}\}^T . \tag{1.12}$$

Let e be the tracking error, i.e.,

$$e \triangleq y - q , \tag{1.13}$$

and denote

$$\mathbf{e} \triangleq \{e_1, e_2, \ldots, e_N\}^T \triangleq \{e, \dot{e}, \ldots, e^{(N-1)}\}^T = \mathbf{x} - \mathbf{q} . \tag{1.14}$$

Our control objective is to have a stable adaptive control algorithm, such that the plant output $y(t)$ follows a predefined bounded reference trajectory $q(t)$, under the constraint that all quantities involved must be bounded.

2.3 INDIRECT ADAPTIVE CONTROL

An indirect adaptive control algorithm in [8] is introduced to achieve the control objective. A DFLS of Eq.(1.8) is used to identify the state $x^{(N)}$,

$$\dot{\hat{y}} = -\alpha\hat{y} + \boldsymbol{\Theta}^T(\mathbf{x}, u) \cdot \overline{\boldsymbol{\Psi}} \ , \tag{1.15}$$

and the identification error is denoted as ξ,

$$\xi \stackrel{\triangle}{=} \hat{y} - x^{(N)} \ . \tag{1.16}$$

Since the function $g(\mathbf{x})$ is of a static nature, a static FLS in the form of Eq.(1.6) is used as its estimator,

$$\hat{g} = \boldsymbol{\Theta}_g^T(\mathbf{x}) \cdot \overline{\boldsymbol{\Psi}}_g \ , \tag{1.17}$$

which will be discussed in more detail later in this section. The main results of the control algorithm are repeated in the following.

2.3.1 Control Law. For the system output to have the desired tracking performance and at the same time always be bounded, a control law consisting of two components, u_c and u_s, was proposed in [8]:

$$u \stackrel{\triangle}{=} u_c - S_u u_s \ , \tag{1.18}$$

where

$$S_u \stackrel{\triangle}{=} \begin{cases} 0 \ , & if \quad \| \mathbf{e} \| < M_{\mathbf{e}} \ , \\ 1 \ , & if \quad \| \mathbf{e} \| \geq M_{\mathbf{e}} \ . \end{cases} \tag{1.19}$$

$M_{\mathbf{e}}$ is a finite positive design parameter that bounds $\| \mathbf{e} \|$.

The first component of the control law, u_c, is the usual control term for tracking performance when the system states are within a desired bound:

$$u_c \stackrel{\triangle}{=} \frac{1}{\hat{g}}[\ \dot{q}_N + \alpha q_N - \boldsymbol{\Theta}^T(\mathbf{x}, 0)\overline{\boldsymbol{\Psi}} - \sum_{k=1}^{N-1} \alpha_k e_k\ ] \ , \tag{1.20}$$

where $\alpha_1, \ \ldots, \ \alpha_{N-1}$, and α are design parameters to be determined, $\hat{g}$ is the estimated value of the unknown function $g(\mathbf{x})$, $\boldsymbol{\Theta}$ and $\overline{\boldsymbol{\Psi}}$ are the quantities of the DFLS identifier Eq.(1.15), and

$$\boldsymbol{\Theta}(\mathbf{x}, 0) \stackrel{\triangle}{=} \boldsymbol{\Theta}(\mathbf{x}, u) \mid_{u \equiv 0} \ . \tag{1.21}$$

Let $\alpha_N \stackrel{\triangle}{=} \alpha$, and

$$\mathbf{A} \triangleq \begin{bmatrix} 0 & 1 & 0 & \dots & 0 \\ 0 & 0 & 1 & \dots & 0 \\ \vdots & \vdots & \vdots & & \vdots \\ 0 & 0 & 0 & \dots & 1 \\ -\alpha_1 & -\alpha_2 & -\alpha_3 & \dots & -\alpha_N \end{bmatrix}.$$

The parameters, $\{\alpha_1, \dots, \alpha_N\}$, are chosen in such a way that $\mathbf{A}$ is a stable matrix, i.e., the roots of the polynomial, $\det[s\mathbf{I} - \mathbf{A}] = s^N + \alpha_N s^{N-1} + \cdots + \alpha_1$, are all located in the open left half of the complex plane.

The second component of the control law, u_s, is a supervisory control term that restrains the system states from drifting beyond a desired bound, and is only activated when system states actually reach this bound:

$$u_s(t) \triangleq sgn(\mathbf{B}^T\mathbf{P}\mathbf{e}) \cdot sgn(g) \cdot \frac{1}{\varepsilon_g} \cdot [\, \alpha \mid x_N \mid + \mid \boldsymbol{\Theta}^T(\mathbf{x}, 0)\overline{\boldsymbol{\Psi}} \mid + M_f + (M_g + \mid \hat{g} \mid) \cdot \mid u_c \mid] \,, \qquad (1.22)$$

where $\mathbf{B}^T \triangleq [\, 0, \dots, 0, 1 \,]_{1\times N}$, M_f is a large positive constant that bounds $f(\mathbf{x})$, that is, $\forall\, \mathbf{x} \in \mathcal{X}$, $\mid f(\mathbf{x}) \mid \leq M_f$, and the sign function, sgn, is defined as

$$sgn(x) \triangleq \begin{cases} 1, & if \quad x \geq 0 \,, \\ -1, & if \quad x < 0 \,, \end{cases} \qquad (1.23)$$

and $\mathbf{P}$ is a positive definite symmetric matrix that satisfies the Lyapunov equation,

$$\mathbf{A}^T\mathbf{P} + \mathbf{P}^T\mathbf{A} = -\mathbf{Q} \,, \qquad (1.24)$$

where $\mathbf{Q}$ is some given positive definite symmetric matrix.

2.3.2 Adaptive Law for Parameter Vector $\overline{\boldsymbol{\Psi}}$. The adaptive law for $\overline{\boldsymbol{\Psi}}$ is obtained by following a Lyapunov synthesis approach [8]:

$$\dot{\overline{\boldsymbol{\Psi}}} = -\mathbf{H}[\boldsymbol{\Theta}(\mathbf{x}, u)h\xi - \boldsymbol{\Theta}(\mathbf{x}, 0)(\mathbf{B}^T\mathbf{P}\mathbf{e})] - S\beta\mathbf{H}\overline{\boldsymbol{\Psi}} \,, \qquad (1.25)$$

where $\mathbf{H}$ is a constant positive definite symmetric matrix, and h is a positive constant that weights the identification error. The last term on the right hand side of Eq.(1.25) represents a projection algorithm modification [2, 3, 11] to avoid circumstances under which the parameter vector $\overline{\boldsymbol{\Psi}}$ becomes too large or even drifts to infinity [11]. The "*on*" or "*off*" status of this term is controlled by the switch S, which is defined as

$$S = \begin{cases} 0\,, & if \quad \| \overline{\Psi} \| < M_{\overline{\Psi}}, \\ & or \quad \| \overline{\Psi} \| = M_{\overline{\Psi}} \; and \\ & \quad \overline{\Psi}^T \mathbf{H}[\Theta(\mathbf{x},u)h\xi - \Theta(\mathbf{x},0)(\mathbf{B}^T\mathbf{P}\mathbf{e})] \geq 0\,; \\ & or \quad \| \overline{\Psi} \| > M_{\overline{\Psi}} \; and \\ & \quad \overline{\Psi}^T \mathbf{H}[\Theta(\mathbf{x},u)h\xi - \Theta(\mathbf{x},0)(\mathbf{B}^T\mathbf{P}\mathbf{e})] > 0\,; \\ 1\,, & otherwise; \end{cases} \tag{1.26}$$

$M_{\overline{\Psi}}$ is a positive constant that bounds $\| \overline{\Psi} \|$. β is a positive design parameter which satisfies

$$\begin{cases} \beta \;\geq\; \dfrac{\overline{\Psi}^T \mathbf{H}[\Theta(\mathbf{x},0)(\mathbf{B}^T\mathbf{P}\mathbf{e}) - \Theta(\mathbf{x},u)h\xi]}{\overline{\Psi}^T \mathbf{H}\overline{\Psi}}\,, \\ if \;\; \| \overline{\Psi} \| = M_{\overline{\Psi}} \; and \; \overline{\Psi}^T \mathbf{H}[\Theta(\mathbf{x},u)h\xi - \Theta(\mathbf{x},0)(\mathbf{B}^T\mathbf{P}\mathbf{e})] < 0\,; \\ \beta \;>\; \dfrac{\overline{\Psi}^T \mathbf{H}[\Theta(\mathbf{x},0)(\mathbf{B}^T\mathbf{P}\mathbf{e}) - \Theta(\mathbf{x},u)h\xi]}{\overline{\Psi}^T \mathbf{H}\overline{\Psi}}\,, \\ if \;\; \| \overline{\Psi} \| > M_{\overline{\Psi}} \; and \; \overline{\Psi}^T \mathbf{H}[\Theta(\mathbf{x},u)h\xi - \Theta(\mathbf{x},0)(\mathbf{B}^T\mathbf{P}\mathbf{e})] \leq 0\,; \end{cases} \tag{1.27}$$

2.3.3 Adaptive Laws for $\hat{g}$. Consider the FLS estimator Eq.(1.17),

$$\hat{g} = \Theta_g^T(\mathbf{x})\overline{\Psi}_g\,, \tag{1.28}$$

where

$$\begin{cases} \Theta_g(\mathbf{x}) & \triangleq & \{\theta_{g1}(\mathbf{x}),\; \ldots,\; \theta_{gJ}(\mathbf{x})\}^T\,, \\ \overline{\Psi}_g & \triangleq & \{\overline{\psi}_{g1},\; \ldots,\; \overline{\psi}_{gJ}\}^T\,, \end{cases} \tag{1.29}$$

"J" is the size of the fuzzy rule base, and θ_{gj} is defined as

$$\theta_{gj}(\mathbf{x}) \triangleq \frac{\prod_{n=1}^{N} \exp[-(\frac{x_n - \overline{x}_{gn}^j}{\sigma_{gn}^j})^2]}{\sum_{j=1}^{J} \prod_{n=1}^{N} \exp[-(\frac{x_n - \overline{x}_{gn}^j}{\sigma_{gn}^j})^2]}\,. \tag{1.30}$$

For the system to be stable it is assumed that $\forall\; \mathbf{x} \in \mathcal{X}$, if $\| \overline{\Psi}_g \| \geq M_{\overline{\Psi}_g}$, then $| \hat{g} | \geq \varepsilon_g$, where $M_{\overline{\Psi}_g}$ is some large positive constant.

Following the Lyapunov synthesis approach, the adaptive law for parameter vector $\overline{\Psi}_g$ is obtained as [8]

$$\dot{\overline{\Psi}}_g = \gamma\Theta_g(\mathbf{x})(\mathbf{B}^T\mathbf{P}\mathbf{e})u_c - S_g\beta_g\gamma\overline{\Psi}_g\,, \tag{1.31}$$

where γ is a positive gain, S_g is a switch to control the "*on*" or "*off*" status of the projection modification term, and β_g is a positive design parameter,

$$\{S_g,\ \beta_g\}^T \triangleq \begin{cases} \{S_g^U,\ \beta_g^U\}^T\ , & if\ \ |\ \hat{g}\ |> \varepsilon_g\ , \\ \{-S_g^L,\ \beta_g^L\}^T\ , & if\ \ |\ \hat{g}\ |= \varepsilon_g\ and \\ & \quad \|\ \overline{\mathbf{\Psi}}_g\ \|< M_{\overline{\mathbf{\Psi}}_g}\ , \end{cases} \tag{1.32}$$

where

$$S_g^U \triangleq \begin{cases} 0\ , & if\ \ \|\ \overline{\mathbf{\Psi}}_g\ \|< M_{\overline{\mathbf{\Psi}}_g}, \\ & or\ \ \|\ \overline{\mathbf{\Psi}}_g\ \|= M_{\overline{\mathbf{\Psi}}_g}\ \ and \\ & \quad \overline{\mathbf{\Psi}}_g^T\mathbf{\Theta}_g(\mathbf{x})(\mathbf{B}^T\mathbf{Pe})u_c \leq 0\ , \\ & or\ \ \|\ \overline{\mathbf{\Psi}}_g\ \|> M_{\overline{\mathbf{\Psi}}_g}\ \ and \\ & \quad \overline{\mathbf{\Psi}}_g^T\mathbf{\Theta}_g(\mathbf{x})(\mathbf{B}^T\mathbf{Pe})u_c < 0\ , \\ 1\ , & \quad otherwise, \end{cases} \tag{1.33}$$

$$S_g^L \triangleq \begin{cases} 0\ , & if\ \hat{g}(\mathbf{B}^T\mathbf{Pe})u_c \geq 0\ , \\ 1\ , & if\ \hat{g}(\mathbf{B}^T\mathbf{Pe})u_c < 0\ , \end{cases} \tag{1.34}$$

$$\begin{cases} \beta_g^U \geq \dfrac{\overline{\mathbf{\Psi}}_g^T\mathbf{\Theta}_g(\mathbf{x})(\mathbf{B}^T\mathbf{Pe})u_c}{\overline{\mathbf{\Psi}}_g^T\overline{\mathbf{\Psi}}_g}\ , & if\ \ \|\ \overline{\mathbf{\Psi}}_g\ \|= M_{\overline{\mathbf{\Psi}}_g}\ \ and \\ & \quad \overline{\mathbf{\Psi}}_g^T\mathbf{\Theta}_g(\mathbf{x})(\mathbf{B}^T\mathbf{Pe})u_c > 0\ ; \\ \beta_g^U > \dfrac{\overline{\mathbf{\Psi}}_g^T\mathbf{\Theta}_g(\mathbf{x})(\mathbf{B}^T\mathbf{Pe})u_c}{\overline{\mathbf{\Psi}}_g^T\overline{\mathbf{\Psi}}_g}\ , & if\ \ \|\ \overline{\mathbf{\Psi}}_g\ \|> M_{\overline{\mathbf{\Psi}}_g}\ \ and \\ & \quad \overline{\mathbf{\Psi}}_g^T\mathbf{\Theta}_g(\mathbf{x})(\mathbf{B}^T\mathbf{Pe})u_c \geq 0\ . \end{cases} \tag{1.35}$$

$$\beta_g^L \geq -\frac{\hat{g}(\mathbf{\Theta}_g^T\mathbf{\Theta}_g)(\mathbf{B}^T\mathbf{Pe})u_c}{\hat{g}^2}\ . \tag{1.36}$$

$M_{\overline{\mathbf{\Psi}}_g}$ is a positive design constant that bounds $\|\ \overline{\mathbf{\Psi}}_g\ \|$ and satisfies

$$M_{\overline{\mathbf{\Psi}}_g} \leq \frac{1}{\sqrt{J}}M_g\ . \tag{1.37}$$

where J is the size of the vectors $\mathbf{\Theta}_g$ and $\overline{\mathbf{\Psi}}_g$ as in Eq.(1.29).

It is possible that $|\ \hat{g}\ |$ exceeds its lower bound, ε_g, during training, as the training is implemented computationally as

$$\hat{g}(t+dt) = \hat{g}(t) + \dot{\hat{g}}(t)dt = \hat{g}(t) + \mathbf{\Theta}_g^T\dot{\overline{\mathbf{\Psi}}}_g(t)dt\ , \tag{1.38}$$

so that a $\mid \hat{g}(t) \mid$ very close to but still within its lower bound, ε_g, produces a $\mid \hat{g}(t+dt) \mid$ which can possibly exceed this prespecified lower bound. In this situation, we can change the value of the gain, γ, and obtain a new value of $\dot{\overline{\mathbf{\Psi}}}_g(t)$, which results in a new value of $\hat{g}(t+dt)$ (such that $\mid \hat{g}(t+dt) \mid \geq \varepsilon_g$). Another choice could simply be to let $\hat{g}(t+dt) \triangleq sgn[\hat{g}(t)]\varepsilon_g$ to avoid this situation, and carry on the adaptive calculation with new samples.

2.4 STABILITY PROPERTIES OF THE ADAPTIVE CONTROL ALGORITHM

The indirect adaptive control scheme described in Section 2.3 satisfies the following stability properties [8]:

1 $\| \overline{\mathbf{\Psi}} \| \leq M_{\overline{\mathbf{\Psi}}}$;

2 $\varepsilon_g \leq \mid \hat{g} \mid \leq M_g$;

3 $\| \mathbf{e} \| \leq M_{\mathbf{e}}$, $\| \mathbf{x} \| \leq M_{\mathbf{e}} + \| \mathbf{q} \|$, and $\mid \xi \mid \leq M_\xi$, where M_ξ is a positive constant;

4 $\mid u(t) \mid \leq \frac{1}{\varepsilon_g} \cdot [\, (2+\frac{M_g+|\hat{g}(t)|}{\varepsilon_g}) \cdot (\sqrt{I} M_{\overline{\mathbf{\Psi}}} + \alpha \mid x_N(t) \mid) + (1+\frac{M_g+|\hat{g}(t)|}{\varepsilon_g}) \cdot (\mid \dot{q}_N(t) \mid + \| \mathbf{a} \| M_{\mathbf{e}}) + M_f \,]$, where $\mathbf{a} \triangleq \{\alpha_1, \ldots, \alpha_N\}^T$, "$I$" is the size of the vector $\mathbf{\Theta}$, and M_f is a positive constant that bounds $\mid f(\mathbf{x}) \mid$;

5 $\int_0^t \| \mathbf{e}(\tau) \|^2 \, d\tau \leq b + \eta \int_0^t [\, r^2(\mathbf{x}, 0, \mathbf{\Theta}, \overline{\mathbf{\Psi}}^*) + r^2(\mathbf{x}, u, \mathbf{\Theta}, \overline{\mathbf{\Psi}}^*) + \Delta_g^2 \,] d\tau$,

$\int_0^t \xi^2(\tau) d\tau \leq b + \eta \int_0^t [\, r^2(\mathbf{x}, 0, \mathbf{\Theta}, \overline{\mathbf{\Psi}}^*) + r^2(\mathbf{x}, u, \mathbf{\Theta}, \overline{\mathbf{\Psi}}^*) + \Delta_g^2 \,] d\tau$,

where b and η are positive constants, $\Delta_g \triangleq \hat{g} - g$, $r(\mathbf{x}, u, \mathbf{\Theta}, \overline{\mathbf{\Psi}}) \triangleq \mathbf{\Theta}^T(\mathbf{x}, u)\overline{\mathbf{\Psi}} - \alpha x_N - f(\mathbf{x}) - g(\mathbf{x})u$, $\overline{\mathbf{\Psi}}^*$ represents an optimal parameter vector which minimizes the static modeling error $r(\mathbf{x}, u, \mathbf{\Theta}, \overline{\mathbf{\Psi}}^*)$;

6 if $r(t) \in \mathcal{L}_2[0, \infty)$ and $\Delta_g \in \mathcal{L}_2[0, \infty)$, i.e., $(\int_0^\infty \| r(\tau) \|^2 \, d\tau)^{\frac{1}{2}} < \infty$ and $(\int_0^\infty \Delta_g^2 d\tau)^{\frac{1}{2}} < \infty$, then, $\lim_{t\to\infty} \| \mathbf{e}(t) \| = 0$, $\lim_{t\to\infty} \mid \xi(t) \mid = 0$.

3. EXPERIMENTAL DEMONSTRATION – TRAJECTORY CONTROL OF A MECHANICAL MANIPULATOR

3.1 DESCRIPTION OF THE MECHANICAL MANIPULATOR

The experimental test bed is a single link manipulator as seen in Figure 1.1. The overall system is schematically illustrated in Figure 1.2. The actuator is a direct drive DC motor with the control to the motor an input voltage in the range of [-10, +10](V). Its angular displacement is measured with an incremental encoder with resolution of 1024000 pulses/rev and angular velocity measured with an analog voltage signal produced by a build-in device. A stainless steel beam (580 × 50 × 1.28 mm) is clamped at one end to a rigid hub mounted directly onto the rotor and can be driven in only the horizontal plane. At the other end, metal blocks (227g) are mounted as the tip load. The tip deflection of the link is measured with an ultrasonic position sensing device with resolution of 0.10 mm. The transmitter of the ultrasonic device is mounted on the hub of the motor, and the receiver fixed at the tip of the link. The link is very flexible in the horizontal plane (1.28 mm thick), and stiff (50 mm wide) in the vertical plane, allowing the assumption of planar motion.

As mentioned in Section 1, this system is characterized by a very flexible link, complex nonlinear joint friction, and an unknown internal speed regulating loop. It is of interest because it not only is challenging for control purposes but also represents an important practical system, since the study of control of robotic systems possessing flexible links has been drawing increasing interest for their potential use in space applications as well as in new generations of industrial robots [1].

Our objective here is to command the tip of the link to follow a desired trajectory $q(t)$ by using the indirect adaptive control algorithm described in the previous section.

3.2 TRAJECTORY CONTROL OF THE MANIPULATOR

The variables of describing the manipulator are illustrated in Figure 1.3, where L is the length of the beam, $v(L,t)$ is the vibrational displacement of the tip of the beam, $\Phi(t)$ is the angular displacement of the rigid body motion of the beam, and $w(t)$ is denoted as the displacement of the tip of the beam, i.e.,

$$w(t) \triangleq L\Phi(t) + v(L,t) \ . \tag{1.39}$$

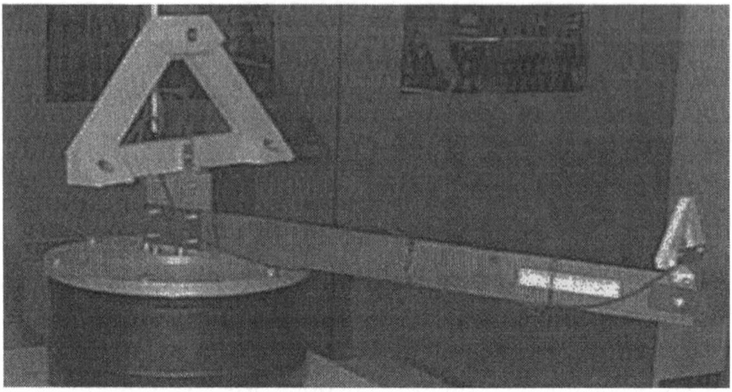

Figure 1.1 Flexible Link Manipulator.

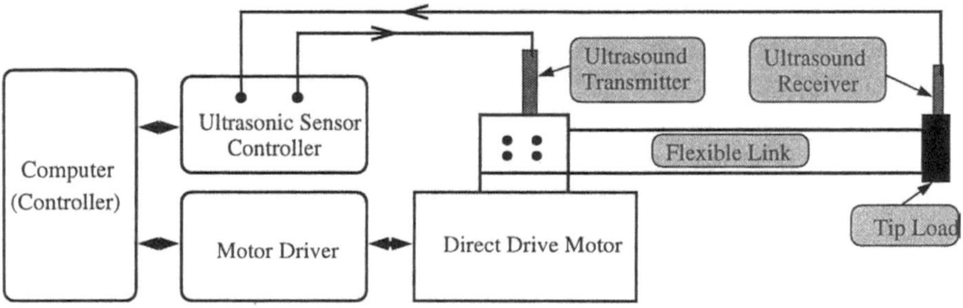

Figure 1.2 Diagram of the Flexible Link Manipulator System.

In the experimental setup, the variables, $\Phi(t)$, $\dot{\Phi}(t)$, and $v(L,t)$, can be directly measured with sensors, while $\dot{v}(L,t)$ is obtained by numerical differentiation of $v(L,t)$, i.e.,

$$\dot{v}(L,t) \triangleq \frac{v(L,t) - v(L,t-T)}{T} . \tag{1.40}$$

As this operation may amplify noise significantly, a low pass filter is used to filter out high frequency noises. A second order low pass digital Butterworth filter is used here.

The manipulator can be modeled as

$$\ddot{w} = f(w,\dot{w}) + g(w,\dot{w})V_m , \tag{1.41}$$

where V_m is the command voltage of the motor, f and g are unknown continuous functions. Let

$$\begin{cases} w_1 & \triangleq & w , \\ w_2 & \triangleq & \dot{w} , \end{cases} \tag{1.42}$$

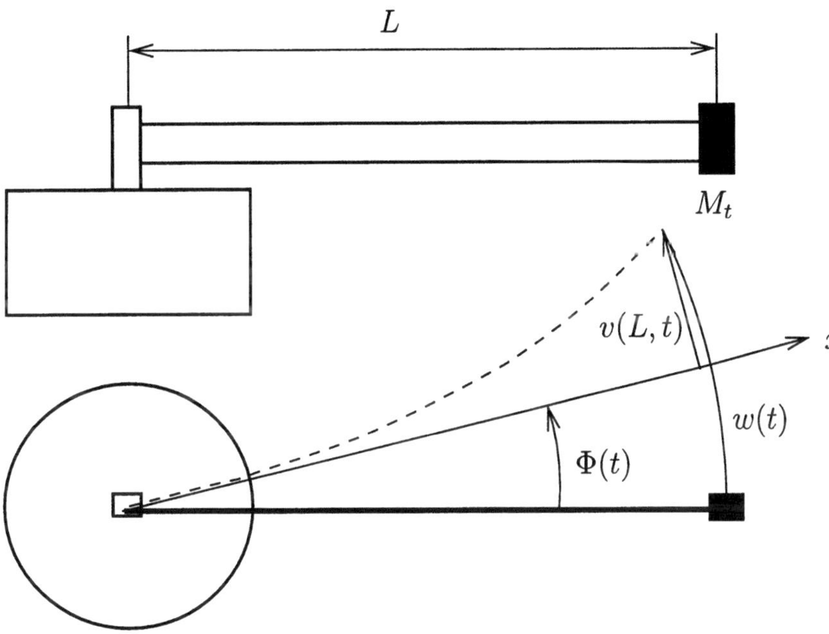

Figure 1.3 Illustration of the Single Link Manipulator.

then

$$\left\{ \begin{array}{lcl} \dot{w}_1 & = & w_2 \, , \\ \dot{w}_2 & = & f(\mathbf{w}) + g(\mathbf{w})V_m \, , \end{array} \right. \tag{1.43}$$

where $\mathbf{w} \triangleq \{w_1, w_2\}^T$.

A DFLS identifier, labeled D, is used to identify the state variable w_2, i.e.,

$$\hat{y} \triangleq D(\mathbf{w}, V_m) \, , \tag{1.44}$$

where $\hat{y}$ represents the identifier's output and is the estimate of w_2. The tracking error, e, and identification error, ξ, are

$$\left\{ \begin{array}{lcl} e & \triangleq & w_1 - q \, , \\ \xi & \triangleq & \hat{y} - w_2 \, . \end{array} \right. \tag{1.45}$$

For conciseness of expression, let

$$\begin{cases} x_1 & \triangleq & \Phi \ , \\ x_2 & \triangleq & \dot{\Phi} \ , \\ x_3 & \triangleq & w_1 \ , \\ x_4 & \triangleq & w_2 \ , \\ x_5 & \triangleq & V_m \ . \end{cases} \tag{1.46}$$

Considering Eq.(1.39), we have

$$\begin{aligned} \hat{y} &= D(\Phi,\ \dot{\Phi},\ v(L,t)\ , \dot{v}(L,t)\ , V_m) = D(\Phi,\ \dot{\Phi},\ w_1\ , w_2\ , V_m) \\ &\triangleq D(\mathbf{z})\ , \end{aligned} \tag{1.47}$$

where $\mathbf{z} \triangleq \{x_1, x_2, x_3, x_4, x_5\}^T$. The identifier, D, has five inputs, $x_p \in \mathcal{X}_p$, $p = 1, \ldots, 5$, and one output, $\hat{y} \in \mathcal{Y}$, where $\mathcal{X}_p$, $\mathcal{Y} \subset R$ are universes of discourse of their respective linguistic variables. In each of $\mathcal{X}_p$, three primary fuzzy sets, A_{pj_p}, $p = 1, \ldots, 5$, $j_p = 1, 2, 3$, are defined. Gaussian membership functions are used for all the primary fuzzy sets,

$$\begin{aligned} \mu_{A_{p1}}(x'_p) &\triangleq \begin{cases} 1\ , & if \quad x'_p < \overline{x}_{p1}\ , \\ \exp[-\frac{1}{2}(\frac{x'_p - \overline{x}_{p1}}{\sigma_{p1}})^2]\ , & if \quad x'_p \geq \overline{x}_{p1}\ , \end{cases} \\ \mu_{A_{p2}}(x'_p) &\triangleq \exp[-\frac{1}{2}(\frac{x'_p - \overline{x}_{p2}}{\sigma_{p2}})^2]\ , \\ \mu_{A_{p3}}(x'_p) &\triangleq \begin{cases} \exp[-\frac{1}{2}(\frac{x'_p - \overline{x}_{p3}}{\sigma_{p3}})^2]\ , & if \quad x'_p < \overline{x}_{p3}\ , \\ 1\ , & if \quad x'_p \geq \overline{x}_{p3}\ . \end{cases} \end{aligned} \tag{1.48}$$

Let the shape parameters of all the primary fuzzy sets be 0.45, i.e.,

$$\sigma_{pj_p} \triangleq 0.45\ , \tag{1.49}$$

and let the position parameters, $\overline{x}_{pj_p}$, be,

$$\{\overline{x}_{p1},\ \overline{x}_{p2},\ \overline{x}_{p3}\}^T \triangleq \{-1,\ 0,\ 1\}^T\ . \tag{1.50}$$

In Eq.(1.47), x'_p represents the filtered value of the DFLS input, x_p, that is, for $p = 1, \ldots, 5$,

$$x'_p \triangleq f_p(x_p)\ . \tag{1.51}$$

The purpose of the prefilters is to transform the input data so as to fit into the desired range.

The output of D, $\hat{y}$, is defined via the DFLS of Eq.(1.15),

$$D(\mathbf{z}'): \quad \dot{\hat{y}} = -\alpha\hat{y} + \mathbf{\Theta}^T(\mathbf{z}')\overline{\mathbf{\Psi}} \,. \tag{1.52}$$

In this equation,

$$\begin{cases} \mathbf{\Theta}(\mathbf{z}') & \triangleq & \{\theta_1(\mathbf{z}'),\ \theta_2(\mathbf{z}'),\ ...,\ \theta_{243}(\mathbf{z}')\}^T \,, \\ \overline{\mathbf{\Psi}} & \triangleq & \{\overline{\psi}_1,\ \overline{\psi}_2,\ ...,\ \overline{\psi}_{243}\}^T \,, \end{cases} \tag{1.53}$$

and

$$\theta_i(\mathbf{z}') = \frac{\prod_{p=1}^{5} \mu_{A_p^i}(x_p')}{\sum_{i=1}^{243} \prod_{p=1}^{5} \mu_{A_p^i}(x_p')} \,, \qquad i = 1, \ldots, 243, \tag{1.54}$$

where, for $p = 1, \ldots, 5$, $i = 1, \ldots, 243$, $A_p^i \in \{A_{pj_p};\ j_p = 1, 2, 3\}$.

$\overline{\mathbf{\Psi}}$ is the free parameter vector to be adaptively adjusted with the training law, Eq.(1.25),

$$\dot{\overline{\mathbf{\Psi}}} = -\mathbf{H}[\mathbf{\Theta}(\mathbf{z}')h\xi - \mathbf{\Theta}(\mathbf{x}', 0)(\mathbf{B}^T\mathbf{P}\mathbf{e})] - S\beta\mathbf{H}\overline{\mathbf{\Psi}} \,. \tag{1.55}$$

where $\mathbf{x}' \triangleq \{x_1', x_2', x_3', x_4'\}^T$. Using a numerical difference to approximate the derivative gives

$$\begin{aligned} \overline{\mathbf{\Psi}}(kT+T) &= \overline{\mathbf{\Psi}}(kT) - T\mathbf{H}[\ \mathbf{\Theta}(\mathbf{z}'(kT))h\xi(kT) \\ &\quad - \mathbf{\Theta}(\mathbf{x}'(kT), 0)(\mathbf{B}^T\mathbf{P}\mathbf{e}(kT)) + S\beta\overline{\mathbf{\Psi}}(kT)\ ], \end{aligned} \tag{1.56}$$

where T is the time incremental step.

The unknown function g is estimated via the FLS of Eq.(1.17),

$$\hat{g} \triangleq \mathbf{\Theta}_g^T(\mathbf{x}')\overline{\mathbf{\Psi}}_g \,, \tag{1.57}$$

where $\hat{g}$ represents the identifier's output and is the estimate of g. The FLS has four inputs, x_1', x_2', x_3', x_4', and one output, $\hat{g}$. Three primary fuzzy sets, B_{pj_p}, $p = 1, 2, 3, 4$, $j_p = 1, 2, 3$, are defined in all the universes of discourse in the input space of the FLS. Gaussian membership functions are used for all the primary fuzzy sets, which are the same as those given in Eqs.(1.47–1.50). The adaptive law for this FLS is given in Eq.(1.31),

$$\dot{\overline{\mathbf{\Psi}}}_g = \gamma\mathbf{\Theta}_g(\mathbf{x}')(\mathbf{B}^T\mathbf{P}\mathbf{e})u_c - S_g\beta_g\gamma\overline{\mathbf{\Psi}}_g \,. \tag{1.58}$$

The command voltage is restricted in [–7V, 7V], i.e.,

$$-7V \leq V_m \leq 7V \,. \tag{1.59}$$

Correspondingly, the supervisory control term is not required in this control system. Let $\alpha = \alpha_1 \stackrel{\triangle}{=} 10$, the time increment T set to 0.026 sec, and the bounds of the parameter vectors, $M_{\overline{\mathbf{\Psi}}}$ and M_g, set to 10^4, and ε_g to 0.01. Let $M_{\overline{\mathbf{\Psi}}_g} \stackrel{\triangle}{=} \frac{1}{9} M_g$, and set the gain matrix, $\mathbf{H}$, to be a diagonal matrix with all diagonal elements equal to 5, and set the gain γ to 5, and the weight parameter h to 2. $\mathbf{Q}$ is defined as a diagonal matrix with all diagonal elements set to 4.

The filtered values of inputs, x'_p, $p = 1, \ldots, 5$, are defined as

$$\left\{ \begin{array}{lllll} x'_1 & = & f_1(x_1) & \stackrel{\triangle}{=} & \Phi(t) \cdot \frac{1}{4} \,, \\ x'_2 & = & f_2(x_2) & \stackrel{\triangle}{=} & \dot{\Phi}(t) \cdot \frac{1}{10} \,, \\ x'_3 & = & f_3(x_3) & \stackrel{\triangle}{=} & w(t) \cdot \frac{1}{2} \,, \\ x'_4 & = & f_3(x_4) & \stackrel{\triangle}{=} & \dot{w}(t) \cdot \frac{1}{5} \,, \\ x'_5 & = & f_3(x_5) & \stackrel{\triangle}{=} & V_m(t) \cdot \frac{1}{5} \,. \end{array} \right. \tag{1.60}$$

It is assumed here that there is no *a priori* expert knowledge on system behavior. This poses more stringent conditions on the fuzzy logic identifier. The initial values of the parameter vectors $\overline{\mathbf{\Psi}}(0)$ and $\overline{\mathbf{\Psi}}_g(0)$ are initially assigned as random numbers. Because of safety concerns, instead of directly applying the adaptive controller to the real mechanical manipulator, simulation experiments were conducted first. That is, the adaptive controller outlined above was first applied in computer simulation to control the trajectory of the manipulator which is represented by a mathematical model. The mathematical model of the manipulator is presented in [5, Section 8.4.3]. The DFLS controller was then applied to command the actual manipulator to follow various sinusoidal trajectories of different frequencies and magnitudes. The final parameter vectors, $\overline{\mathbf{\Psi}}$ and $\overline{\mathbf{\Psi}}_g$, obtained in those simulation experiments were used as initial values in the real experiments, and further adjusted on-line. The mechanical system was initially at rest, i.e., $\mathbf{w}(0) = \{0, 0\}^T$.

For a desired trajectory,

$$q(t) = 0.4 \sin(0.8\pi t + \frac{\pi}{2}) - 0.4 \,, \tag{1.61}$$

the command voltage, V_m, and the controlled trajectory, w, are shown in Figure 1.4(a–b), the estimations for variable $\dot{w}$ and for function g are shown in Figure 1.5(a–b). The adaptation process stops at $t = 10$ sec. It is seen that the controlled trajectory converges very fast and follows the desired trajectory very well.

For a more complicated desired trajectory,

$$q(t) = 0.3 \sin(0.6\pi t) + 0.3 \sin(0.8\pi t) + 0.6 \,, \tag{1.62}$$

the command voltage, V_m, and the controlled trajectory, w, are shown in Figure 1.6(a–b), and the estimations for variable $\dot{w}$ and for function g are shown in Figure 1.7(a–b). The adaptation process also stops at $t = 10$ sec. Again, the controlled trajectory converges very fast and follows the desired trajectory very well.

These experiments demonstrate the effectiveness of the DFLS adaptive control algorithm in dealing with nonlinear, ill-defined systems.

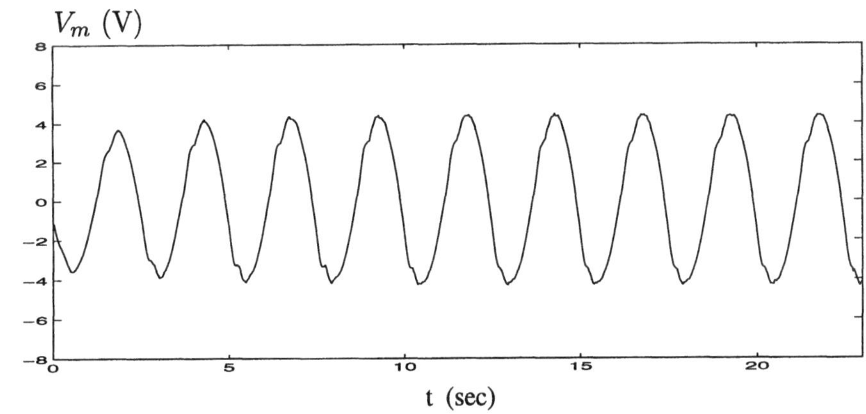

(a) Controller Output – Command Voltage

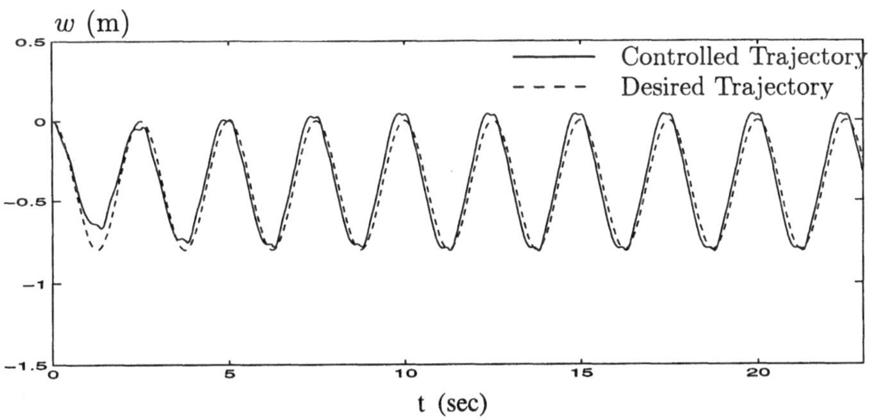

(b) Controlled Link Tip Trajectory – w
(Adaptation Terminates at t=10 sec)

Figure 1.4 Trajectory Control of a Mechanical Manipulator — Experiment 1.

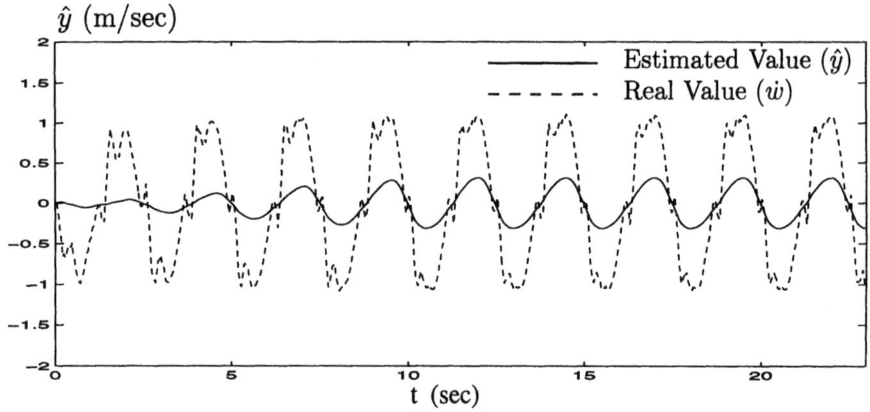

(a) Identifier Output – Link Tip Velocity

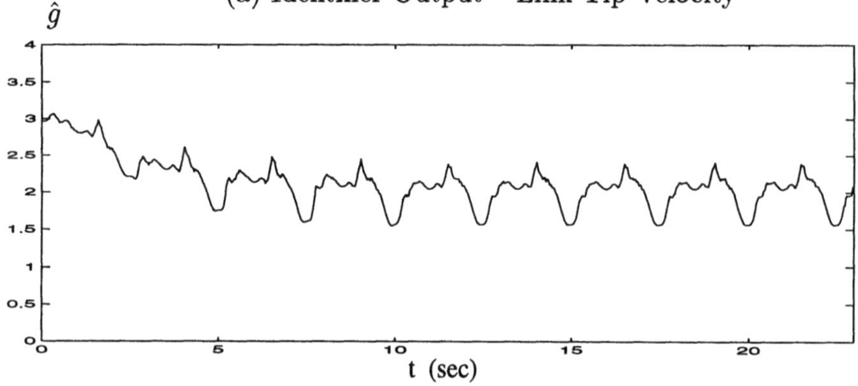

(b) Estimation for g
(Adaptation Terminates at t=10 sec)

Figure 1.5 Estimation of $\dot{w}$ and g — Experiment 1.

4. CONCLUDING REMARKS

- Experiments on trajectory control of a flexible link manipulator were conducted, where a DFLS based indirect adaptive control algorithm was applied. The manipulator features a very flexible link, complex nonlinear joint friction, and an unknown internal speed regulating loop, posing a significant challenge for automatic control. Satisfactory results were demonstrated for two different desired trajectories. This indicates that the DFLS based indirect adaptive control algorithm is very effective in dealing with complex, nonlinear systems, and can be useful in practical applications.

- In fuzzy logic control system design, if some expert knowledge of system behavior is available the initial values of the parameter

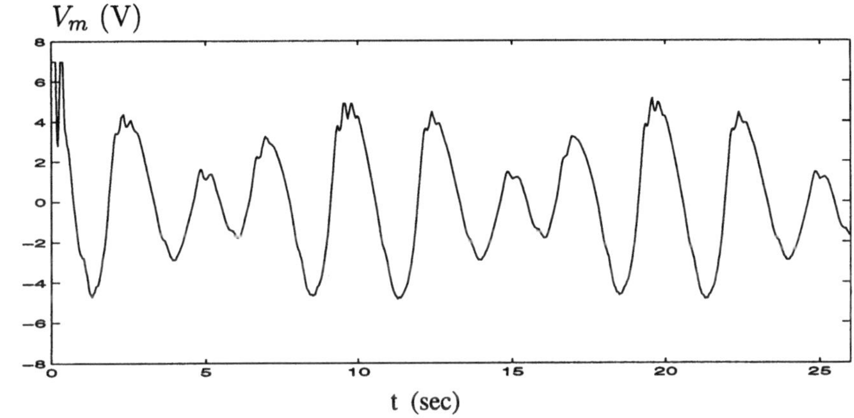

(a) Controller Output – Command Voltage

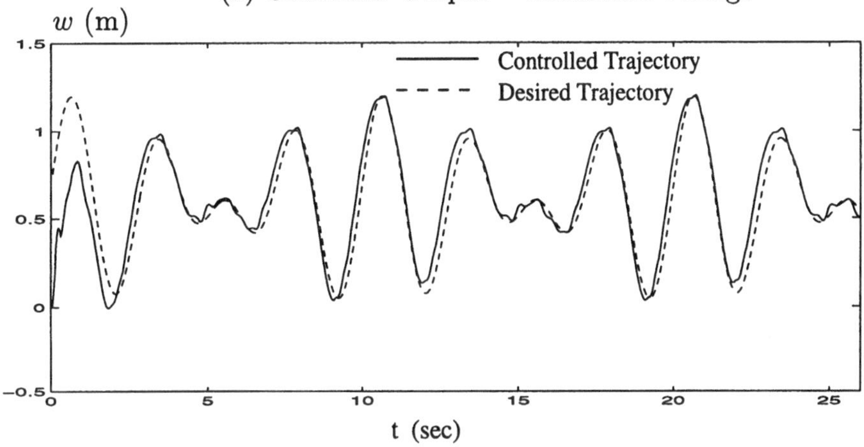

(b) Controlled Link Tip Trajectory – w
(Adaptation Terminates at t=10 sec)

Figure 1.6 Trajectory Control of a Mechanical Manipulator — Experiment 2.

vectors to be trained, $\overline{\mathbf{\Psi}}$ and $\overline{\mathbf{\Psi}}_g$ in this chapter, can be assigned accordingly. These initial values are generally better than those randomly assigned, and in turn, the system may require less training and display improved behavior as demonstrated in [6]. In the absence of human expertise, there remains no alternative to random selection, and more stringent conditions are imposed on the fuzzy logic identifiers, which is the situation in our experiments. Nevertheless, closed loop system performance will be as good as those properties specified in Section 2.4.

- In our fuzzy logic controller design the off-line design parameters were selected quite arbitrarily, without a great deal of effort, and

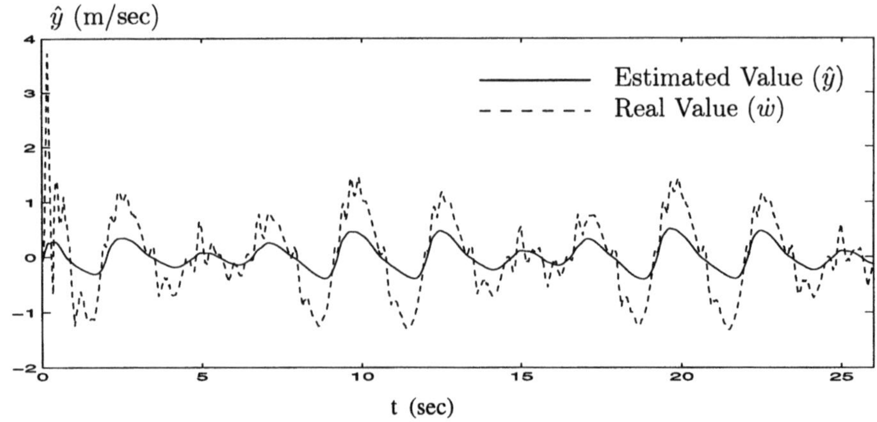

(a) Identifier Output – Link Tip Velocity

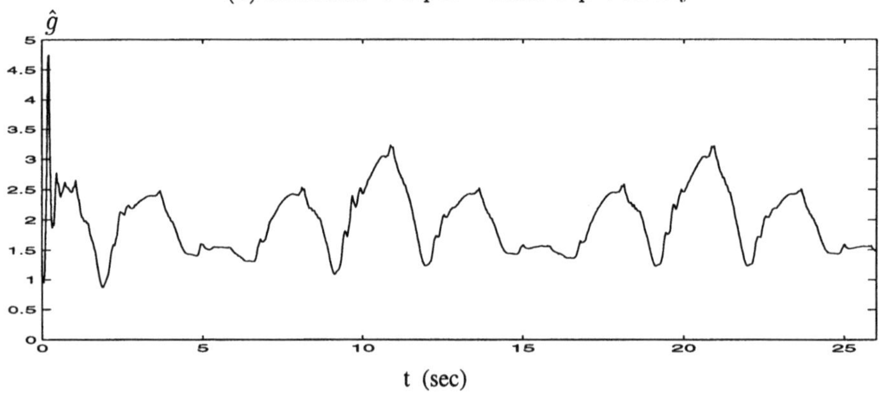

(b) Estimation for g

(Adaptation Terminates at t=10 sec)

Figure 1.7 Estimation of $\dot{w}$ and g — Experiment 2.

they are by no means considered to be optimal. This also poses more stringent demands on the identification and control algorithms.

References

[1] Alasty, A. "Force Control of Flexible Manipulators," *Ph.D. Dissertation.* Department of Mechanical and Aerospace Engineering, Carleton University, Ottawa, Canada, 1997.

[2] Goodwin, G.C., and Mayne, D.Q., "A Parameter Estimation Perspective of Continuous Time Model Reference Adaptive Control," *Automatica*, Volume 23, 1987, pp. 57–70.

[3] Ioannou, P.A., and Datta, A., "Robust Adaptive Control: Design, Analysis and Robustness Bounds," in *Foundations of Adaptive Control,* edited by Kokotovic, P.V., Berlin: Springer-Verlag, 1991, pp. 71–152.

[4] Lee, C. C., "Fuzzy Logic in Control Systems, Fuzzy Logic Controller–Part I and II," *IEEE Transactions on Systems, Man, and Cybernetics.* Volume 20, No. 2, March/April 1990, pp. 404–435

[5] Lee, J.X., "On Fuzzy Logic Systems: Nonlinear System Identification, and Adaptive Control", *Ph.D. Dissertation*, Carleton University, Ottawa, Canada, 1997.

[6] Lee, J.X., and Vukovich, G., "The Dynamic Fuzzy Logic System: Nonlinear System Identification and Application to Robotic Manipulators," *Journal of Robotic Systems*, 14(6), 1997, pp.391–405.

[7] Vukovich, G., and Lee, J.X., "Stable Identification and Adaptive Control – A Dynamic Fuzzy Logic System Approach," in *Fuzzy Evolutionary Computation*, Pedrycz, W. ed., Boston, London, Dordrecht: Kluwer Academic Publishers, 1997, pp.223–248.

[8] Lee, J.X., and Vukovich, G. "On Adaptive Fuzzy Logic Control of Nonlinear Systems – Synthesis and Analysis," in *Analytical Issues in Fuzzy Control – Synthesis and Analysis*, Farinwata, S.S., Filev, D., and Langari, R., eds., England: John Wiley and Sons, Inc., 1998. (to appear)

[9] Mamdani, E.H., "Advances in the Linguistic Synthesis of Fuzzy Controllers," In *Fuzzy Reasoning and its Applications*, Edited by Mamdani, E.H. and Gaines, B.R., Academic Press, 1981, pp.325–334.

[10] Mamdani, E.H. and Assilian, S., "An Experiment in Linguistic Synthesis with a Fuzzy Logic Controller," *International Journal of Man, Machine Studies*, Volume 7, 1975, pp. 1–13.

[11] Polycarpou, M.M., and Ioannou, P.A., "Stable Nonlinear System Identification Using Neural Network Models," in *Neural Networks in Robotics* ed by Bekey, G. and Goldberg, K., Norwell, MA: Kluwer Academic Publishers, 1993, pp.147–164.

[12] Spooner, J.T. and Passino, K.M., "Stable Adaptive Control Using Fuzzy Systems and Neural Networks," *IEEE Transactions on Fuzzy Systems*, Vol.4, No.3, Aug. 1996, pp.339–359.

[13] Su, C.-Y., and Stepanenko, Y., "Adaptive Control of a Class of Nonlinear Systems with Fuzzy Logic," *IEEE Transactions on Fuzzy Systems*, Vol. 2, No. 4, 1994, pp.285–294.

[14] Sugeno, M., Edited *Industrial Applications of Fuzzy Control,* Amsterdam, North-Holland: Elsevier Science Publishers B.V., 1985.

[15] Wang, L.X., "Stable Adaptive Control of Nonlinear Systems," *IEEE Transactions on Fuzzy Systems*, Volume 1, No. 2, 1993, pp. 146–155.

[16] Wang, L.X., *Adaptive Fuzzy Systems and Control,* Englewood Cliffs, NJ: PTR Prentice Hall, 1994.

[17] Wang, L.-X., *A Course in Fuzzy Systems and Control,* Upper Saddle River, NJ 07458: Prentice Hall PTR, 1997.

[18] Wang, L.-X., "Combining Mathematical Model and Heuristics into Controllers: An Adaptive Fuzzy Control Approach," *Fuzzy Sets and Systems*, 89, 1997, pp.151–156.

[19] Yager, R.R. and Filev, D.P., *Essentials of Fuzzy Modeling and Control,* New York, Chichester, Brisbane, Toronto, Singapore: John Wiley & Sons, Inc., 1994.

Index

MIX
Papier aus verantwortungsvollen Quellen
Paper from responsible sources
FSC® C105338
www.fsc.org

If you have any concerns about our products,
you can contact us on
ProductSafety@springernature.com

In case Publisher is established outside the EU,
the EU authorized representative is:
Springer Nature Customer Service Center GmbH
Europaplatz 3, 69115 Heidelberg, Germany

Printed by Libri Plureos GmbH
in Hamburg, Germany